LABORATORY MANUAL
for
Human Biology

THIRD EDITION

Bert Atsma
UNION COUNTY COLLEGE

Sandra Hsu
SKYLINE COLLEGE

with contributions by

Robin McFarland
CABRILLO COLLEGE

PEARSON
Benjamin Cummings

San Francisco Boston New York
Cape Town Hong Kong London Madrid Mexico City
Montreal Munich Paris Singapore Sydney Tokyo Toronto

Publisher: Daryl Fox
Senior Acquisitions Editor: Deirdre Espinoza
Assistant Editor: Alison Rodal
Managing Editor: Wendy Earl
Production Editor: David Novak
Copyeditor: Lori Cavanaugh
Proofreader: Joella Krall
Text Design: The Left Coast Group, Inc.
Compositor: The Left Coast Group, Inc.
Cover Design: Frances Baca
Senior Manufacturing Buyer: Stacey Weinberger
Senior Marketing Manager: Sandra Lindelof
Executive Marketing Manager: Lauren Harp

PEARSON
Benjamin Cummings

ISBN 0-8053-7198-2
2 3 4 5 6 7 8 9 10—CRK—09 08 07 06 05
www.aw-bc.com

CONTENTS

PREFACE

This laboratory manual was written for use with *Human Biology: Concepts and Current Issues,* Third Edition, by Michael Johnson, a wonderful textbook written for students who may not have a strong background in science and, indeed, may not be science majors at all. Similarly, the goal of our laboratory book is to provide a meaningful laboratory experience for students, regardless of their level of preparation for a human biology course. We firmly believe that providing the nonmajor with a solid foundation in the science of human biology is an appropriate goal and that a laboratory experience is an important part of that foundation.

Features

Although it is possible that students with little or no laboratory experience may sometimes find the science laboratory an intimidating environment, much can be done to make students more comfortable with laboratory learning. This laboratory manual contributes to a more comfortable learning environment for the nonmajor by using a simple, consistent format.

- **Objectives** are listed first in each exercise so that students immediately understand what they are supposed to get out of the laboratory experience.
- **Materials** needed to perform the laboratory activities are set forth, both at the beginning of the exercise and at the start of each individual activity. This placement is convenient for the instructor or staff in preparing the laboratory and for the student in executing the activity.
- A short **introduction** is provided for each topic so that the student can learn the information relevant to the activity he or she will be performing, regardless of whether he or she has read the matching textbook section or covered it in the lecture portion of the course.
- **Activities** are clearly called out with distinct headings so that students are alerted to places where they should be performing lab work. The activities are set forth in a clear, step-by-step format and presented in a way that assumes the student is doing this kind of work for the first time.
- **Embedded questions** are asked frequently throughout each exercise to keep students focused on what they should be learning while they are doing their lab work.
- **Review questions** at the end of each exercise may be used by students to practice and reinforce the concepts they have covered, and these questions may also be used by instructors to evaluate student learning.
- **Answers** to the questions in this book are available online at http://suppscentral.aw.com/. Instructors should contact their local Benjamin Cummings sales representative for the username and password.

Organization

The initial exercises of this lab manual introduce students to the lab environment. Exercise 1 introduces the scientific method and lab protocols. Exercise 2 covers the use and care of the microscope. The remaining exercises closely follow the organization of the Johnson textbook, *Human Biology.* Exercises 3 and 4 concern cell anatomy and physiology, and Exercise 5 provides the link between cells, tissues, and organs. In Exercise 6, students are introduced to the terminology used to characterize and discuss anatomical structures and regions. Exercises 7–19 cover the 10 body systems, while Exercise 20 provides a fun introduction to genetics. Exercise 21 is new to this edition, and introduces students to the fascinating field of DNA Technology. Finally, Exercise 22 covers evolution, and Exercise 23 invites students to address ecological issues, such as the impact of various lifestyles and population growth upon the environment and human interaction.

Approach

There are so many fascinating things to be learned about the human body that it is not surprising that human biology course instructors often have varying approaches to its presentation. As with many things in life, there is more than one right way to organize a human biology course, and our somewhat ambitious table of contents reflects an awareness of this. It is expected that most instructors, with 2 to 3 hours of lab time per week, will pick and choose from these numerous exercises and activities to match their goals and methods of presentation. Some of the shorter exercises were written with the idea in mind that in courses with longer laboratory periods, they would be combined with all or parts of other exercises.

Some instructors disagree with the use of dissection work for nonmajors, and others feel that dissection is an integral part of learning how organs and structures look *and* feel. We tend to believe that a certain amount of dissection work is meaningful, even for the nonmajor, in many places. For example, the fetal pig does much to convey the flimsy, sacklike nature of the gallbladder, and the sheep brain demonstrates the toughness of the dura mater around the brain. A few minutes with an actual specimen makes a lasting impression, perhaps in more ways than one. Nonetheless, there are certainly legitimate reasons why instructors do not wish to include dissection in their human biology courses. Understanding this preference, we provided an opportunity in most exercises that call for dissection to learn much of the same material by using widely available models.

Acknowledgments

Many people contributed their efforts to this book. First, we would like to acknowledge the contributions of Robin McFarland of Cabrillo College, who teaches both human anatomy and general biology and is actively involved in bringing her special interest in evolution and ecology to the classroom. Exercises 22 and 23, which feature timely and relevant activities, bear the mark of Robin's expertise.

We thank Kay Ueno, Managing Editor for the first edition, who took on the extraordinarily difficult task of working with two authors who were separated by 3,000 miles, two mountain ranges, and occasionally different approaches. Her dedication and inestimable contribution to the successful completion of the first edition are sincerely appreciated.

For the 2nd edition, we would again like to thank Daryl Fox for his leadership, and David Novak for his strenuous efforts in getting everything just right. We thank Ziki Dekel, whose skillful management and editing during the work on the second edition made an enormous contribution to the success of the revision process.

For this third edition, we would like to thank Alison Rodal and Marie Beaugureau, our highly talented and hard-working editors. We again thank David Novak for his outstanding efforts in the production of this book. And, we thank our reviewers for this edition, as their valuable insights were enormously helpful to us: Keith Overbaugh (Northwestern Michigan College), Marilynn Bartels (Black Hawk College), Nancy O'Keefe (Purdue University Calumet), Aimee Lee (University of Southern Mississippi), Roberta Meehan (Troy State University), Marie Gabbard (Boise State University), Peter King (Francis Madison University, South Carolina), Todd Scarlett (University of South Carolina, Lancaster), Sara Van Orden (Quinebang Valley Community College).

Finally, we thank our past students for teaching *us* what works best through their questions and exertions in the laboratory. And, of course, we thank our families for their cooperation while daddy or mommy was writing.

BERT ATSMA, Union County College
SANDRA HSU, Skyline College

TO THE STUDENT

Many students dread a science course, particularly its laboratory component. You have had reading, writing, and arithmetic, even a little history, and possibly a full year or two of high school science, but your classwork may not have prepared you for a science laboratory. For most non-majors, college-level science in general, and biology specifically, may be unfamiliar territory to explore. But exploring new territory can be exciting, especially with good maps and a good guide. With the aid of your instructor, we hope that this manual helps you successfully navigate the laboratory portion of your human biology course.

This laboratory manual was designed to prepare you for a positive laboratory experience, regardless of your level of preparation for a course in human biology. We have used a simple, consistent format to make laboratory learning a more comfortable and enjoyable experience. The **objectives** alert you to what you are supposed to learn from each exercise. For your convenience, the **materials** required are listed both at the beginning of each exercise and each activity. A short **introduction** begins each exercise providing the background information necessary to fully appreciate the activities you will perform. The **activities** are spelled out in a clear, step-by-step format, helping you to more easily follow the directions, even if this is your first laboratory experience. **Questions** are asked frequently throughout the activities and at the end of each exercise to ensure that you are "learning while doing" your work.

To get the most out of your laboratory experience and perform your best, we recommend that you read the exercise you will be performing *before* you come to class. This really does work. The introductory information will prepare you for new terms and concepts that you will encounter in the lab. Being familiar with the procedures will help you perform the activities more smoothly and more safely.

Speaking of safety, you will find safety guidelines printed on the inside front cover of this lab manual. Please take the precautionary statements in these guidelines seriously, as well as any additional safety information provided by your instructor. We have made an effort to reduce the risk of personal injury by recommending the use of the safest chemicals and procedures appropriate to each activity. Careless use of chemicals and equipment is something that only you can prevent. Your instructor will advise you of any protective equipment that you should use prior to each activity. In particular, we feel it is a good idea to wear eye protection when handling any chemicals, especially if you wear contact lenses. Remember, safety first!

Finally, be prepared for your instructor to modify some of the activities in this laboratory manual, as there is more than one way to learn about the marvels of the human body. If you wish to make comments on this lab manual or offer suggestions for improvements in future editions, please write us at the publisher's address below. We sincerely hope that your human biology laboratory experience will be positive and memorable, and we wish you great success!

BERT ATSMA and SANDRA HSU
Benjamin Cummings
Applied Sciences
1301 Sansome Street
San Francisco, CA 94111

LIST OF SUPPLY HOUSES

This is a partial list of suppliers of equipment, animals, and chemicals, and it should not be considered a recommendation for these companies. Many supply companies have regional addresses. Only one address is listed for each company.

Aldrich Chemical Company, Inc.*
1001 W. Saint Paul Avenue
Milwaukee, WI 53233
1-800-558-9160
www.sigma-aldrich.com

Carolina Biological Supply Company
2700 York Road
Burlington, NC 27215
1-800-334-5551
www.carolina.com

Fisher Scientific†
2170 Margin Avenue
Santa Clara, CA 95050
1-800-766-7000
www.fishersci.com

ICN Biochemicals
1263 South Chillicothe Road
Aurora, OH 44202
1-800-854-0530
www.icnbiomed.com

Intelitool (Phipps & Bird)
P.O. Box 7475
Richmond, VA 23221
1-800-955-7621
www.intelitool.com

LabChem
200 William Pitt Way
Pittsburgh, PA 15238
412-826-5230

Nasco
901 Janesville Avenue
P.O. Box 901
Fort Atkinson, WI 53538-0901
1-800-558-9595
www.enasco.com

Sigma Chemical Company*
P.O. Box 14508
St. Louis, MO 63178-9974
1-800-325-3010
www.sigma-aldrich.com

VWR Scientific
3745 Bayshore Boulevard
Brisbane, CA 94005
1-800-932-5000
www.vwrsp.com

Ward's
P.O. Box 92912
Rochester, NY 14692
1-800-962-2660
www.wardsci.com

* The Aldrich and Sigma brands are both sold through Sigma-Aldrich, Inc.
† For orders outside the United States, contact Fisher Scientific Worldwide: One Liberty Lane, Hampton, NH 03842, USA; 603-926-5911; www.fisherscientific.com

EXERCISE

1

SCIENTIFIC METHOD AND LABORATORY PROTOCOL

Objectives

After completing this exercise, you should be able to

1. Understand the concept of discovery.
2. Explain the concept of scientific method and its steps.
3. Understand the metric system, and calculate metric conversions.
4. Convert percentages, decimals, and fractions.
5. Implement proper dissection and experimental protocols.
6. Implement proper lab safety protocols.

Materials for Lab Preparation

Safety Equipment

- ❑ Hood
- ❑ Safety shower
- ❑ Eyewash
- ❑ Fire extinguisher
- ❑ Fire blanket
- ❑ Gloves
- ❑ Goggles

Dissection Tools

- ❑ Scalpel or knife
- ❑ Probe
- ❑ Forceps
- ❑ Scissors
- ❑ Dissection tray (waxed or unwaxed)
- ❑ Dissection pins or twine

Supplies

- ❑ Coin
- ❑ Cleaning disinfectant (e.g., Sanisol in spray bottle)
- ❑ Autoclave bag for biologically hazardous material
- ❑ Bleach jar for soaking glassware
- ❑ Timer

ntroduction

All activities in this exercise are intended to serve as an introduction to the scientific laboratory. They also may be useful to you as a reference when embarking on later activities in this laboratory manual.

The first three activities will acquaint you with the scientific process designed for testing ideas and making discoveries. The remaining activities introduce you to some of the scientific tools of the trade: the use of metric measurements and commonly used dissection tools. Last, but not least, you will review safety protocol in the laboratory.

Scientific Method of Discovery

Even before the word "science" was around, people observed the world around them, formulated questions, tested their ideas, and communicated their conclusions to their friends and neighbors. The tried and tested ideas eventually became part of our general body of knowledge. Today, whether in our homes, in school, or in business, we continue to add or make adjustments to that body of knowledge, without realizing that to do so, we employ some of the same elements of the discovery process (albeit informally or unconsciously) employed by the scientist in the laboratory.

In popular literature and media—particularly science fiction movies—the scientist is often portrayed against the back drop of a laboratory, frequently "mad as a hatter," hovering over a complex test tube apparatus or microscope. The everyday reality is seldom so dramatic, although the eventual outcomes of scientific work are no less exciting. What is missing from both our everyday process of discovery and the ever-popular movie version of scientific discovery is the systematic, step-by-step process over weeks, months, and years that characterizes a scientist's work (Figure 1.1).

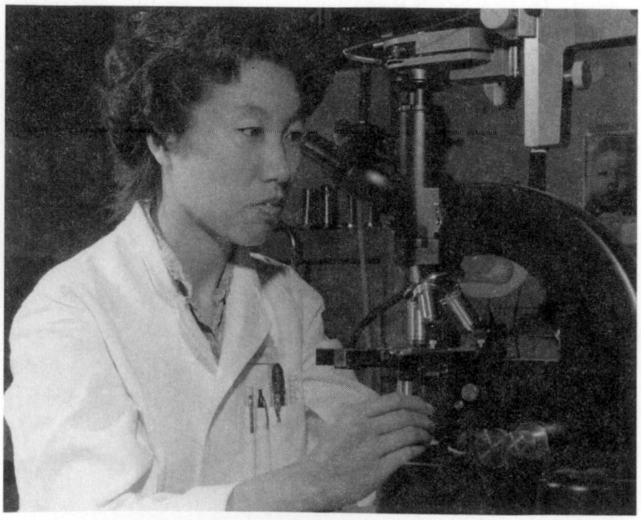

FIGURE 1.1 Dr. Wang, a National Cancer Institute research scientist, at work.

Like all of us, the scientist is interested in getting dependable answers to questions—answers with which the members of our community, scientific or otherwise, can agree. It is not the scope of the question that distinguishes the scientific discovery from ordinary, personal discoveries; what makes scientific discovery distinct and compelling is the process by which questions are answered.

ACTIVITY 1

Getting Acquainted with Scientific Method

Let's look at the key steps in the process called **scientific method.** This process allows scientists to obtain knowledge and develop scientific theories that explain occurrences in the natural world. This activity will introduce you to the steps of the scientific method, which you will implement in your own experiment later in this lab. In this activity, we will apply the steps of the scientific method to the common cold.

Step 1: Observe and Generalize

Decide on a phenomenon you would like to study, in this case, the common cold. Observe different characteristics of this phenomenon and then generalize about it. Take into consideration what is already known on the subject, varying opinions, past mistakes, and new aspects that you may not have foreseen. This process might go as follows:

Observation 1: When my friends get a cold, I sometimes get one too after spending time with them.

Observation 2: People with colds cough and sneeze a lot.

Observation 3: People with colds do not always cover their nose or mouth when coughing or sneezing, or wash their hands.

Generalization: Colds can be passed from person to person.

Step 2: Formulate a Hypothesis

A **hypothesis** is a tentative, unproven statement about the natural world, which attempts to explain a specific phenomenon. Formulate a hypothesis by making an explicit statement about the problem, question, or idea that you intend to study. Based on the observations and generalizations about the common cold you made in Step 1, the hypothesis you formulate might go as follows:

Hypothesis: Colds are caused by something that is physically transmitted from person to person.

Step 3: Make a Testable Prediction

A testable prediction is a test based on the hypothesis. Predictions are often in the form of an "if . . . then" statement, in which the "if" part of the statement is the hypothesis.

Prediction: If healthy people are near people with colds, then they will catch a cold.

Step 4: Design an Experiment

An experiment is a carefully planned and executed manipulation of the natural world that has been designed to test your prediction. You must design an experiment which accounts for all factors that may vary during the course of the experiment, or **variables,** except for the **controlled variable,** which is the factor or element under study.

Let's design an experiment to study how colds may be transmitted. In this case, the controlled variable is the common cold. We will have two groups: the **control group,** which will not be exposed to the controlled variable (the common cold), and the **experimental group,** which will be exposed to the controlled variable. The experiment we design may go as follows:

Experiment Step 1: Select a large group of people without colds.

Experiment Step 2: Randomly divide this first group into the control group, and the experimental group.

Experiment Step 3: Select a second group of people to be their companions. Half of them should have cold symptoms; half of them should be free of cold symptoms.

Experiment Step 4: The individuals of the control group will get a companion without a cold; the individuals of the experimental group will get a companion with a cold. For 2 weeks, 1–2 hours per day, each couple will be in continuous close contact. This might be in the form of sitting together for the same lecture or lab for 2 weeks, eating lunch together, or studying together during the weekend.

Step 5: Analyze the Data and Modify Your Hypothesis

The data must be gathered and evaluated in an accurate manner, without prejudice. Do not change "bad data"; look at it clearly and honestly even if it does not support your hypothesis. If the data is numerical, organize it into tables and graphs. This makes it easier for you and other people to quickly grasp the main points.

Once you've analyzed your data, determine whether it supports the hypothesis or not. It may be necessary to modify the hypothesis in order to develop a sound conclusion, based on your evaluation of the data.

Conclusion: If the experimental group's cold symptoms are found to be statistically higher, then your prediction is verified, and your hypothesis is supported.

Hypothesis Modification: If your experimental group does not exhibit a higher incidence of cold symptoms than the control group, you will have to modify your hypothesis to fit these new findings and develop a new experiment. Perhaps the cold symptoms would be more evident if you allowed more contact, such as sharing the same dishes or writing utensils, or increase the experiment to 4 weeks.

Step 6: Developing a Conclusion and Sharing Your Findings

A well-tested hypothesis becomes a **theory,** or a hypothesis that has been extensively tested and supported over time.

Once you have modified your hypothesis, develop a theory based on your new understanding. It should stand up to repetitive tests, for now at least, though it may change as time goes on, as theories often do. Your theory should be more broadly defined than your hypothesis.

Theory: A micro-organism is transmitted by people with colds.

Now that you've got a solid theory, make the findings known! The ultimate goal of a scientist is to advance public knowledge, and publication of findings is the process by which our universal body of knowledge grows and benefits the general public. The traditional method of disseminating information is through publication in scientific journals. Journals are subject to the scrutiny of review by the scientist's peers, and, once published, findings often contribute to an overall body of knowledge on a given topic. How could you share your findings on the results of the cold experiment with your peers?

Sharing Your Findings: Submit an article to the school paper called *Experiment on Colds.* ■

ACTIVITY 2

Building and Testing a Hypothesis (Coin Toss)

Materials for this Activity

Coin

Let's apply the steps of the scientific method to a common situation, the coin toss.

Step 1: Observe and Generalize

Reflect on the times you've observed a coin toss. Why is this a favorite method for deciding on a course of action? What are the factors which make it a "fair" method? After considering these questions, write your observations and generalization.

Observation 1: One side represents only one symbol

Observation 2: ______________________

Generalization: ______________________

Step 2: Formulate a Hypothesis

Propose a hypothesis that investigates the probability of the coin landing on either heads or tails, and the usefulness of this method in deciding on a course of action.

Hypothesis: Heads will be more.

Step 3: Make a Testable Prediction

Predict the possible outcome of your experiment, using your hypothesis as a basis. Formulate the prediction in an "if . . then" statement. Be as specific as possible.

Prediction: If the coin is tossed 24 times heads will be the majority.

Step 4: Design an Experiment

Design an experiment to test your prediction. Remember Activity 1. You'll want to do everything possible to avoid experimental error when designing your experiment. You will also want to define the different variables. Be specific as you work out the details of how you will carry out your experiment. Consider issues such as, how you will toss the coin or how many tosses are adequate to test your hypothesis (see Figure 1.2). Write down your procedures below or on another sheet of paper if you need more space.

Experiment Step 1: Flip the penny 24 times.

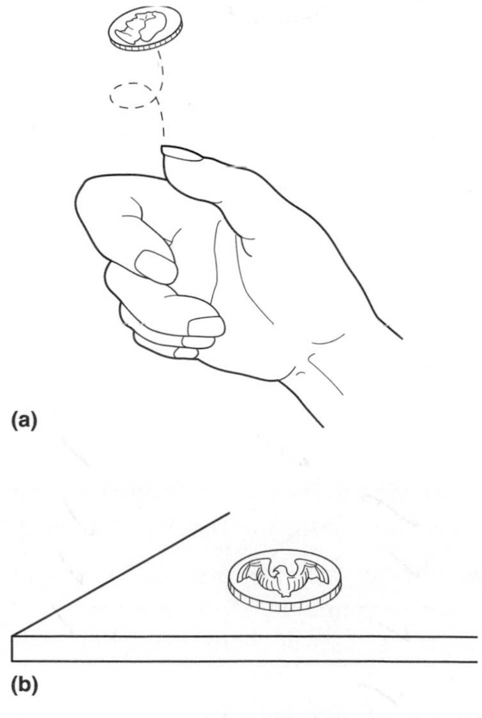

FIGURE 1.2 Building and testing a hypothesis with a simple coin-toss experiment.

Experiment Step 2: record results in chart.

Experiment Step 3: ______________________

Step 5: Analyze the Data and Modify Your Hypothesis

Present your data in a way that is meaningful to the evaluation of your hypothesis. You must record your data in an organized and a thorough way. Remember that you will be evaluating this data to prove your hypothesis. Use Box 1.1 to record your data during this experiment. It is a good example of how to record and present your data.

After performing the experiment, how will you evaluate the data? What simple arithmetic technique will you use to analyze your data? Write a brief explanation of how you will evaluate your data.

According to the result, tails was the minority with the ratio of 8:24. Heads was the majority with a ratio of 16:24 = (4:7)

Once you've analyzed the data, compare the results to your prediction, and suggest any modifications to your hypothesis. If your prediction was supported by the data, write a conclusion for your hypothesis. If your prediction was not supported, develop a modified hypothesis. Consider the various factors that could affect your results. For example, could the size of your sample have had an effect on results? This is always a significant factor when dealing with statistical data. Explain in the conclusion space below, any errors that you think might have affected your outcome. One of the final steps in discussing data is explaining experimental error. For example, is the amount of error significant? Is there something about your error(s) that may help you to redesign a better experiment? Explain any experimental errors in the space below for your conclusions. If necessary, write a modified hypothesis.

Conclusion: This experiment did support our group's hypothesis. The hypothesis that was ~~was~~ that stated was heads will be the majority.

Modified Hypothesis: —

Box 1.1 Coin Toss Data

Trial Number	Heads	Tails	Trial Number	Heads	Tails	Trial Number	Heads	Tails	Trial Number	Heads	Tails
1		✓	7	✓		13		✓	19	✓	
2	✓		8	✓		14	✓		20		✓
3	✓		9	✓		15		✓	21	✓	
4	✓		10	✓		16	✓		22		✓
5	✓		11		✓	17		✓	23		✓
6	✓		12	✓		18	✓		24	✓	

Congratulations! You have completed a short tour of the scientific method. Now that you have been through the process for a simple hypothesis, imagine developing a hypothesis and designing its test for a more complex biological issue such as, "does acid rain affect the growth of plants?" While you will not be able to test such an issue in this course, you can form a small group and brainstorm some ideas for testing this hypothesis. Spend 10 minutes discussing this topic with your group and summarize your ideas in the space provided. Your instructor can facilitate sharing each group's summary with the class.

Metric Measurements

In all fields of science, the physical dimensions of a phenomenon must be measured. At this time, all nations use the **metric system** for scientifically measuring length, mass, volume, and temperature. The metric system has two advantages:

1. Standard references. This allows measurements to be checked independent of the language of any single counting system.
2. Conversion. Measurements can be converted between different units in a universally agreed-upon manner (Figure 1.3).

Standard References

Without a standard reference, such as the metric system, we could not know that a gram (g) in Japan weighs exactly the same as a gram in the United States, which would severely hinder the process of scientific discovery. In both countries, the weight of a gram is measured by weighing the mass of 1 cubic centimeter of water at the temperature of its maximum density. Water, of course, behaves the same in

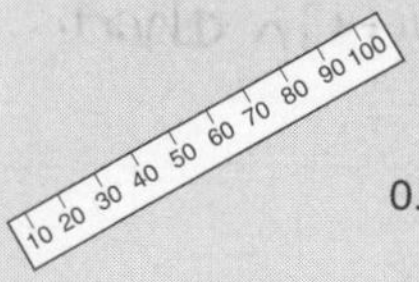

Length

Basic unit: meter (m)

0.000001 meter = micrometer (µm)
0.001 meter = millimeter (mm)
0.01 meter = centimeter (cm)
1000 meters = kilometer (km)

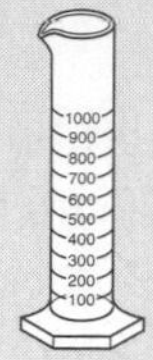

Volume

Basic unit: liter (l or L)

0.001 liter = milliliter (ml)*
0.01 liter = centiliter (cl)
0.1 liter = deciliter (dl)
**milliters = cubic centimeters (cc)*

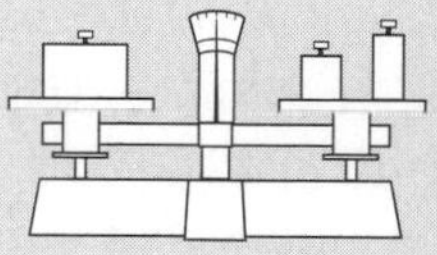

Mass

Basic unit: gram (g)

0.001 gram = milligram (mg)
0.01 gram = centigram (cg)
1000 grams = kilogram (kg)

Pressure

Basic unit: millimeters of mercury (mm Hg)

No alternate units are commonly used.

Time

Basic unit: second (sec)

0.000001 second = microsecond (µsec)
0.001 second = millisecond (msec)
60 seconds = minute (min)
3600 seconds = hour (hr)

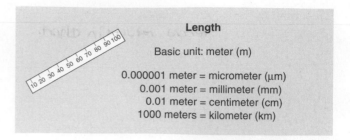

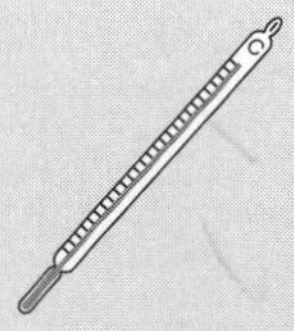

Temperature

Basic unit: degrees Celsius (°C)

No alternate units are commonly used.

FIGURE 1.3 Metric measurements.

all countries, so by basing the standard reference for mass on water, all countries will achieve the same standards for a gram.

The standard reference for a meter is more complicated. A meter (m) represents a unit of length equal to 1,650,763.73 wavelengths in a vacuum of the orange-red line of the spectrum of krypton 86. One degree Celsius (C) represents one-hundredth of the temperature scale, of which 0° is the freezing point of water and 100° is the boiling point.

You should memorize the basic units for each physical dimension (see Figure 1.3), as well as the most common metric units, for example centi- (one-hundredth), milli- (one-thousandth), micro- (one-millionth), and nano- (one-billionth). Note that there is no period after a metric unit. For example, 1 m (no period) is equivalent to writing and saying "one meter."

Conversions

Making conversions easily is the second advantage of using the metric system. Metric conversions are based on powers of 10 (Table 1.1). Thus, different orders of magnitude are easily converted by moving the decimal point to the right or left. For example, a common question is: How many cm are there in 5 m? The answer is 500 cm, which was determined by moving the decimal two places to the right. Use the acronym **LSR,** which stands for **L**arge **S**mall **R**ight, to help you remember this rule. You can also rely on common sense to determine the answer, if you keep in mind, in this instance, that a centimeter (cm) is smaller than a meter (m). So when you are trying to convert a larger unit (m) to a smaller unit (cm), the equivalent numerical value you seek will be bigger.

Table 1.1 Metric Prefixes and Conversions

Metric Prefixes

Prefix	Symbol	Term	Exponent	Multiplication factor
deci	d	one-tenth	10^{-1}	× 0.1
centi	c	one-hundredth	10^{-2}	× 0.01
milli	m	one-thousandth	10^{-3}	× 0.001
micro	μ	one-millionth	10^{-6}	× 0.000,001
nano	n	one-billionth	10^{-9}	× 0.000,000,001

Metric Conversions

To convert from	To	Move decimal point	Factor
meter (liter, gram)	decimeter	1 place to right	× 10
meter (liter, gram)	centimeter	2 places to right	× 100
meter (liter, gram)	millimeter	3 places to right	× 1,000
meter (liter, gram)	micrometer	6 places to right	× 1,000,000
meter (liter, gram)	nanometer	9 places to right	× 1,000,000,000
decimeter	meter (liter, gram)	1 place to left	÷ 10
centimeter	meter (liter, gram)	2 places to left	÷ 100
millimeter	meter (liter, gram)	3 places to left	÷ 1,000
micrometer	meter (liter, gram)	6 places to left	÷ 1,000,000
nanometer	meter (liter, gram)	9 places to left	÷ 1,000,000,000

Let's calculate some metric conversions. Suppose that you want to convert 0.4 m to millimeters:

$$0.4 \text{ m} \times 1{,}000 \text{ mm/m} = 400 \text{ mm}$$

Or, because we are moving from larger to smaller units, let's apply the LSR rule and move the decimal to the *right* three places:

$$0.4 \text{ m} \rightarrow 400 \text{ mm}$$

Let's convert 5 μg to grams:

$$5 \text{ μg} \times 1\text{g}/1{,}000{,}000 \text{ μg} = 0.000{,}005 \text{ g}$$

The opposite of the LSR rule applies because we are moving from smaller to larger units. Let's move the decimal to the *left* six places.

$$5 \text{ μg} \rightarrow 0.000{,}005 \text{ g}$$

For a complete list of metric conversions, including temperature conversions between Celsius and Fahrenheit, please refer to Appendix A on page 283.

Converting Percentages to Fractions

Many of our eyeball estimates involve percentages or fractions. For example, let's talk about length. Your fingers are longer than your toes by about 2½ times, or 250%, or by a factor of 2.5. Another way of saying the same thing is the length of your toes is about 40%, or two-fifths, of your fingers. Converting between numbers in decimals, percentages, and fractions is a useful skill, not only in biology, but in everyday life.

Let's calculate the percentage represented by the fraction two-fifths.

Step 1: $2 \div 5 = 0.4$

Step 2: $0.4 \times 100 = 40$

Step 3: Attach the %.

Answer: 40%

Let's calculate the fraction represented by the percentage 40%:

Step 1: Set up a fraction: $\frac{40}{100}$

Remember, in percentages the denominator is always 100.

Step 2: Divide the numerator and denominator by the largest possible number.

In this case, divide both numbers by 20:

$$\frac{40}{100} \div \frac{20}{20} = \frac{2}{5}$$

It is recommended that you make a table of the most common fractions and percentages. Then treat it like the multiplication table. Review it again and again until you know it by heart, so that you will "automatically" know the answer. Doing so will give you a head start in the following lab report. Remember that if you're not sure, you can always calculate the answer.

Review of Dissection and Experimental Protocols

In human biology, the lab exercises will include conducting anatomical dissections and physiological experiments. Some animals are similar in structure and function to humans, so dissection provides a useful means for understanding your own physical structure. Proper dissection requires patience, skill, and an understanding of commonly used dissection tools. To acquaint you with the dissection tools that you will encounter in future exercises, try the next activity.

ACTIVITY 3

Identifying Dissection Tools

Materials for this Activity

Scalpel or knife
Probe
Forceps
Scissors
Dissection tray (waxed or unwaxed)
Dissection pins or twine

1. Pick out the dissection tools listed above from your supply area and bring them to your workstation. Describe the purpose of each tool in Box 1.2. Don't worry about getting it "right"; just make your best guess, and keep in mind the old adage, "Form follows function."

2. Match each of your tools with those illustrated in Figure 1.4 and enter the appropriate name for each in the space provided.
3. Now refer to Figure 1.5, which depicts selected steps in a rat dissection. In the space provided under each photograph, name the tool or tools that you think would be used to perform the specific dissection task portrayed, and explain why. Keep in mind that a dissection task may require more than one tool, and that a tool may be used for more than one purpose.

Review of Lab Safety Protocols

Conducting an experiment involving glassware, flames, acids, and other people also requires skill and patience. The most important thing is for you to

Box 1.2 Dissection Tools and Their Purpose

Dissection Tools	**Purpose**
Scalpel or knife	
Probe	
Forceps	
Scissors	
Dissection tray (waxed or unwaxed)	
Dissection pins or twine	

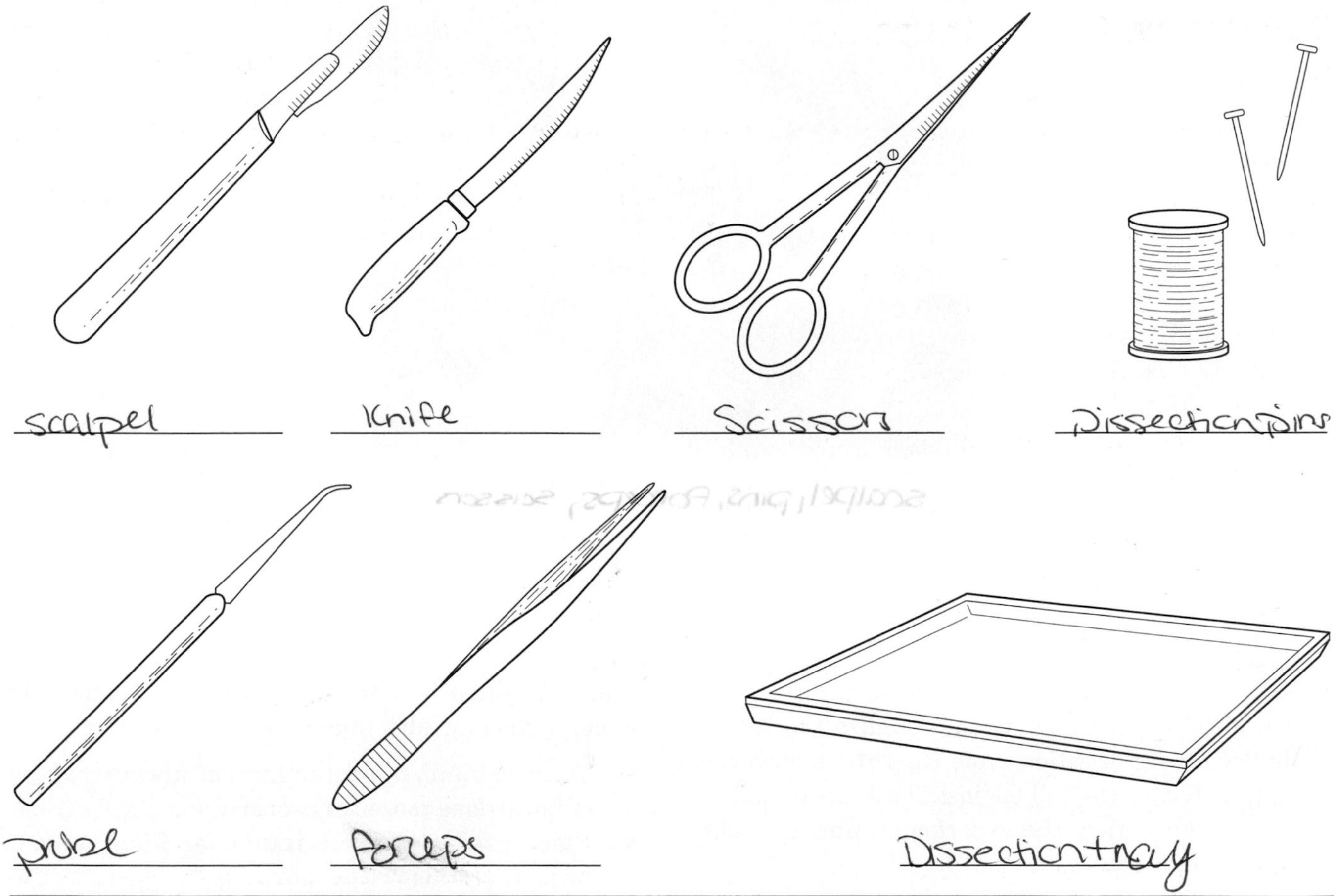

FIGURE 1.4 Commonly used dissection tools.

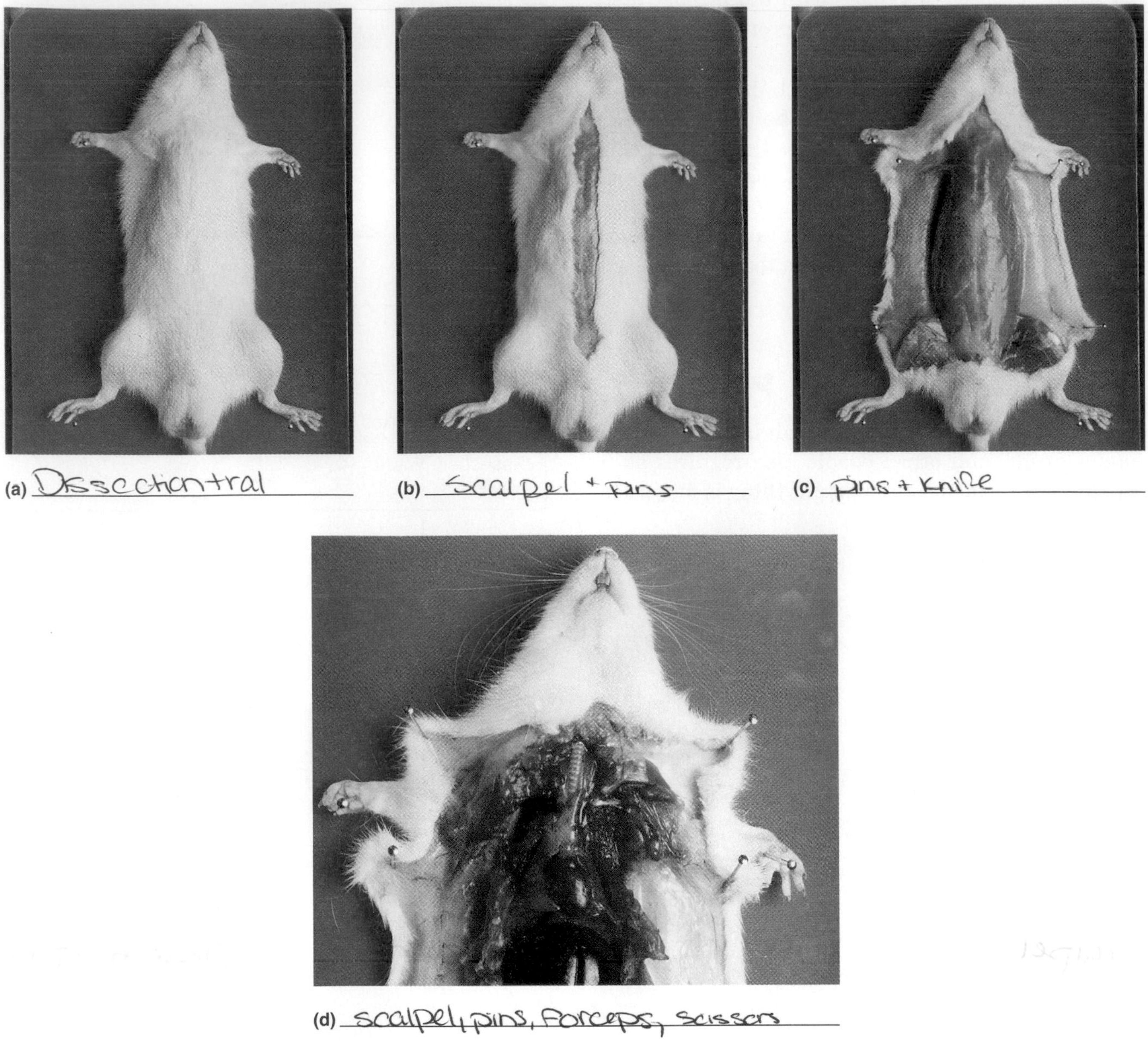

FIGURE 1.5 Selected steps in a rat dissection.

be fully prepared. Always read ahead carefully before beginning an experiment, visualize the steps, write down your questions, and clear them with the instructor before the lab begins. Think about safety issues as you review the experiment, and consider how you might address them.

Every laboratory contains some safety equipment. Although the equipment maintained from one laboratory to another may vary according to needs, Figure 1.6 illustrates the equipment most commonly found in biology labs. Remember the following:

- To avoid injury or contamination, always dispose of hazardous material properly.
- Place used glassware into a jar filled with a bleach solution.
- Always use gloves before beginning a dissection as a precaution when handling freshly killed animals, which sometimes harbor parasites. When

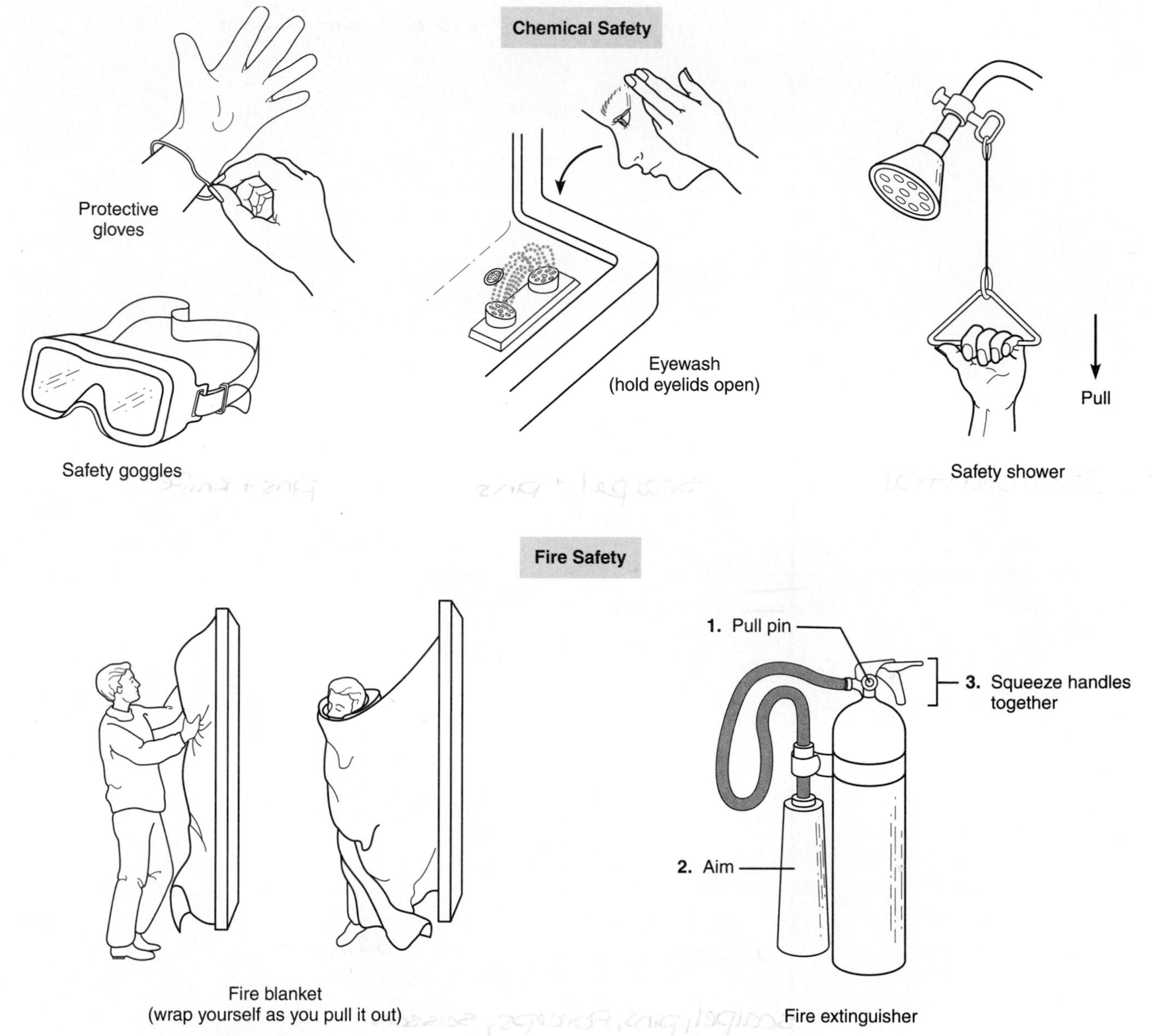

FIGURE 1.6 Laboratory safety equipment.

handling preserved specimens, gloves may reduce the possibility of skin irritation.

- Place biological material into an autoclave bag.
- Clean, dry, and return all equipment to their proper locations. This is especially important for microscopes and small machines such as the centrifuge, calorimeter, and pH meter.
- After you complete your experiment, always wipe down the counter with a disinfectant so that the next student has a clean surface to use.
- If any glassware or equipment is broken, notify your instructor immediately.
- Wash your hands thoroughly upon entering and leaving a biological laboratory.

Human biology can be a fascinating course. We hope you have an enjoyable learning experience!

EXERCISE

2

THE MICROSCOPE

Objectives

After completing this exercise, you should be able to

1. Care properly for microscopes.
2. Understand magnification and calculate the total magnification.
3. Identify the parts of the microscope and list their functions.
4. Focus with a microscope.
5. Compare and describe the effects of scanning-, low-, and high-power lenses.
6. Determine the depth of focus.
7. Prepare a wet mount slide.
8. Properly clean up and store the equipment.

Materials for Lab Preparation

Equipment and Supplies

- ❑ Compound light microscopes
- ❑ Millimeter ruler (plastic)
- ❑ Lens paper and cleaner
- ❑ Autoclave bag
- ❑ Jar of 10% bleach for soaking glassware
- ❑ Disinfectant

Prepared Slides

- ❑ Letter "e"
- ❑ Three colored threads

Supplies for Wet Mount Slide Preparation

- ❑ Clean slides
- ❑ Clean coverslips
- ❑ Flat toothpicks
- ❑ Paper towels
- ❑ Methylene blue or Wright's stain in dropper bottle

Introduction

In a human biology laboratory, the microscope is probably the most widely used instrument. Since the 17th century, it has enabled scientists to extend their senses into the microscopic world, to literally see beyond the naked eye. Blood looks like red ink to the naked eye, but under the microscope it becomes an amber fluid filled with hundreds of thousands of tiny red pancakes and tiny clear bags containing funny-shaped purple structures. A hangnail, under a microscope, becomes a cylindrical specimen composed of flat cells arranged in a whorl-like pattern, while a drop of clear pond water may become a large body of water alive with small, swimming organisms covered with yellow and green flecks.

This exercise will introduce you to the use and care of the standard compound light microscope.

Care of Microscopes

Remember that the microscope is a delicate and expensive instrument. A good microscope should last many, many years. Accidents and damage can be avoided by using the following guidelines:

1. Carry a microscope with two hands. *Do not swing it!*
2. Clean the lenses only with lens paper and cleaner. *Do not use paper towels!*
3. Do not force anything—lenses, knobs, levers—ask for help.
4. Never leave a slide on the microscope stage once you have completed your observations and are preparing the microscope for storage.
5. Store the microscope with the lowest objective lens in place.
6. Store the microscope with the lamp cord around the arm, not the objective lens.
7. Report microscope problems, then use another microscope. *Do not store damaged microscopes!*

Microscopes and Magnifications

The microscopes commonly used in this course are compound light microscopes. Light is produced from a lamp, passed through the condenser, iris diaphragm, stage, and finally through the specimen. The image is magnified as it passes through two lenses—first, the objective lens, and then the ocular lens. The image that you see, depending on the objective lens you are using to view your specimen, will have been magnified from 40 to 400 times, which is far greater than the typical magnifying glass.

You might think that the more magnification, the better. This is only correct to a certain extent. There are three reasons that magnification in a light microscope cannot or should not be increased indefinitely:

1. The maximum magnification for a light microscope is approximately 1000 times. Beyond this, the image is magnified but no longer clear.
2. Remember that the more you magnify, the smaller the area you are observing. In other words, you're seeing a smaller and smaller part of the specimen, being blown up more and more.
3. It's more important (for beginning students) to see the whole specimen than to see a very small part of the specimen well. This is comparable to "seeing the forest, rather than the leaves." Remember that whenever you work with microscopes, you will eventually need to correlate all of these separate images and relate them back to the original whole specimen.

Because not all microscope models are alike, your objective lenses may differ. In this exercise, the power of a lens will be referred to as follows:

Scanning power (about 4×)
- Used to locate specimen; low power should be used if the microscope does not have scanning power

Low power (about 10×)
- Used to see the whole or large portions of the specimen

High power (about 40×)
- Also called the **high-dry lens,** as no oil is required for clear viewing
- Used to see small, detailed parts of the specimen

Oil immersion (about 100×)
- Used to see very small specimens (e.g., bacteria)
- Always immersed in oil to improve resolution
- Usually not used in beginning biology labs

The job of the microscope is to magnify small images. The compound light microscope accomplishes this task with two sets of lenses: the ocular lenses, which provide a constant magnification factor, and various objective lenses, which provide a variable magnification factor. It is important to know, and to be able to communicate, the total magnification of an image. In biology books, you will notice that photographs of slides usually indicate how much magnification has occurred. For example, 106× means the image has been magnified 106 times.

Total magnification (TM) is determined by multiplying the power of the ocular lens (OcL) by the power of the objective lens (ObL).

$$TM = OcL \times ObL$$

Work through Activity 1 to acquaint yourself with your microscope, then calculate the total magnification of the lens combinations of your microscope and record the results in Box 2.1.

ACTIVITY 1

Identifying the Parts of the Microscope

As you read the description of the parts of the microscope, begin to label the structures shown in Figure 2.1. We will start at the top and generally move downward. Identify all bold face items.

Ocular Lens

- Also known as the eyepiece
- May be one only or a set of two
- Lens closest to your eye, usually the highest part of the microscope
- Includes a magnification factor engraved on the barrel of the ocular lens; for example, 10× indicates the image is magnified 10 times
- Should be cleaned *only* with lens cleaner and lens paper

Nosepiece

- Revolving circular mechanism that holds the different objective lenses
- Rotation of the nosepiece (*not* pushing on the barrel of the objective lens) is used to change the objective lens

Objective Lenses

- Individual lenses attached to the nosepiece
- Includes a magnification factor engraved on the barrel of each objective lens; for example, 4×, 10×, 40×, or 400× indicates the magnification
- Should be changed by rotating the nosepiece, *not* by pushing the barrel of the objective lens directly

Arm

- Upright bar that connects the ocular lenses to the rest of the microscope
- Provides the safest way to hold a microscope: Use two hands—one hand on the arm, the other under the base of the microscope

Stage

- Also called the mechanical stage
- A surface that supports and secures the slide, with the help of **stage clips**

Stage Controls

- Usually located on the side or on top of the stage
- Front/back controls: move the slide front to back, and vice versa
- Side/side controls: move the slide from side to side

Condenser

- Located under the stage
- Focuses the light through the hole in the stage and onto the specimen
- Adjusts the quality and amount of light passing through the specimen

Box 2.1 Calculating Total Magnification for the Microscope

	Scanning	Low	High	Oil immersion
Objective lens (ObL)	______	______	______	______
Ocular lens (OcL)	______	______	______	______
Total magnification (TM)	______	______	______	______

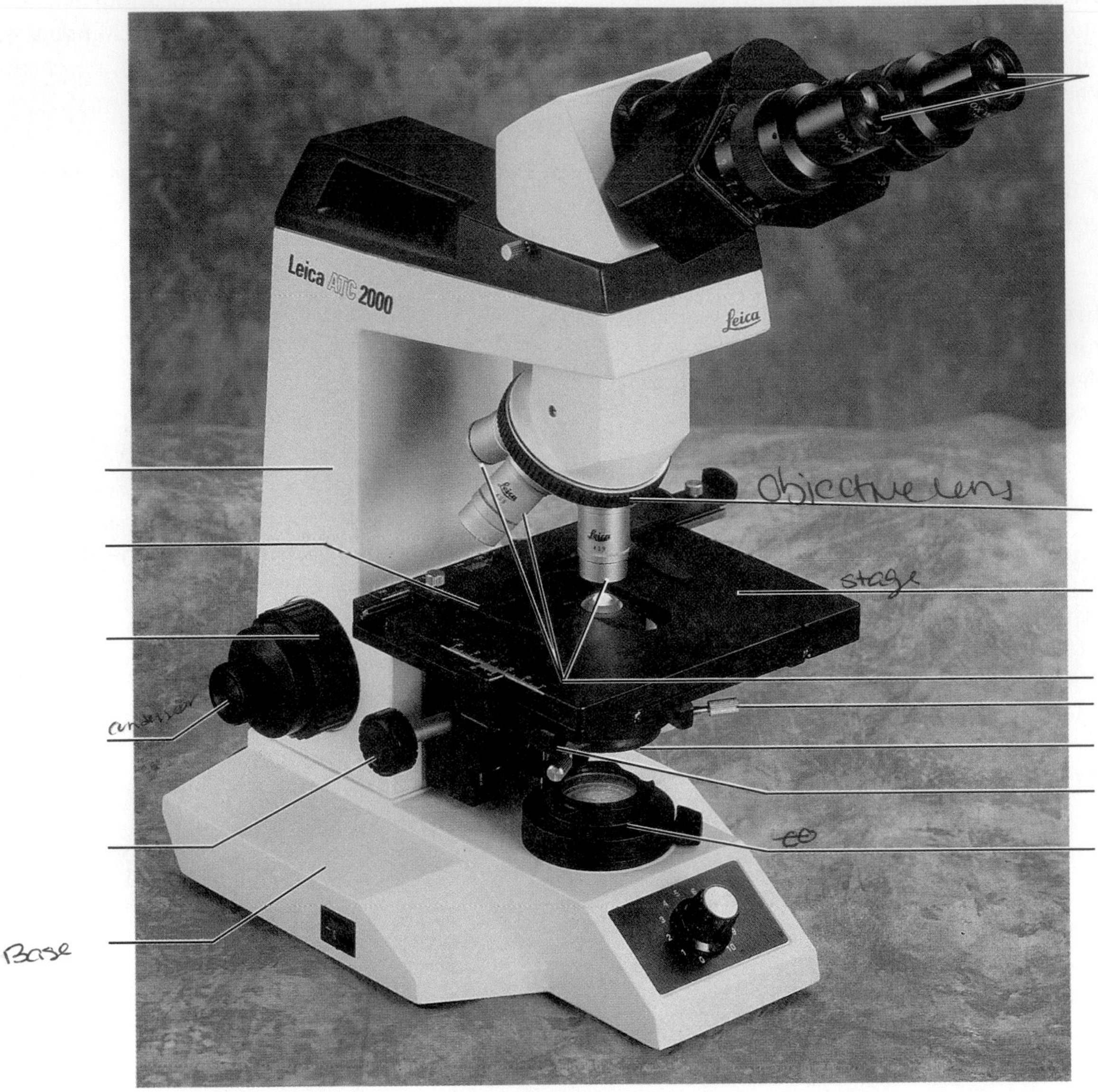

FIGURE 2.1 Parts of a compound light microscope.

- May be raised or lowered with the **condenser-adjustment knob**

Iris Diaphragm

- Located under the condenser
- Adjusts the intensity of light passing through the specimen
- Opens or closes as the **iris diaphragm lever** is moved

Coarse-Adjustment Knob

- Large knob located on the arm
- Adjusts, in large increments, the distance between the stage and objective lens
- Used initially to bring the specimen into general focus; it is dangerous to use this knob when the objective lens is already near the slide
- Should be turned very slowly to avoid breaking the slide

Fine-Adjustment Knob

- Small knob attached to the coarse-adjustment knob, on the arm
- Adjusts, in small increments, the distance between the stage and objective lens
- Typically used after the objective lens is already near the slide and the specimen is almost in focus
- Should be turned very slowly to avoid breaking the slide

Lamp

- Small light source located under the condenser
- May be turned on or off with a switch located on the base

Base

- Square- or horseshoe-shaped support for the whole microscope
- Usually quite heavy to prevent tilting ■

ACTIVITY 2

Focusing with the Microscope

Materials for this Activity

Prepared slide of the letter "e"

Obtain a letter "e" slide, and place it on the **stage.** Looking from the side of the microscope, use the **stage controls** to center the letter "e" over the hole in the center of the stage. Secure the slide with the **stage clips.** Rotate the **nosepiece** to bring the **scanning-power lens** (or the lowest power objective lens if your microscope does not have a scanning power lens) into place; listen for the "click," which indicates the objective lens is in position. To prevent damage to the lens, remember to look at the objective lens from the side while turning the nosepiece. Do not look through the ocular lens yet.

Again put your head to the side, and watch while you slowly turn the **coarse-adjustment knob** forward (clockwise), until the stage is ½ to 1 inch from the objective lens. Do not touch the slide. Observe how the stage moves upward, bringing the slide closer to the objective lens. Now put your eye to the **ocular lens.** You should see the blurry shape of the letter "e."

Sharpen the focus further in two stages: Begin by slowly using the **coarse-adjustment knob.** Finish by slowly using the **fine adjustment knob.** Notice that only a few turns of the fine adjustment knob are necessary.

Light control is another important factor to focusing an image. For example, when a slide specimen is quite thick, you may need to increase the amount of light. This may be the case if the letter "e" was printed on thick paper. If the specimen is very thin, you may find that reducing the light will produce a better contrast, and thus a better focus.

Put your head to the side and find the **iris diaphragm** lever, which is usually located on the condenser, underneath the stage. Place your finger on the lever and look at the slide. See what happens as you move the lever. In one direction, light will flood through the slide; in the opposite direction, the light will begin to dim. Later, when viewing biological specimens, you will have to play with the iris diaphragm lever to determine the amount of light needed to obtain a clear focus.

Draw the letter "e" in the circle of Box 2.2 as you observe it through the eyepiece.

Box 2.2 Drawing of Letter "e" Using Scanning (or Low) Power

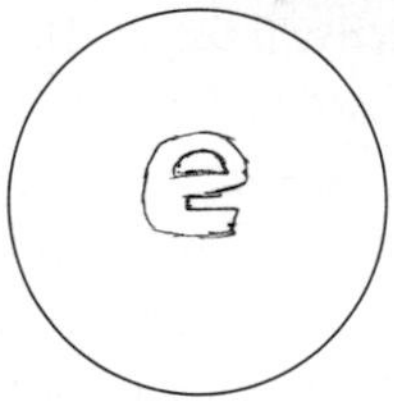

TM = 4x

1. Compare the image of the letter "e" as it is placed on the stage to its image as seen through the microscope. How does the orientation differ? Is the top side still on the top? Is the right side still on the right side?

____

____

____

2. If the light was too bright or too dim, how did you adjust it?

____

____

3. Let's say that you never focused on the letter "e." You just saw streaks and blurs. What are the possible reasons for this? What can you do differently?

____

____

Now switch to the low-power lens, and then to the high-power lens. Most microscopes are **parfocal,** which means that the image remains focused when you change objective lenses. Some minor adjustment may be necessary using the fine adjustment knob. Putting your head to the side, observe that the distance between the objective lens and the slide gets closer and closer as the power of the lens increases. When the image is focused, the **working distance** between the slide and the scanning-power lens is about ½ to 1 inch. With the low-power lens, it's about ¼ to ½ inch, and with the high-power lens, it's about ⅛ inch. In other words, at the highest magnification, there is very little distance between the lens and the slide. Slides are easily broken at this stage. *Never adjust the coarse focus when using high power!*

Sketch the entire field in Box 2.3, as observed using the three powers. Try to indicate the actual proportions of the letter "e" in relationship to the whole field. Observe how much of the letter "e" is visible at each magnification. As much as possible, draw in details like the blotchiness of the ink, the increasing amount of space around the ink, the jagged edges of the letters, and the strands and small objects of the paper itself.

Compare the three drawings.

1. At what TM was the ink thickest and most uniform?

   10

2. At what TM was the ink splotchy and inconsistent?

   40

Box 2.3 Letter "e" at Different Magnifications

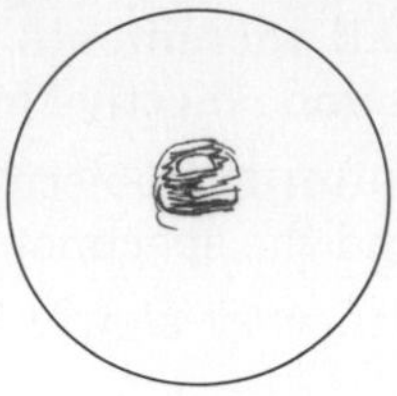

Scanning power

TM = 4

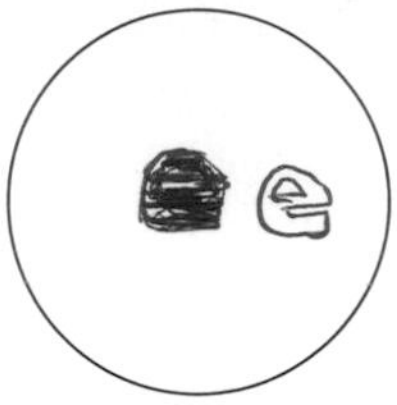

Low power

TM = 10

High power

TM = 40

3. Which power provided the clearest image?

   10

4. Was the orientation of the letter "e" the same at all powers?

   no

5. Did you have to adjust the light at each objective lens? Were substantial adjustments necessary?

   ~~The magnification lens~~)

   coarse adjustment knob

ACTIVITY 3

Determining the Depth of Focus

Materials for this Activity

Prepared slide of colored threads

It is possible to focus on different depths in the same specimen. Put your head to the side of the microscope, and observe what happens as you turn the coarse-adjustment knob *forward*, or clockwise. Generally, the *stage moves upward, which brings the slide closer to the objective lens*. This means that if you have focused the lenses on the top level of the specimen, turning the adjustment knob forward, or clockwise, enables you to focus on successively lower sections of the specimen until you have finally focused on the bottom surface of the slide. The opposite occurs when you turn the adjustment knob backward, or counterclockwise. The stage moves downward, which moves the slide away from the objective lens. This means that if you have focused on the top level of the specimen, turning the adjustment knob backward will bring the focus to successively higher levels of the slide, until you are finally focused on the top surface of the slide (i.e., the glass of the coverslip).

Obtain a prepared slide of three crossed strands of colored thread, usually red, blue, and yellow. All three strands cross at or near the same point. Center and secure the slide, and focus with the scanning-power lens.

Turn the fine-adjustment knob backward, or counterclockwise, and note the order of the colored threads as they come into focus. When the specimen goes out of focus, you are at the top surface of the slide. Now, turn the fine-adjustment knob forward, or clockwise, and note the threads' order of appearance. The order should be reversed. Record the sequence and repeat the process.

When this exercise is done too quickly and too often, you may experience some nausea or vertigo. Simply rest with your eyes closed.

Try these steps using the low-power lens, then the high-power lens. Note the additional sharpness of the colors and individual strands of the thread.

1. Which lens was the most effective?

   Low

   ___

2. If the high-power lens was not the most effective, why not?

   The picture wasn't "clear"

   ___

3. What was the high-power lens useful for?

   Observe the individual strings

   ___

4. What is the order of the threads, from top to bottom?

   Blue, red, yellow

   ___

5. What was the most confusing part of this exercise?

   Adjusting on high power to

   clear the blurriness

   ___

   ___ ■

ACTIVITY 4

Preparing a Wet Mount Slide

Materials for this Activity

Paper towel

Clean slide and coverslip

Methylene blue stain or Wright's stain in dropper bottle

Flat toothpick

1. Place a small drop of methylene blue or Wright's stain on the clean slide (Figure 2.2a).
2. Gently scrape the inner surface of your cheek with the flat part of the toothpick. (Be careful not to draw blood. If blood is drawn, tell the instructor.) Transfer the cells to the slide by carefully stirring the end of the toothpick with the cheek cells in the drop of stain (Figure 2.2b). The stain serves to darken the nuclei and edges of the cheek cells, thus substantially improving the contrast between different parts of the cheek cells. Be careful to use no more than a small drop of stain, or the cells will appear too dark.

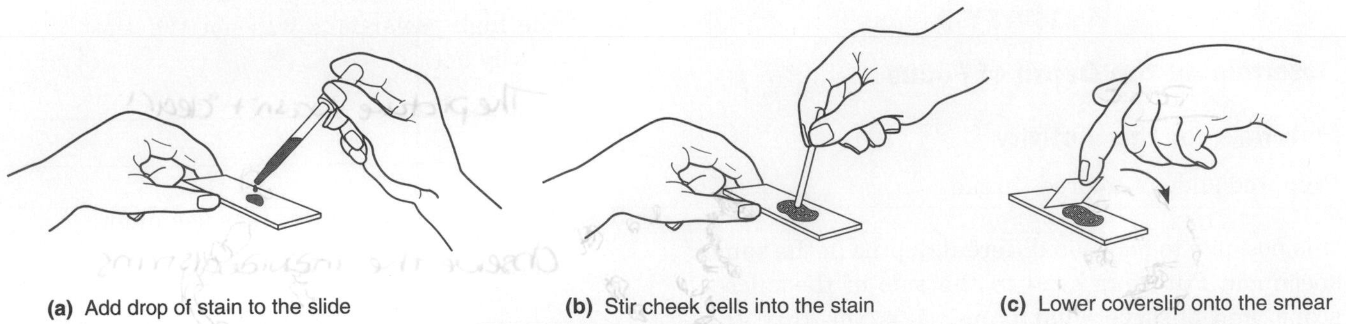

(a) Add drop of stain to the slide
(b) Stir cheek cells into the stain
(c) Lower coverslip onto the smear

FIGURE 2.2 Preparation of a wet mount slide.

3. Immediately dispose of the toothpick in the autoclave bag.
4. Hold the coverslip by the sides to avoid getting fingerprints on it. Place one edge of the coverslip next to one edge of the smear, then lower it slowly into the smear (Figure 2.2c). By lowering it slowly you will avoid forming air bubbles. Absorb any excess fluid around the edges of the coverslip and on the slide with a twist of paper towel. Dispose of the paper towel in the autoclave bag.
5. Place the cheek smear slide onto the stage of your microscope. Focus first with the scanning-power lens. Scan for some individual cheek cells with the scanning-power lens. Look for cells that are separated from the other cells and are not folded at the edges. Notice that each cell has a distinctive dark center called the nucleus. Draw one of these cells in Box 2.4, under *Individual Cheek Cells*, in the circle labeled *Scanning power.* Be sure to include the proportion of the cell to the whole visual field, including details like the shape of the cell and nucleus and if the cytoplasm appears clear, cloudy, or grainy.
6. Change to the low-power lens, and draw the cell in the appropriate circle in Box 2.4. Remember this is a parfocal microscope, and the lens change should not involve any major adjustments of the microscope. Finally, move on to the high-power lens, and draw the cell again in the appropriate circle.
7. Return to the scanning-power lens, and scan for a cluster of cheek cells. Again, look for a cluster that is separate from the other clusters and that is not crumpled or squeezed at the edges. The nucleus of each cell should be approximately in the center. Draw the cluster of cheek cells in Box 2.4, under *Grouped Cheek Cells* and the appropriate magnification.
8. Repeat this process for the low- and high-power lenses.

After completing the table on cheek cells, study it for a moment. Compare the view and details at the different magnifications. What are the changes? Look at the drawings of the grouped cells. How do they hold together? At increased magnifications, is this question answered?

Imagine how the cheek cells are arranged inside your mouth. Does it make sense that they form the mouth lining or that they can be scraped off so easily with a toothpick? When you chew an apple or a piece of lean meat, how do the cheek cells protect the mouth?

1. What is the shape of the individual ~~cheek cells~~ Pond water?

   no particular shape

2. Do you think ~~cheek cells~~ pond are thick or thin?

   thick

3. What does the thickness of a ~~cheek cell~~ pond tell you about its function?

   Better protections

4. Was the nucleus centrally or peripherally located?

   peripherally

5. When the cells are spread out, how many layers did you observe?

   two-three

Box 2.4 Cheek Cells at Different Magnifications

Individual ~~Cheek C~~ells Pond

Scanning power **Low power** **High power**

~~Grouped Cheek Cel~~ls

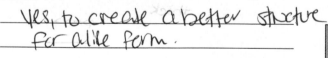

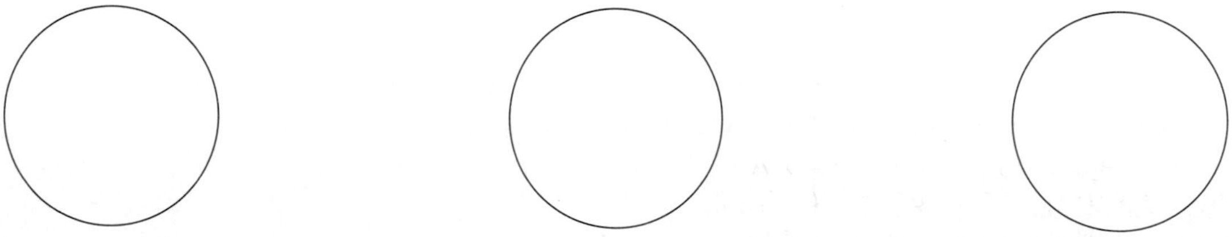

Scanning power **Low power** **High power**

6. Do you think the natural ~~cheek structur~~e pond contains one or more layers of cells? Give a reason for your answer.

   no because the more organism, the more layers are formed

7. How do you think the ~~cheek cel~~ls pond cell connect to each other?

   Yes, to create a better structure for alike form.

Cleanup and Storage of Equipment

After you are satisfied with your drawings, begin to clean up your work area. The toothpicks and soiled paper towels should have been placed into the autoclave bag at the instructor station. The glassware (slides and coverslips) now should be placed into a jar containing a bleach solution.

Look at the microscope, and wipe down the stage, lens, and arm with lens paper and cleaner. *Do not use paper towels!* Make sure that all traces of saliva and stain are removed. Prepare the microscope for storage (Box 2.5). Using two hands, carry and store the microscope in the appropriate cubicle.

Finally, wipe down your counter with a disinfectant and paper towels in order to leave a clean work surface for the next student.

Box 2.5 Checklist for Proper Microscope Storage

- ❑ Store the microscope with the lowest objective lens in place.
- ❑ Store the microscope with the lamp cord around the arm, not the objective lens.
- ❑ Report microscope problems, then use another microscope.

DO NOT STORE DAMAGED MICROSCOPES!

EXERCISE

3

THE ANATOMY AND DIVERSITY OF CELLS

Objectives

After completing this exercise, you should be able to

1. Explain why cells are the basic unit of life.

2. Identify (on a model or in a diagram) the major cellular organelles and describe their functions.

3. Explain why cellular diversity is necessary for the human body.

Materials for Lab Preparation

Equipment and Supplies

- ❑ Compound light microscope

Prepared Slides

- ❑ Kidney
- ❑ Sperm
- ❑ Skeletal muscle
- ❑ Multipolar neuron

Introduction

The basic units of life are the **cells,** which separate the disordered, random chemistry of *nonliving* things from the ordered biochemistry of *living* things. Although many primitive cells such as bacteria do not have a large central **nucleus,** all cells have a **plasma membrane** to encase their special biochemistry. Cell membranes keep their complex chemistry discrete from contamination by the chemistry of the nonliving world.

Organelles

Within the plasma membrane are various **cytoplasmic organelles** that are responsible for particular life functions (Figure 3.1). **Organelle** is the general name given to cellular structures made of complex macromolecules. Each has a job essential to the life of complex cells, such as producing energy or directing cell division. The basic study of cells involves the study of membranes and organelles.

There are two basic categories of organelles. **Membranous organelles**, as the name implies, are cellular structures wrapped in phospholipid bilayers. The **nonmembranous organelles** are typically made of complex proteins. See Table 3.1 for a summary of the most important organelles.

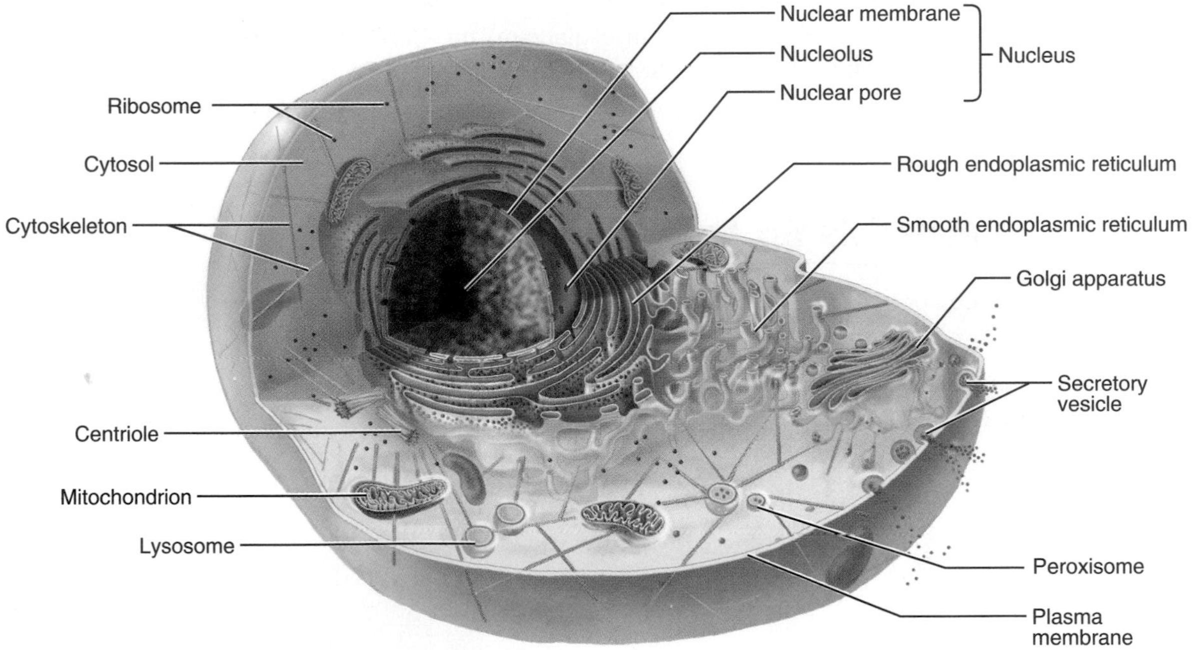

FIGURE 3.1 Generalized animal cell.

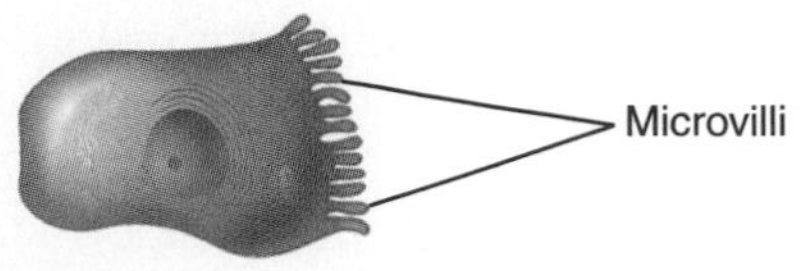

FIGURE 3.2 A single cell with its outer membrane of microvilli.

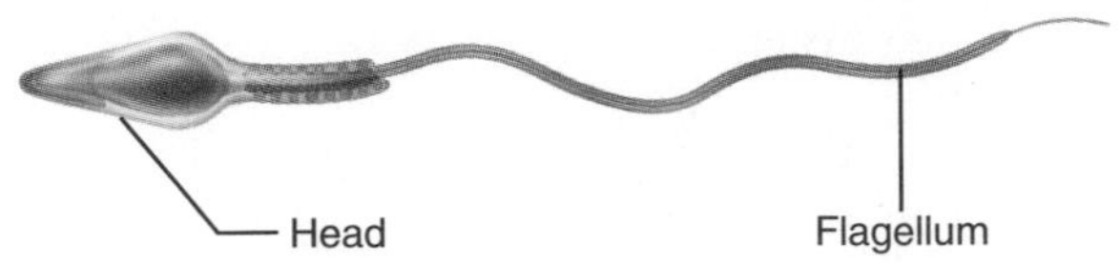

FIGURE 3.3 A mature sperm.

Table 3.1 Important Cell Structures and Organelles

Organelle	Description
Membranous	
Nucleus	Usually a spherical membrane shell that houses DNA and makes RNA for use in directing protein synthesis. By controlling what proteins are made, the nucleus controls most cellular functions.
Endoplasmic reticulum (ER)	"Mazelike" network of membrane often spanning out from the nucleus. The rough ER is dotted with ribosomes and is often an industrial complex for protein production. The smooth ER is a site where some of the proteins produced do their jobs and also where detoxification, materials processing, and lipid synthesis occur.
Golgi apparatus	Site for packaging and special processing of molecules (mainly proteins) for export out of the cell.
Mitochondria	Nicknamed "the powerhouse of the cell," these membranous energy transducers convert molecular energy from one molecule to another (usually from sugars, lipids, and amino acids to ATP).
Vacuole	Typically a large bubble of membrane used for storage inside a cell.
Vesicle	Small membrane spheres with various storage or transport functions.
Lysosome/peroxisome	Vesicle or small vacuole containing digestive enzymes.
Nonmembranous	
Cytoskeleton	Network of protein fibers and tubules that supports and moves the cell.
Flagella and cilia	Whip-like projections composed of contractile proteins; important for moving a cell (e.g., sperm) or moving other substances in a multicellular organism (e.g., ciliated epithelium of the human respiratory tract).
Ribosomes	Mixture of rRNA and proteins; worksites for protein synthesis.
Centrioles	Part of the larger centrosome, responsible for directing many aspects of cell division in most animal cells.
Microvilli	Folds in the plasma membrane to increase surface area (e.g., for absorption, transport).

ACTIVITY 1

Identifying Organelles

Using Figures 3.1, 3.2, 3.3, and Table 3.1, match the following organelles to their descriptions.

______ Mitochondria	a.	vesicle containing digestive enzymes
______ Lysosome	b.	converts molecular energy from sugars to ATP
______ Golgi apparatus	c.	contains centrioles; directs parts of cell division
______ Centrosome	d.	packages materials for export out of the cell
______ Smooth ER	e.	composed of contractile proteins that move a cell
______ Rough ER	f.	network of protein fibers/tubules that supports the cell
______ Flagella	g.	membranous organelle associated with protein synthesis
______ Cytoskeleton	h.	membranous organelle associated with lipid synthesis ■

Diversity of Human Cells

Although all human cells come from the original, generic fertilized egg, human development eventually produces a wide variety of cells. Oocytes are very large, and sperm are very small. Muscle cells are long and slender, and neurons are highly branched. The reason why each of the cells you will be viewing is so different from the others is that each cell has been modified to its own specialized task. You will learn more about these cells and what they do in later exercises.

ACTIVITY 2

Observing Cellular Diversity

Materials for this Activity

Compound light microscope

Prepared Slides

Kidney

Sperm

Skeletal muscle

Multipolar neurons

1. Obtain microscope slides of the kidney, sperm, skeletal muscle, and multipolar neurons.
2. View each slide using the high-power (high-dry) magnification on your microscope.
3. Sketch a few of the representative cells from each slide that you view in the blank boxes provided with Figures 3.4, 3.5, 3.6, and 3.7. Remember to record the magnification used to view the slide in the space provided.
4. For each slide you observed, note in the appropriate section of Box 3.1 a description of it, for example, its shape, size, arrangement, or other characteristics.

   Keeping in mind the notion that "form follows function," choose each cell's job from the following list and enter it in the function column of Box 3.1.

   Movement through fluid

   Contraction

   Communication

   Absorption and secretion ■

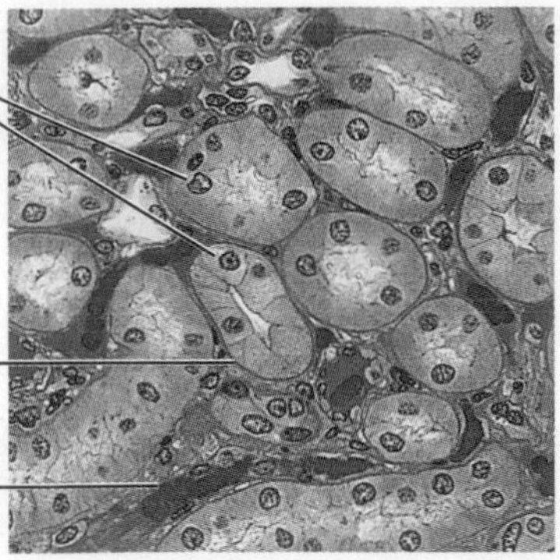

Magnification

__________ ×

FIGURE 3.4 Simple cuboidal epithelium in kidney tubules (270×).

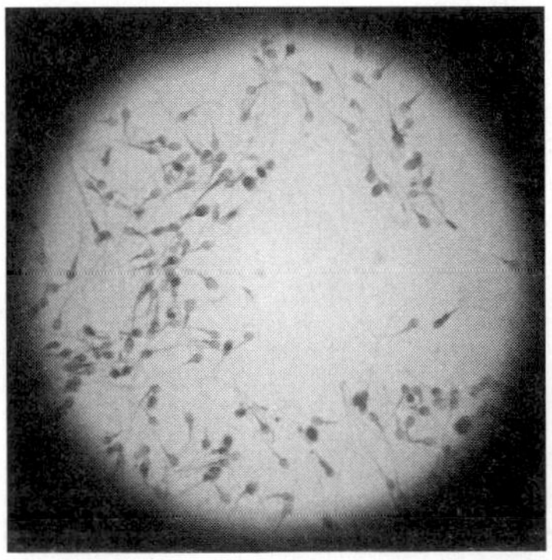

Magnification

__________ ×

FIGURE 3.5 Sperm cells from a sperm smear (430×).

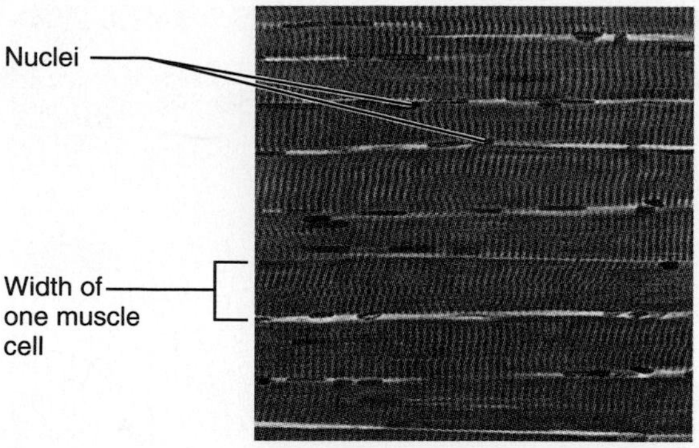

Magnification
__________ ×

FIGURE 3.6 Skeletal muscle (~100×).

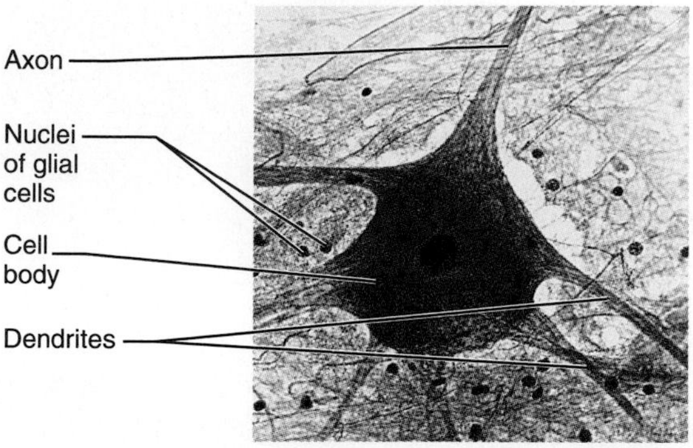

Magnification
__________ ×

FIGURE 3.7 Multipolar neuron (170×).

Box 3.1 Cellular Diversity

Cell	Brief description of characteristics	Function
Kidney		
Sperm		
Skeletal muscle		
Neuron		

Section ______________________ Name ______________________

Date ______________________ Instructor ______________________

EXERCISE 3

THE ANATOMY AND DIVERSITY OF CELLS

Review Questions

1. Why is the plasma membrane so important to cells and to life in general?

2. List four membranous organelles found in cells, and briefly describe them.

3. List four nonmembranous organelles found in cells, and briefly describe them.

4. List four organelles that produce or process materials for the cell, and briefly describe them.

5. Do you think all cells have roughly the same number of mitochondria? Explain your answer.

6. Why is cellular diversity important in building complex organisms such as the human body?

____

____

____

____

7. Choose two of the cells you examined in this lab, and describe how each cell's shape and/or size fits its function.

____

____

____

____

EXERCISE

4

CELL PHYSIOLOGY

Objectives

After completing this exercise, you should be able to

1. Describe the basic functions of cell membranes.
2. Understand how substances move across membranes.
3. Explain the concepts of diffusion and osmosis.

Materials for Lab Preparation

Supplies

- ❑ Petri dish with agar gel
- ❑ Methylene blue crystals
- ❑ Potassium permanganate crystals
- ❑ Two sets of forceps
- ❑ China marker
- ❑ Ruler
- ❑ Precut, presoaked dialysis tubing (4-inch segments)
- ❑ Thread
- ❑ 200-ml beakers
- ❑ Clinistix or other suitable sugar testing strips
- ❑ Small test tube
- ❑ Fresh potato
- ❑ Knife
- ❑ Paper towels
- ❑ Sample of whole blood (ox blood recommended)
- ❑ Clean microscope slides and coverslips
- ❑ Medicine dropper
- ❑ Biohazard container

Equipment

- ❑ Triple-beam balance or other scale with sensitivity of at least 0.1 g
- ❑ Compound light microscope

Solutions

- ❑ Mixture of 5% glucose and 1% boiled starch solution in squeeze bottle
- ❑ Iodine solution in dropper bottle
- ❑ Distilled or deionized water
- ❑ 2.5% saline
- ❑ Isotonic (0.9%) saline

Introduction

As you learned in the previous exercise, cells are the basic units of life, and their **plasma membranes** are the structures that keep their complex chemistry discrete from contamination by the chemistry of the nonliving world. But there are times when substances need to move in and out of cells, and the plasma membrane regulates that process as well.

The Plasma Membrane

The plasma membrane is the most essential component of a cell because it encases and protects the complex chemistry of life inside the cell. Its main component is a molecule called a **phospholipid.** Phospholipids are unusual "hybrid molecules" that are made up of a polar "head" and a nonpolar "tail." The polar head is attracted to water, while the neutral tails are repelled by water. In the plasma membrane, these molecules form two layers, where the polar heads form the outside layers and the hydrophobic tails make up the interior. This structure, which allows certain molecules to pass through while keeping others out, is the **phospholipid bilayer.** As such, the plasma membrane is **selectively permeable.**

Cells can modify their plasma membranes by making protein channels that span the membrane. These protein channels allow for the movement, or **diffusion,** of specific items, such as ions and polar molecules, from areas of high concentration to low concentration. Without these channels, the ions and polar molecules would not be able to pass across the plasma membrane (Figure 4.1).

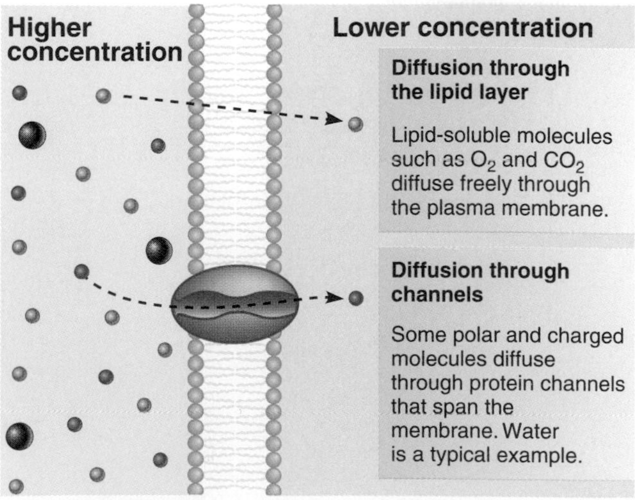

Higher concentration
Lower concentration
Diffusion through the lipid layer
Lipid-soluble molecules such as O2 and CO2 diffuse freely through the plasma membrane.
Diffusion through channels
Some polar and charged molecules diffuse through protein channels that span the membrane. Water is a typical example.

FIGURE 4.1 Two forms of passive transport. Both involve transport down a concentration gradient without the expenditure of additional energy.

Passive transport is the name for the process of simple diffusion across the plasma membrane. Lipid-soluble items pass right through the phospholipid bilayer, while items that are not lipid soluble (ions and polar molecules) must utilize the protein channels, or carriers (if present) to pass through. Molecules of different sizes tend to diffuse at different rates, as you will see in the following activity.

ACTIVITY 1

Measuring Diffusion Rates of Dye Through an Agar Gel

Materials for this Activity

Petri dish with agar gel
Methylene blue crystals
Potassium permanganate crystals
Two sets of forceps
China marker
Ruler

1. Obtain a Petri dish containing a layer of agar gel. With your forceps, carefully place a small amount of potassium permanganate on the agar gel near one edge of the dish (Figure 4.2). *Note:* Use a toothpick or similar tool to create a well in the agar gel. This will prevent the crystals from scattering.
2. Using a different set of forceps, place a small amount of methylene blue on the agar gel near the opposite edge.

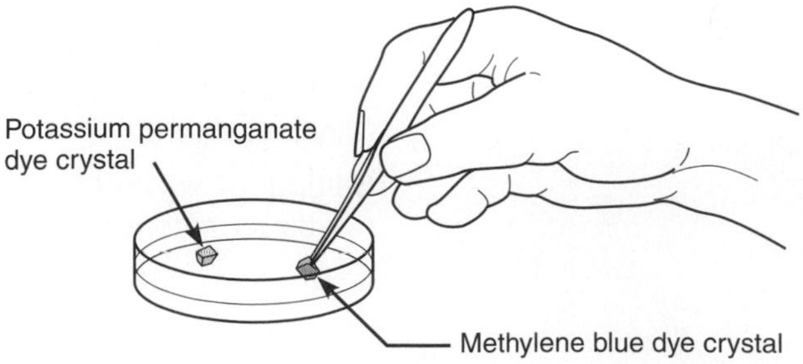

FIGURE 4.2 Setup for measuring diffusion rates using methylene blue and potassium permanganate dyes through an agar gel.

3. Use a china marker to put a circle on the underside of the Petri dish (without turning the petri dish upside down) to mark the border of your placement of these materials.
4. At 15-minute intervals, measure the distance the dye has moved from your mark at time zero. Enter the data in Box 4.1.

   Which dye diffused through the agar at a faster rate?

   ____

   Compare the diffusion rates. What is the approximate diffusion distance ratio of methylene blue to potassium permanganate (e.g., 2:1, 3:1, 2:3)?

   ____

   Given that potassium permanganate has a molecular weight approximately half that of methylene blue, what is the connection between molecular size/weight and diffusion rate?

   ____

   ______________________________ ■

Selectively Permeable Membranes

Now that you understand diffusion, it should be clear that only the wall of a solid container stops most molecules once they head toward the area of low concentration. But what if a porous wall or selectively permeable membrane were placed in the way of these molecules? Would they pass through the barriers or bounce off them? Common sense suggests that passing through will depend at least partly upon the size (which is proportional to the molecular weight) of the molecule versus the size of the pores in the membrane. Visualize straining cooked spaghetti. Water passes through the holes in the strainer/colander, but the spaghetti noodles are trapped on one side. The colander is like a selectively permeable membrane.

There are factors other than size (such as the polarity of the molecules) that determine whether molecules move through a membrane. But because those factors are more difficult to test in the laboratory, we will focus on the molecular size issue for the following experiment.

Box 4.1 Distance Dye Moved Over Time

Time	Methylene blue	Potassium permanganate
15 minutes	10 ml	20 ml
30 minutes	12ml	21ml
45 minutes	12m	21m
60 minutes	13 ml	22 ml

ACTIVITY 2

Observing Movement Through a Selectively Permeable Membrane

Materials for this Activity

Precut, presoaked dialysis tubing (4-inch segments)
Thread
200-ml beaker
Mixture of 5% glucose and 1% boiled starch solution in squeeze bottle
Clinistix or other suitable sugar testing strip
Iodine solution in dropper bottle
Small test tube

1. Obtain a piece of presoaked dialysis tubing about 4 inches long. Fold about 1 inch of the bottom end of the tube. Use thread to tie the middle of the folded area. Tie it as tightly as you can, and then tie this same area with another piece of thread. This duplication will ensure that you have made your tube into a bag that will not leak.
2. Open the top of the bag and insert the spout of a squeeze bottle of glucose/starch solution into the open end of your bag. Add the solution to the bag until it is about two-thirds full. Twist the bag closed about 1 inch from the top of the bag, and tie it with thread as you did for the bottom. You may find that a helping hand from a partner will be useful during this step.
3. Rinse the outside of the bag with tap water, and gently blot it dry on a paper towel. Place the bag in a 200-ml beaker that is about half full of tap water. Figure 4.3 shows how your setup should look.
4. Wait 40 minutes, and then test the water using the following procedures:

Sugar test: Obtain a Clinistix strip or other suitable testing strip for sugar. Dip the strip into the water according to the directions on the product's container.

Is there sugar in the water? Yes

Starch test: Pour some of the water from the beaker into a clean test tube. About 1 inch of water in the test tube should be sufficient. Obtain a dropper bottle of iodine solution, and add several drops to the water in the test tube. If the liquid in the test tube is yellow-orange, there is no starch in the water (a negative starch test). If the liquid turns a dark color, usually navy blue or black, there is starch in the water (a positive starch test).

Is there starch in the water? NO

Which molecules, sugar or starch, were small enough to pass through the membrane (the dialysis tubing bag)? Which molecules were too large to pass through the membrane?

sugar. starches are too large.

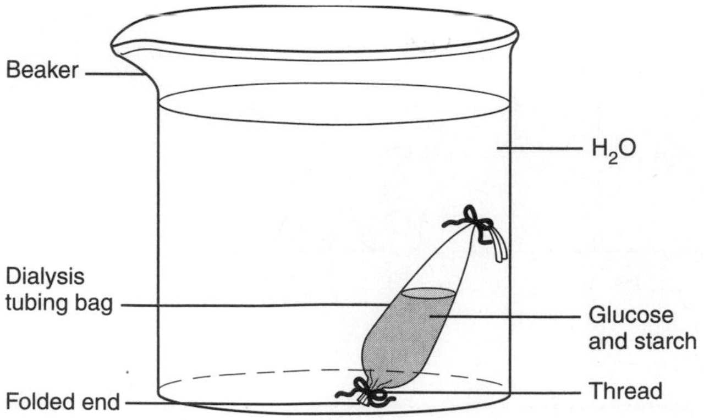

FIGURE 4.3 Setup for observing diffusion rates through a nonliving selectively permeable membrane.

Osmosis—Part One

Water easily passes through the pores in a selectively permeable membrane. However, as you have already discovered, other molecules do not always do so. Could these other molecules, which are trapped on one side of the membrane, affect the movement of water molecules across the membrane? The net diffusion of water across a selectively permeable membrane, known as **osmosis,** is driven by solutes affecting the concentration of water. (Consult your textbook for a more complete discussion of osmosis.) Where there is more solute, there is less water available. Like all molecules, water will move toward its area of lower concentration.

Tonicity is the term used to describe the solute concentration in the fluid surrounding a cell. An **isotonic** fluid has the same solute concentration as the cell. **Hypotonic** fluids have lower solute concentrations, and **hypertonic** fluids have higher solute concentrations. Keeping in mind that more solute means less water, which fluid would have more water, the hypertonic or hypotonic solution?

ACTIVITY 3

Observing Osmosis in Potato Cubes

Materials for this Activity

Fresh potato

Knife

Distilled or deionized water

200-ml beakers

Forceps

China marker

Triple-beam balance or other scale with sensitivity of at least 0.1 g

Solutions

2.5% saline

Isotonic saline

1. Peel and cut the fresh potato into three cubes that are approximately 1 inch in size.

2. Obtain three 200-ml beakers, and label each with a china marker as follows: the first as 0.9% saline (isotonic), the second as distilled water (hypotonic), and the third as 2.5% saline (hypertonic). Add the appropriate solution to a level of about 1 inch.

3. With your forceps, dip one potato cube into the first beaker, remove promptly, pat dry with a paper towel, and weigh as accurately as possible. Record this weight as the initial weight in Box 4.2.

Box 4.2 Weight of Potato Cubes

Beaker solution	Initial weight	Final weight	
0.9% saline (isotonic)	~~7.14 g~~ 8.80 g	~~8.76~~ 7.57	1.23
Distilled water (hypotonic)	~~11.43~~ ~~11.08 g~~ 10.80	~~10.49~~ 10.03	.79
2.5% saline (hypertonic)	~~10.85~~ 9.74 g	~~9.66~~ 9.45 g	.29

4. Repeat Step 3 for the other two cubes and beakers. (*Note:* this step is to reduce experimental error regarding dry and wet cubes.)

5. Place the potato cubes back in their respective beakers for 1 hour. Remove, pat dry, and weigh them as accurately as possible. Record this weight in Box 4.2.

 Which potato cube gained weight? _none_

 Which potato cube lost weight? _all_

 Which potato cube remained closest to its original weight? _saline_

6. In Figure 4.4, draw an arrow indicating the direction in which diffusion of water is occurring: into the cube, out of the cube, or equally in both directions. Explain your results below.

 all three lost weight

Isotonic Distilled water 2.5% Saline

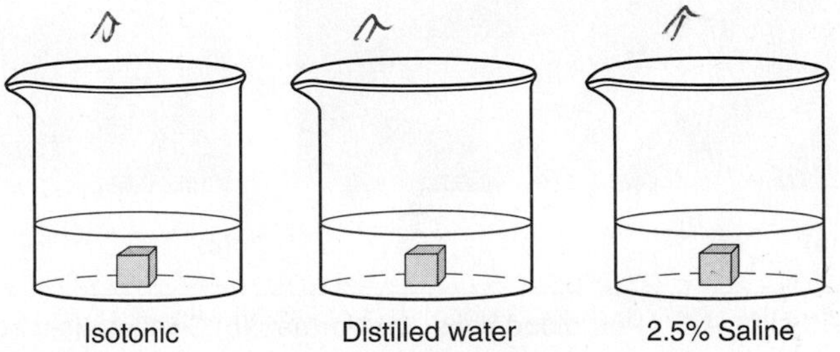

FIGURE 4.4 Osmosis in potato cubes.

Osmosis—Part Two

In the previous experiment, you established that water was moving either into or out of the cells in the potato cube depending upon the solute concentration in the water that surrounded it. But what of human cells? It is true that plant cells respond to the changes driven by osmosis differently than animal cells. For example, the cell wall of a plant cell prevents the cell from bursting when excess water enters from the surrounding environment, but animal cells have no cell walls and would burst. The same overall process occurs in all cells. Imagine what would happen to trillions of cells in your body if your body fluids were permitted to become too dilute. Keeping cells from shriveling up or bursting is a major reason behind the mechanisms the human body has in place for carefully controlling water and solute level (e.g., kidney function). In the next activity, we will demonstrate this principle with red blood cells.

ACTIVITY 4

Observing Osmosis in Red Blood Cells

Materials for this Activity

Sample of whole blood (ox blood recommended)
Distilled or deionized water
Clean microscope slides and coverslips
China marker
Medicine dropper
Compound light microscope
Biohazard container or other suitable disposal vessel for the slides

Solutions

2.5% saline
Isotonic saline

1. Set three clean, dry slides on your work area, and label them near one edge as "A," "B," and "C," respectively, with a china marker.
2. On slide "A" place a drop of isotonic saline solution. On slide "B" place a drop of distilled water. On slide "C" place a drop of 2.5% saline solution.
3. Using the medicine dropper, add a small drop of blood to each slide, and rock each slide gently to mix the solutions. Place a coverslip over the liquid on each slide. Allow an immersion time of 5 minutes before examining slides "A," "B," and "C" using your microscope.
4. Now use your microscope to look at slide "A," and carefully focus the image of the red blood cells using your high-power objective lens. The cells should have a normal "biconcave" shape, like a jelly doughnut slightly pushed in at the center (Figure 4.5a). Summarize the results in Box 4.3.

   Review what you did in the potato cube experiment when you placed the potato cube in the isotonic saline solution. Look at the arrow(s) you were asked to draw in Figure 4.4 for the potato cube in the isotonic saline solution. Keeping in mind that the same *process* is occurring here, briefly describe what is happening to the cells on slide "A".

   ____

   ____

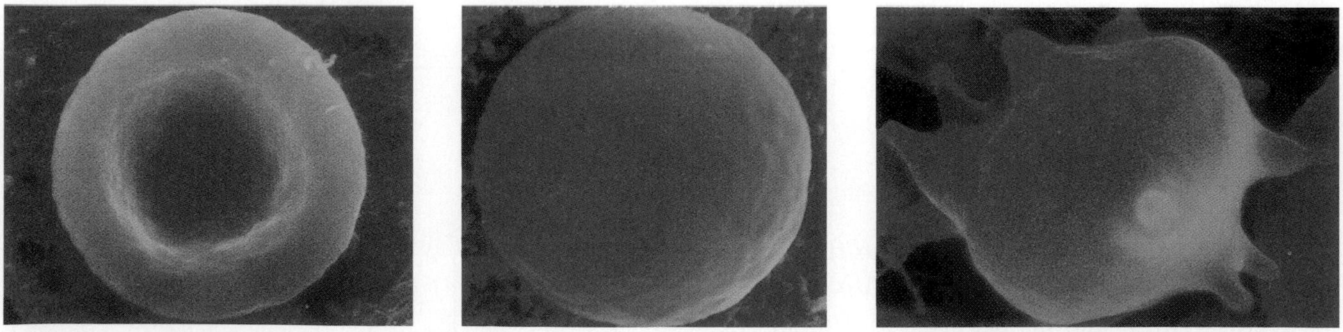

(a) (b) (c)

FIGURE 4.5 Red blood cells. (a) Normal. (b) Swollen. (c) Crenated.

5. Next, use your microscope to look at slide "B," and carefully focus the image of the red blood cells using your high-power objective lens. The cells should have taken on a somewhat swollen shape (Figure 4.5b). The bursting of these cells that normally follows is referred to as **hemolysis.** Let the slide sit for another 15 minutes, and then look at it a second time. Compare the two views in terms of estimating the percentage of size change, and describe the shape change. Summarize the results in Box 4.3.

   Review what you did in the potato cube experiment when you placed the potato cube in distilled water. Look at the arrow(s) you were asked to draw in Figure 4.4 for the potato cube in distilled water. Keeping in mind that the same *process* is occurring here, briefly describe what is happening to the cells on slide "B".

6. While waiting to look at slide "B" a second time, look at slide "C" using your microscope, and carefully focus the image of the red blood cells using your high-power objective lens. Now let this slide sit for another 15 minutes, and then look at it a second time. Compare the two views in terms of estimating the percentage of size change and describing the shape change. The cells should have taken on a somewhat wrinkled shape (Figure 4.5c). This wrinkling is referred to as **crenation.** Summarize the results in Box 4.3.

   Review what you did in the potato cube experiment when you placed the potato cube in the 2.5% saline solution. Look at the arrow(s) you were asked to draw in Figure 4.4 for the potato cube in the 2.5% saline solution. Once again, keeping in mind that the same *process* is occurring here, briefly describe what is happening to the cells on this slide.

Box 4.3 Results of Osmosis in Red Blood Cells

Solution added	Appearance of cells	Net movement of water
0.9% saline (isotonic)	Clear	
Distilled water (hypotonic)	Cloudy	
2.5% saline (hypertonic)	Cloudy	

EXERCISE

5

TISSUES

Objectives

After completing this exercise, you should be able to

1. Explain why tissues are a crucial link between cells and organs.
2. List the four basic tissue types and describe them.
3. List the five major epithelial tissues and describe them.
4. List the four major connective tissues and describe them.
5. List the three kinds of muscle tissue and describe them.
6. List the two categories of nervous tissue and describe them.

Materials for Lab Preparation

Equipment and Supplies

- ❑ Compound light microsope

Prepared Slides

- ❑ Simple squamous epithelium/lung
- ❑ Stratified squamous epithelium/esophagus
- ❑ Simple cuboidal epithelium/kidney
- ❑ Simple columnar epithelium/small intestine
- ❑ Pseudostratified ciliated columnar epithelium/trachea
- ❑ Areolar connective tissue
- ❑ Dense white fibrous connective tissue
- ❑ Hyaline cartilage
- ❑ Adipose tissue
- ❑ Skeletal muscle
- ❑ Cardiac muscle
- ❑ Smooth muscle
- ❑ Multipolar neuron smear

Introduction

As you learned in the cell exercise, there is quite a bit of specialization and diversity among the cells of the human body. But one lonely cell cannot do much for an organism the size of the average human. **Tissues,** or groups of similar cells working together to accomplish a basic function, form a crucial link between cells and organs.

There are four primary types of tissues in the human body. **Epithelial tissues** form linings. **Connective tissues** connect, support, protect, and store things for the human body. **Muscle tissue** contracts and produces movement. Finally, **nervous tissue** conducts electrochemical impulses important for communication.

ACTIVITY 1

Making the Link Between Cells, Tissues, and Organs

In Box 5.1, choose any six organs and list them, one in each box, under *Organ*. Under *Tissues*, list two tissues found in each organ. Under *Function of Tissue*, briefly describe what that tissue does for the organ and system to which it belongs. You may find your textbook or later chapters in this lab manual helpful as references for this exercise, but first try to reason it out based upon the description in the introduction. ■

Epithelial Tissues

Epithelial tissues form the linings of the human body. This is why we typically find them on the inside and/or outside of an organ. They are made of tightly packed cells, seldom having room for blood vessels, and are described as **avascular** (without blood vessels). Epithelial cells are attached to the tissue they are lining via a thin layer of connective tissue called the **basement membrane.**

The names of epithelial tissues may seem daunting, but the naming system is fairly simple. They are first named for their degree of layering: one layer = **simple,** two or more layers = **stratified.** Then they are named for the shape of the cell: flat = **squamous,** cube-shaped to spherical = **cuboidal,** and tall = **columnar.** So, for example, an epithelial tissue made of a single layer of flat cells would be called simple squamous epithelium.

Box 5.1 Organs, Tissues, and Functions

Organ	Tissue	Function of Tissue
Kidney	1. 2.	1. 2.
	1. 2.	1. 2.
	1. 2.	1. 2.
	1. 2.	1. 2.
	1. 2.	1. 2.
	1. 2.	1. 2.

ACTIVITY 2

Viewing and Learning About Selected Epithelial Tissues

Materials for this Activity

Compound light microscope

Prepared Slides

Simple squamous epithelium/lung

Stratified squamous epithelium/esophagus

Simple cuboidal epithelium/kidney

Simple columnar epithelium/small intestine

Pseudostratified ciliated columnar epithelium/trachea

View the various epithelial tissues described in the following paragraphs. While examining each tissue carefully, be on the lookout for characteristics that may be clues to the tissue's function.

Simple Squamous Epithelium

Simple squamous epithelium is made from a single layer of flat cells. As such, it provides the thinnest possible layer for diffusion and filtration. It is important for the air sacs of the lungs and blood capillaries to have this tissue so that molecules such as oxygen can easily diffuse through this layer of cells.

View the slide of this tissue, using your microscope, and sketch the image in the blank box provided with Figure 5.1. Remember to record the magnification. How do your observations compare with the photomicrograph in the figure?

____

____

Stratified Squamous Epithelium

Stratified squamous epithelium is made from many layers of flat cells, although the mitotic cells near the basement membrane may appear more rounded. This tissue provides protection from friction and is generally a good barrier. That is why this tissue is found lining the mouth, esophagus, vagina, and epidermis of the skin.

View a slide of this tissue with your microscope, and sketch the image in the blank box provided with Figure 5.2. Remember to record the magnification.

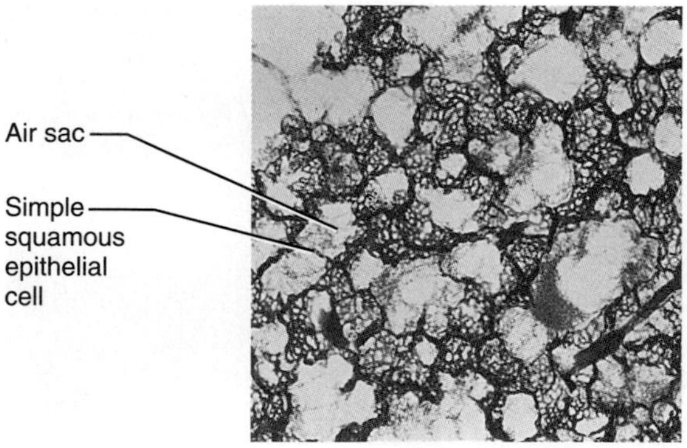

Magnification ________ ×

FIGURE 5.1 Simple squamous epithelium in normal lung alveoli (400×).

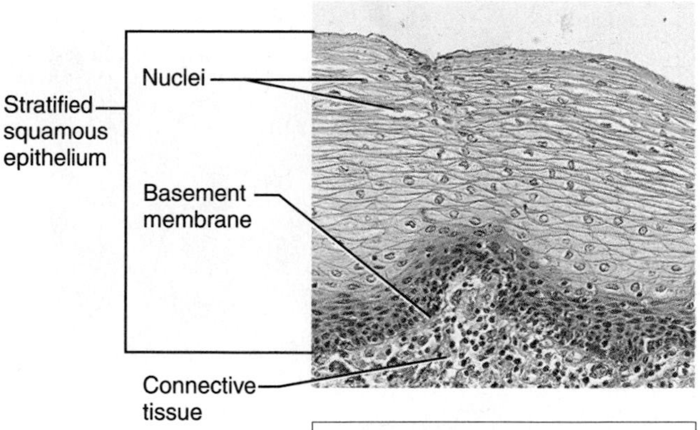

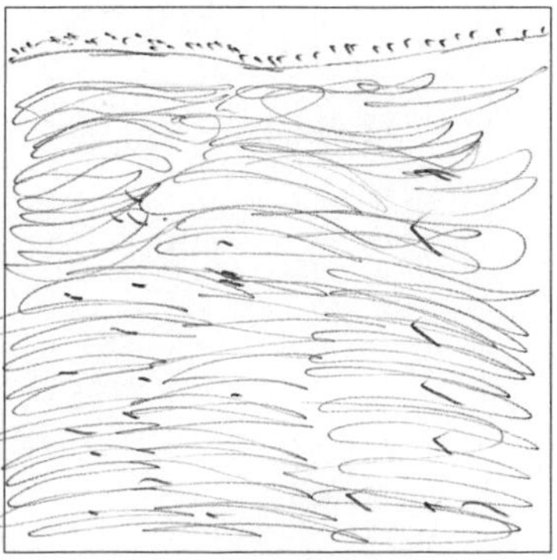

Magnification 40 ×

FIGURE 5.2 Stratified squamous epithelium in lining of esophagus (151×).

How do your observations compare with the photomicrograph in the figure?

Simple Cuboidal Epithelium

Simple cuboidal epithelium is made from one layer of rounded or cube-shaped cells. Although poorly stained membranes sometimes make it difficult to determine the exact shape of the cell, the mostly round nuclei of the cells in this tissue are easy to spot. This tissue provides an absorptive or secretory layer in organs such as the kidney and is also found in many of the glands of the human body, such as the thyroid.

View a slide of this tissue with your microscope, and sketch the image in the blank box provided with Figure 5.3. Remember to record the magnification. How do your observations compare to the photomicrograph in the figure?

Simple Columnar Epithelium

Simple columnar epithelium is made from a single layer of tall cells with oval nuclei. This tissue performs absorption, secretion, and sometimes glandular functions for the human body. The stomach, intestines, and a few miscellaneous ducts are places where this tissue is located.

View a slide of this tissue with your microscope, and sketch the image in the blank box provided with Figure 5.4. Remember to record the magnification. How do your observations compare to the micrograph in the figure?

Pseudostratified Ciliated Columnar Epithelium

Pseudostratified ciliated columnar epithelium has the appearance of several layers of tall cells, but all of these cells touch the basement membrane. So, it is technically one layer of taller and shorter cells, thus the name *pseudo*stratified. This epithelium is a

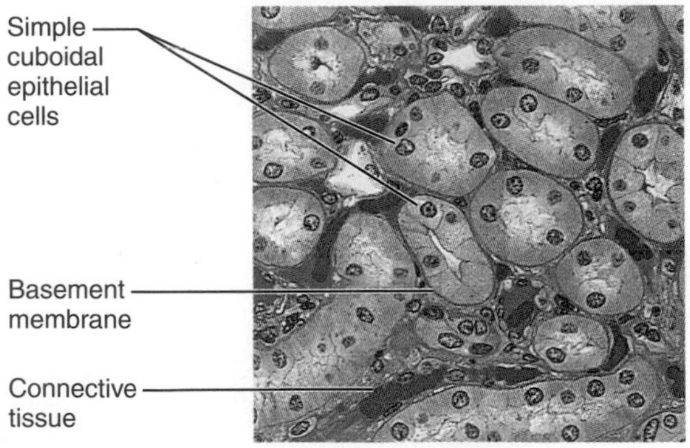

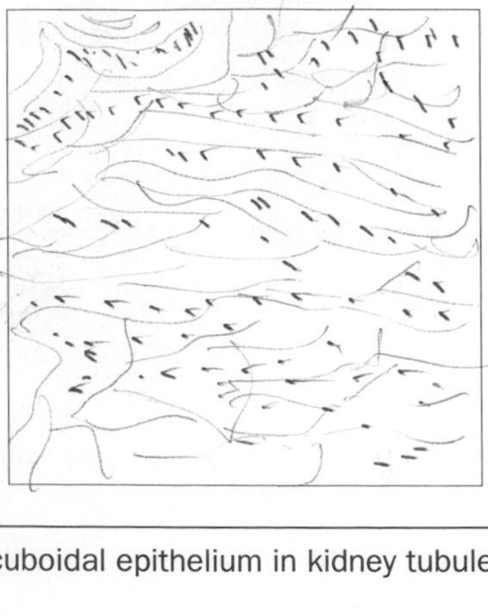

Magnification 40 ×

FIGURE 5.3 Simple cuboidal epithelium in kidney tubules (270×).

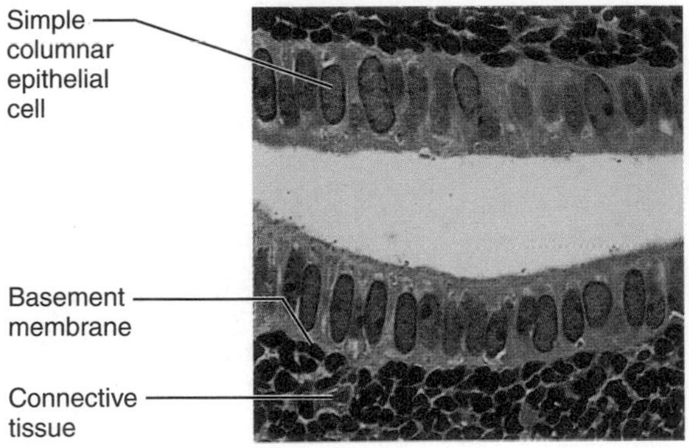

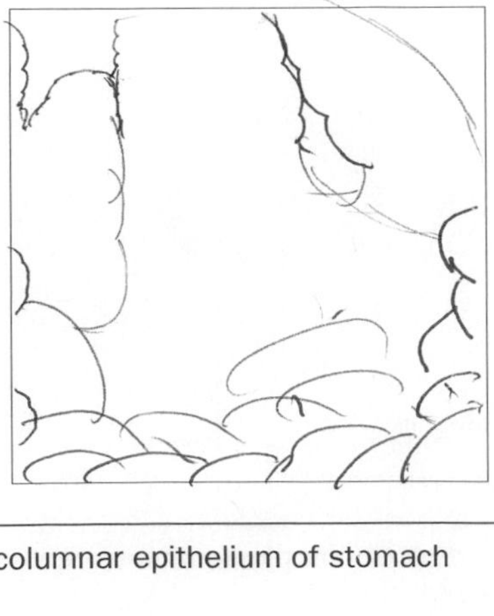

Magnification 40 ×

FIGURE 5.4 Simple columnar epithelium of stomach mucosa (500×).

specialized tissue found throughout most of the respiratory tract, where it protects the lungs by removing dirt particles and mucus.

View a slide of this tissue with your microscope, and sketch the image in the blank box provided with Figure 5.5. Remember to record the magnification. How do your observations compare to the photomicrograph in the figure?

____

______________________________ ■

Connective Tissues

Connective tissues perform a wide variety of functions for the human body. They connect, protect, store, and provide the substrate in which some structures in the body may be embedded (such as sweat glands embedded in the dermis of the skin). In many ways, connective tissues are the opposite of epithelial tissues. That is, they tend to contain comparatively few cells; instead they have a great deal of nonliving material called the **matrix** that is secreted by the cells. There are two components in the matrix: ground substance (which varies in composition from very hard to liquid) and protein fibers, such as collagen or elastin. Collagen gives strength and elastin gives flexibility.

Although many connective tissues contain protein fibers, others, such as adipose tissue, qualify for this tissue category because of their protective role as a tissue that cushions. Connective tissues are the largest and the most diverse of the four tissue groups. In the following activity, you will explore selected connective tissues. Bone, which is a classic example of a connective tissue because of its protective role, will be covered in the skeletal system. Blood, which chemically connects all parts of the body, will be covered in the circulatory system.

When viewing the different kinds of connective tissue listed in the following activity, note the diverse matrix elements found as you progress from one type to another.

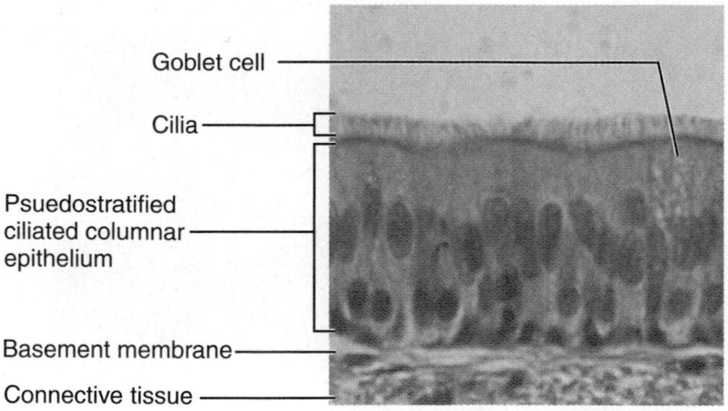

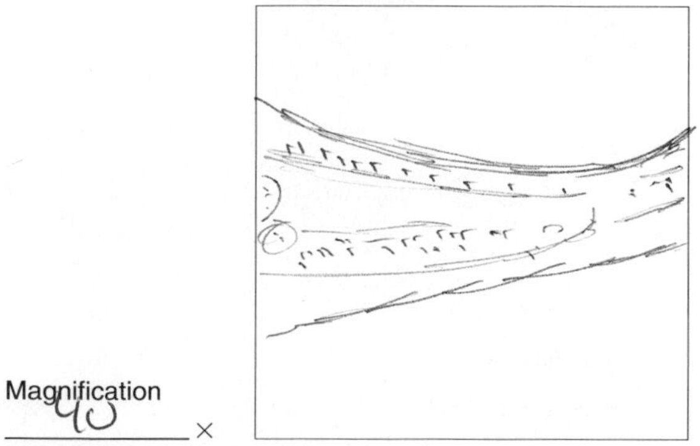

FIGURE 5.5 Pseudostratified ciliated columnar epithelium lining the human trachea (800×).

ACTIVITY 3

Viewing and Learning About Selected Connective Tissues

Materials for this Activity

Compound light microscope

Prepared Slides

Areolar connective tissue
Dense white fibrous connective tissue
Hyaline cartilage
Adipose tissue

Areolar Connective Tissue

Areolar connective tissue is made of an irregular arrangement of fibers with **fibroblasts** (fiber-producing cells) and other cells sprinkled in between them. As such, it permits flexible attachment and makes a good, loose packing material. It is found below the skin, between muscles, and around various organs.

View a slide of this tissue with your microscope, and sketch the image in the blank box provided with Figure 5.6. Remember to record the magnification. How do your observations compare to the photomicrograph in the figure?

____

____

Dense White Fibrous Connective Tissue

Dense white fibrous connective tissue is made of densely packed fibers, running mostly in the same direction, with fibroblasts wedged in between the fibers. This tissue provides strength of attachment, especially when the mechanical stress will be along a predictable direction. This is the case with ligaments, tendons, and capsules around various organs, where we find dense fibrous connective tissue.

View a slide of this tissue with your microscope, and sketch the image in the blank box provided with Figure 5.7. Remember to record the magnification. How do your observations compare to the photomicrograph in the figure?

____

____

____

Hyaline Cartilage

Hyaline cartilage is made of thick, tough collagen fibers "glued" and packed so densely that the matrix appears smooth. These fibers produce a tissue that is strong yet somewhat flexible. Sites where the ribs attach to the sternum and the ends of bones attach to movable joints are examples of places where hyaline cartilage is found. Instead of fibroblasts, hyaline cartilage has **chondrocytes** (mature cartilage-producing cells) in little holes called **lacunae.**

View a slide of this tissue with your microscope, and sketch the image in the blank box provided with Figure 5.8. Remember to record the magnification. How do your observations compare to the photomicrograph in the figure?

____

____

____

Adipose Tissue

Adipose tissue is made of large, rounded irregular cells that look empty but are really filled with a large fat vacuole. These **adipocytes** do more than store energy. They also form tissue that cushions

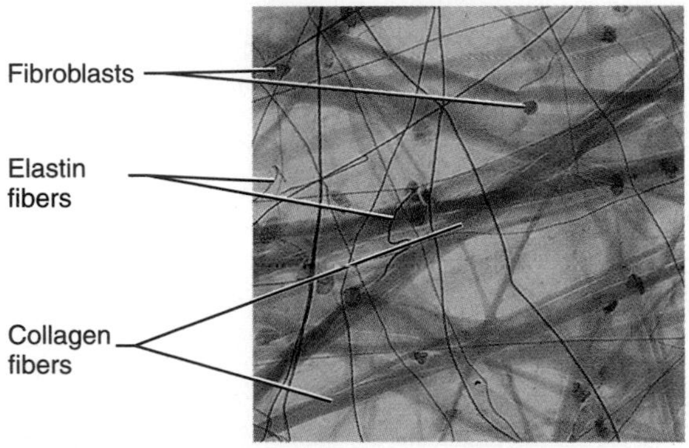

Magnification ____________×

FIGURE 5.6 Loose areolar connective tissue (160×).

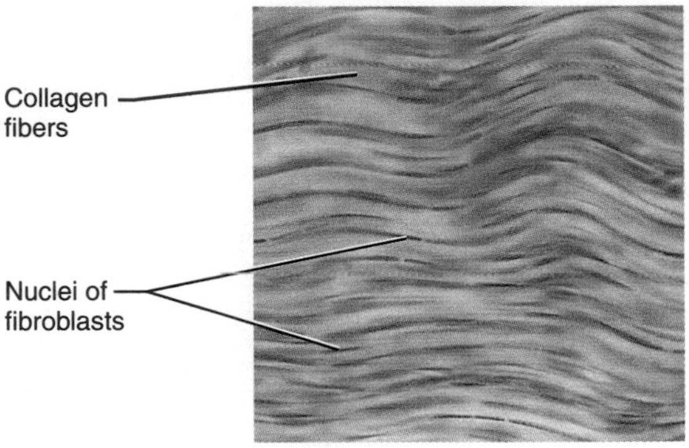

Magnification ____________×

FIGURE 5.7 Dense white fibrous connective tissue (160×).

and insulates. That is why adipose tissue is located below the skin and around various organs.

View a slide of this tissue with your microscope, and sketch the image in the blank box provided with Figure 5.9. Remember to record the magnification. How do your observations compare to the photomicrograph in the figure?

____

____

____ ■

Muscle Tissue

Muscle tissue is unique in its ability to contract and cause movement. Like nervous tissue, it is also conductive. The **contractile proteins** (actin and myosin), and the structure of the cell, are modified into three different categories of muscle tissue: skeletal, cardiac, and smooth muscle.

ACTIVITY 4

Viewing and Learning About Selected Muscle Tissues

Materials for this Activity

Compound light microscope

Prepared Slides

Skeletal muscle
Cardiac muscle
Smooth muscle

Skeletal Muscle

Skeletal muscle, as the name implies, is the muscle typically attached to the skeleton for mostly voluntary movement. The contractile proteins are packed so tightly into alternating patterns of light and dark areas that the cells take on the appearance of having cross-stripes, or **striations.** If this tissue looks

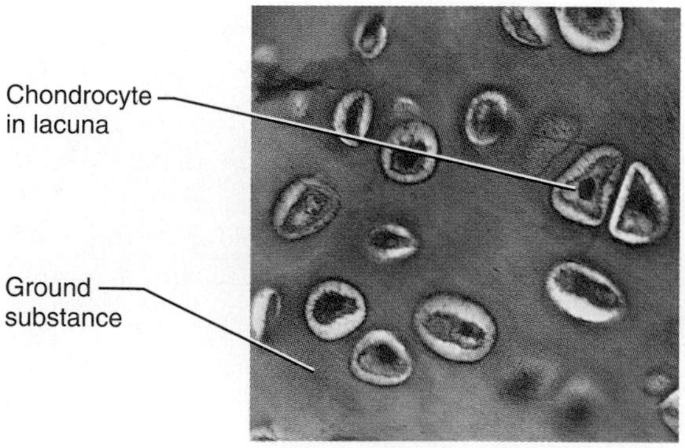

Magnification

__________ ×

FIGURE 5.8 Hyaline cartilage from the trachea (300×).

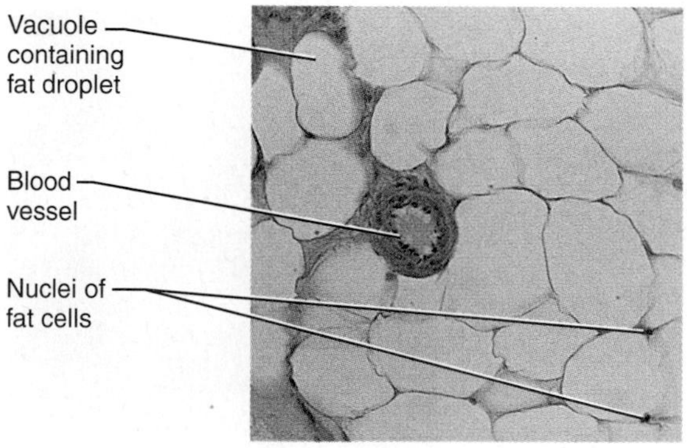

Magnification

__________ ×

FIGURE 5.9 Adipose tissue from the subcutaneous layer under skin (140×).

familiar, it is because the multinucleated skeletal muscle cell slide was the one chosen in the cellular diversity activity in Exercise 3.

View a slide of this tissue with your microscope, and sketch the image in the blank box provided with Figure 5.10. Remember to record the magnification. How do your observations compare to the photomicrograph in the figure?

____

____

____

Cardiac Muscle

Cardiac muscle is a highly fatigue-resistant muscle tissue found in the heart. Like skeletal muscle, cardiac muscle is also striated. Its branched appearance and dark areas of connection between cells, called **intercalated discs,** help to differentiate this tissue from skeletal muscle.

View a slide of this tissue with your microscope, and sketch the image in the blank box provided with Figure 5.11. Remember to record the magnification. How do your observations compare to the photomicrograph in the figure?

____

____

____

____

Smooth Muscle

Smooth muscle is the involuntary muscle found in many internal organs such as the stomach. The cells lack striations and are shorter than those of skeletal muscle.

View a slide of this tissue, and make a sketch in the blank box provided with Figure 5.12. Remember to record the magnification. How do your observations compare to the photomicrograph in the figure?

____

____

____

____

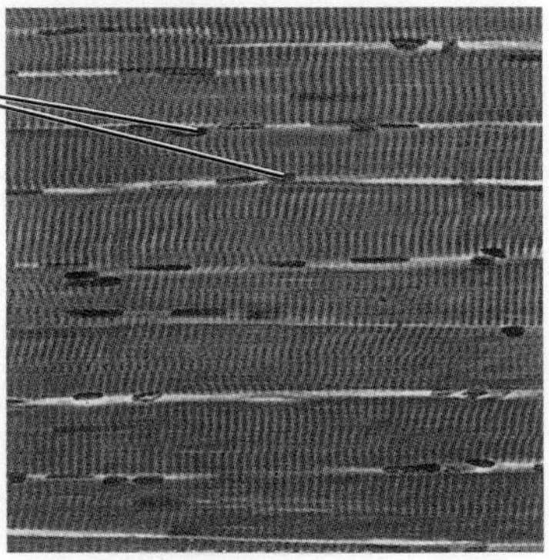

Magnification

__________ ×

FIGURE 5.10 Skeletal muscle (100×).

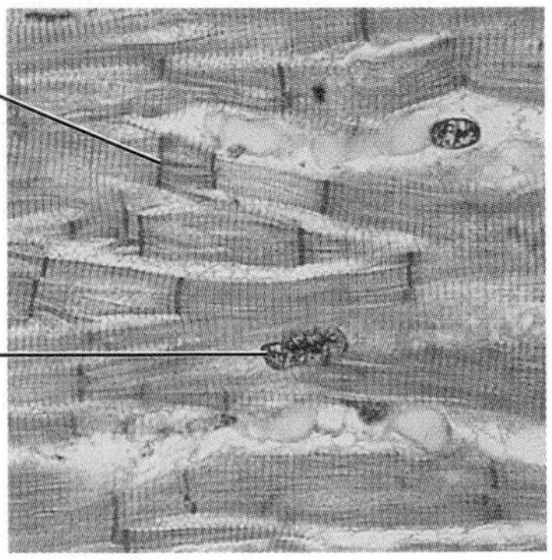

Magnification

__________ ×

FIGURE 5.11 Cardiac muscle (225×).

Nervous Tissue

Nervous, or neural, tissue falls into two categories: **neurons,** which are the conductive cells, and **neuroglia,** which are the nonconductive supporting cells of the nervous system. For this exercise, we will examine only the neuron.

ACTIVITY 5

Viewing and Learning About Selected Nervous Tissues

Materials for this Activity

Compound light microscope

Prepared slide of a multipolar neuron smear

Neurons are usually highly branched cells with a large spherical area containing the nucleus. The long branches pick up or carry information to other neurons (and other cells such as muscle cells), sometimes over long distances.

View a slide of a multipolar neuron smear, and sketch the image in the blank box provided with Figure 5.13. Remember to record the magnification. How do your observations compare to the photomicrograph in the figure?

____

____ ■

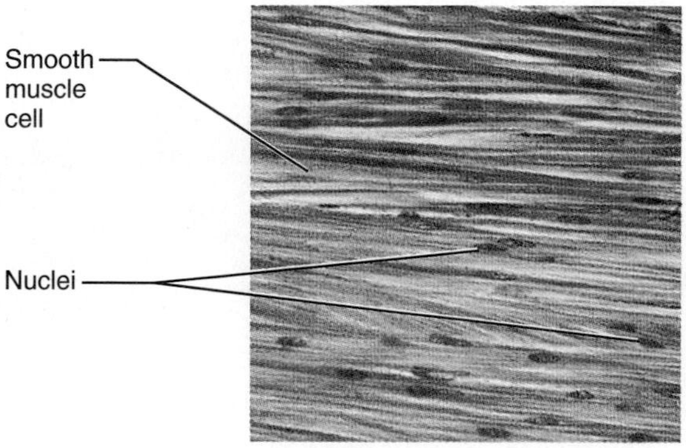

Magnification

____________ ×

FIGURE 5.12 Smooth muscle (250×).

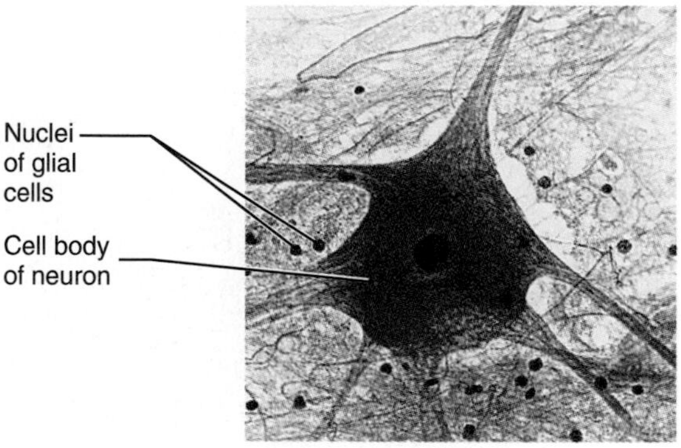

Magnification

____________ ×

FIGURE 5.13 Multipolar neuron (170×).

EXERCISE

6

ORIENTATIONS TO THE HUMAN BODY

Objectives

After completing this exercise, you should be able to

1. Understand and use anatomical terminology.
2. Describe the standard anatomical position (SAP).
3. Understand and use directional terms.
4. Identify anterior and posterior surface regions.
5. Explain the results of three anatomical planes.

Materials for Lab Preparation

All Human Models Available in the Lab

- ❑ Torsos, heads, organs
- ❑ Dissectible models

Introduction

Let's say that you're going to a party in a city nearby. You've never been there before, but knowing *how* to get there seems to be a minor issue. This peace of mind comes from your possession of two navigational tools: directions and a map. Directions provide a way to orient yourself and navigate the terrain from point A to point B. Maps are images that represent the terrain; they can be geographic (land) or anatomical (body). But whether we are talking about getting from one city to another or explaining the human body, directions and maps are useful only if you understand how to use them, which includes recognizing landmarks and using the right terminology. The task of this laboratory exercise is to teach you how to look at and describe the human body in a clear and accurate manner, using terminology that is used universally by scientists and the medical profession at large.

Standard Anatomical Position

In biology courses and in clinical settings, precise terminology is used to describe body parts and position. For that terminology to be meaningful, there must be a universally agreed-upon reference point.

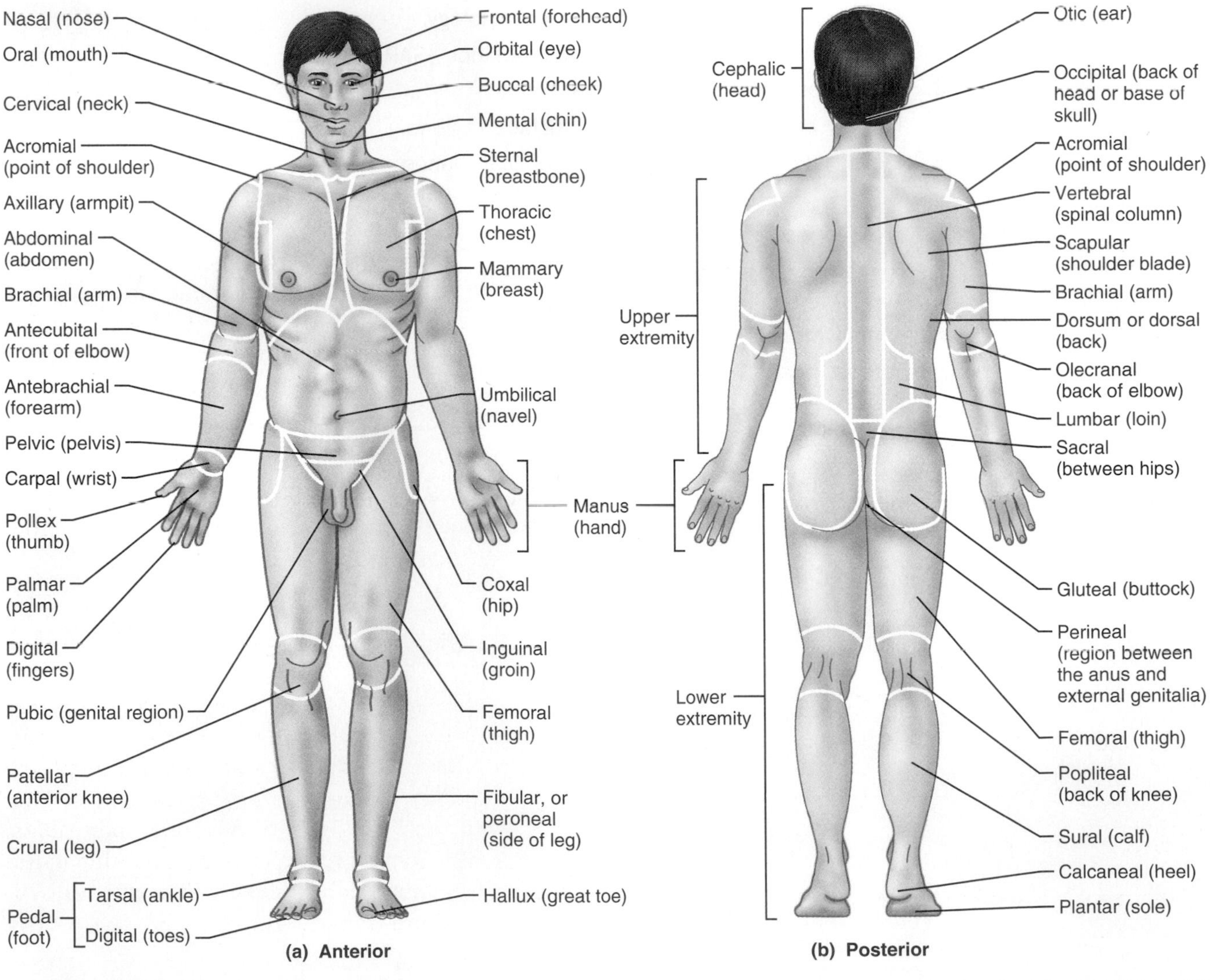

FIGURE 6.1 The standard anatomical position with surface landmarks. (a) Anterior view. (b) Posterior view.

Whose body should be used? Yours? Mine? A third person's? How should this body be positioned? Standing, sitting, lying down?

All anatomical terminology assumes an anatomical reference point, which is the human body in the **standard anatomical position (SAP)** (Figure 6.1). The standard anatomical position includes both the **anterior** (front) and **posterior** (back) views of the human body. Whether you are describing exterior landmarks or internal ones, always assume you're referring to the SAP.

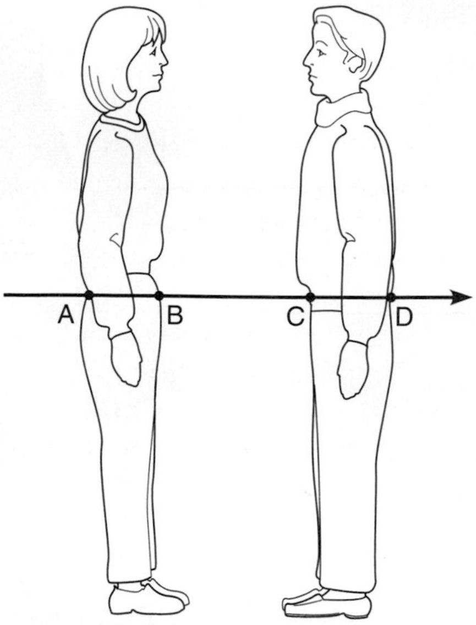

FIGURE 6.2 Two facing bodies pierced with an arrow. A, B, C, and D indicate entry and exit points.

ACTIVITY 1

Agreeing upon an Anatomical Reference Point

Standing face to face with a lab partner, raise your right hand, and ask your partner to do the same. From *your* perspective, your partner will have raised his or her left hand. However, custom and experience teaches us very early to adapt to this apparent anomaly and to understand in situations like this that the person we are facing indeed has raised his or her right hand. What we have done is simply shift our perspective from our own to theirs. Now let's make a trickier point: "Your front is your partner's back."

1. Stand facing your lab partner. Imagine the same arrow piercing through both of your bodies. Compare your situation to the one in Figure 6.2. What are the entry and exit points in the path of the arrow from your point of view?

   First entry point

   ____

   First exit point

   ____

   Second entry point

   ____

   Second exit point

   ____

   The point of this exercise is to make you aware of how important perspective can be when trying to accurately describe body parts and position. In anatomy, it is *not* the observer that is the reference point, but rather it is the body (or a lab model or a photograph of a body) being viewed.

2. Test your understanding of perspective and reference by filling in the blank lines in Figure 6.3, using the terms *right* or *left*. ■

There are three useful orientation tools: **directional terms, surface regions** (also called landmarks or body regions), and **planes.**

Directional Terms

Just as you use directional terms in your everyday conversations—such as north, south, up, down, outside, inside, front, and back—anatomists use corresponding terms that are more precise and universally agreed upon to describe body parts. Where we might say, "The eyes are above the mouth," the anatomist would say, "The eyes are superior and lateral to the mouth."

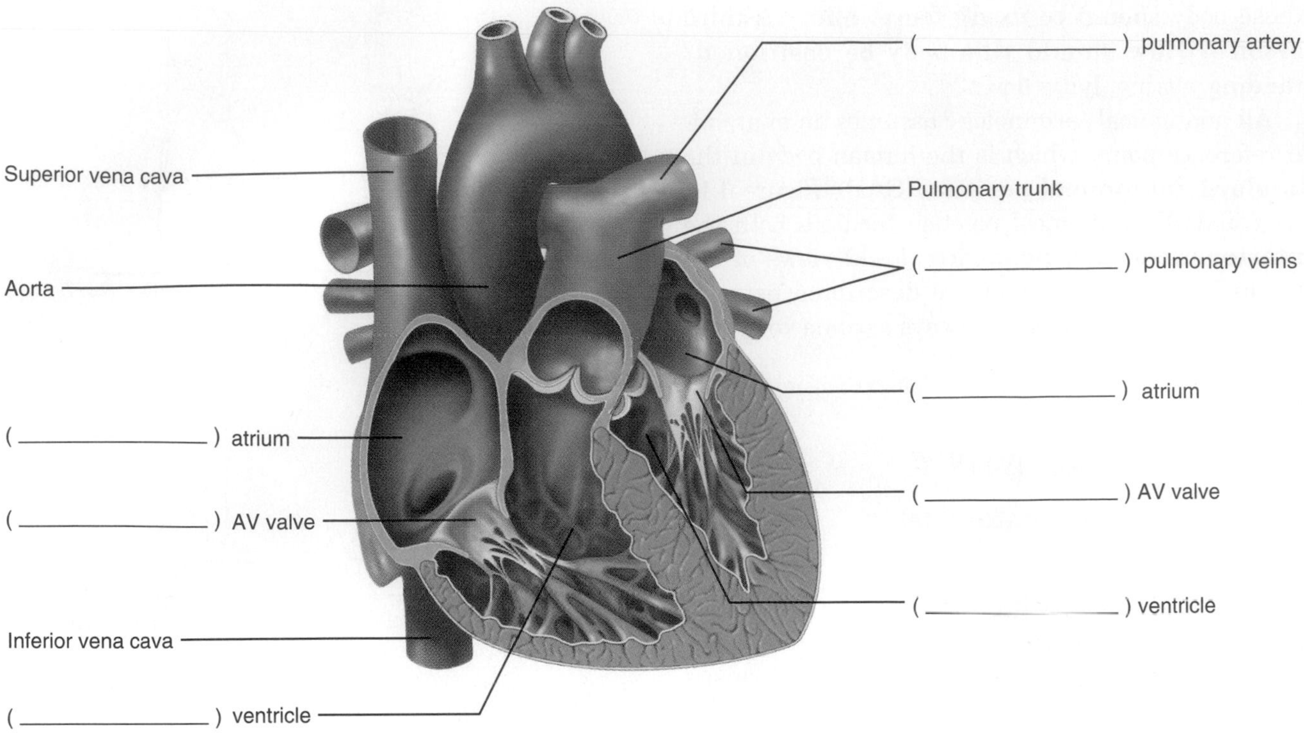

FIGURE 6.3 Gross anatomy of heart. Frontal section, showing major blood vessels, chambers, and valves.

ACTIVITY 2

Using Directional Terms

Review the directional terms and definitions in Table 6.1. Note the direction of the arrows, and imagine how you would describe the relationship between common features of your body, for example, the position of the head relative to the abdomen. In the far right column you will see examples of the use of each term.

1. Describe the relationship between the following parts of your body:

   Shoulders to neck

   ____

   Lips to teeth

   ____

   Ear to shoulder

   ____

   Foot to knee

   ____

   Belly button to buttocks

   ____

2. Refer to Figure 6.3. Use the directional terms from Table 6.1 to describe the relationships of the following structures of the heart.

   Superior vena cava to pulmonary trunk

   ____

   Interior vena cava to aorta

   ____

   Aorta to right atrium

   ____

   Pulmonary trunk to aorta

   ____

Table 6.1 Orientation and Directional Terms

Term	Definition	Example	
Superior (cranial)	Toward the head end or upper part of a structure or the body; above	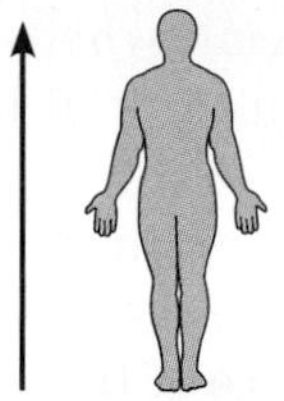	The head is superior to the abdomen
Inferior (caudal)	Away from the head end or toward the lower part of a structure or the body; below	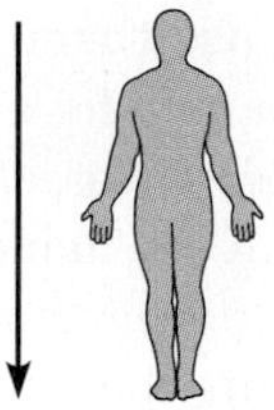	The navel is inferior to the chin
Anterior (ventral)*	Toward or at the front of the body; in front of	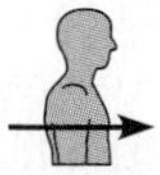	The breastbone is anterior to the spine
Posterior (dorsal)*	Toward or at the back of the body; behind	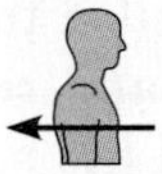	The heart is posterior to the breastbone
Medial	Toward or at the midline of the body; on the inner side of	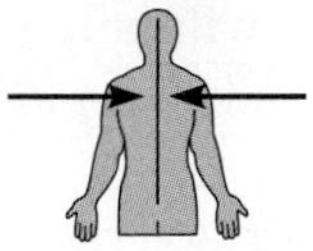	The heart is medial to the arm
Lateral	Away from the midline of the body; on the outer side of	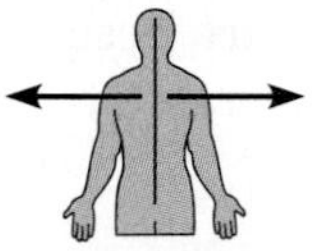	The arms are lateral to the chest
Intermediate	Between a more medial and a more lateral structure	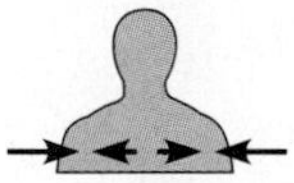	The collarbone is intermediate between the breastbone and shoulder
Proximal	Closer to the origin of the body part or the point of attachment of a limb to the body trunk	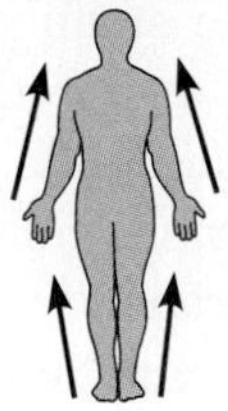	The elbow is proximal to the wrist
Distal	Farther from the origin of a body part or the point of attachment of a limb to the body trunk	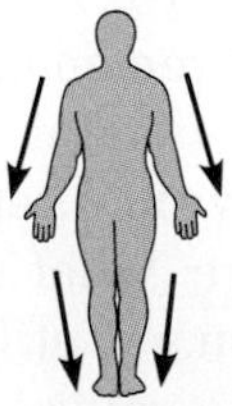	The knee is distal to the thigh
Superficial (external)	Toward or at the body surface	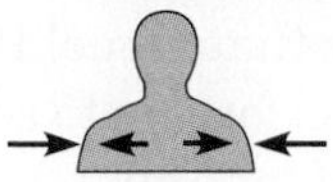	The skin is superficial to the skeletal muscles
Deep (internal)	Away from the body surface; more internal	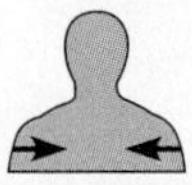	The lungs are deep to the skin

*Whereas the terms *ventral* and *anterior* are synonymous in humans, this is not the case in four-legged animals. *Ventral* specifically refers to the "belly" of a vertebrate animal and thus is the inferior surface of four-legged animals. Likewise, although the dorsal and posterior surfaces are the same in humans, the term *dorsal* specifically refers to an animal's back. Thus, the dorsal surface of four-legged animals is their superior surface.

Surface Regions

Remember our discussion on the importance of local landmarks when providing directions. In anatomy, recognizing landmarks is equally crucial, whether looking at the surface of the body, at bones, or at internal organs.

On the human body, these landmarks are called **surface** or **body regions.** We use surface regions to represent specific locations on the surface of the human body, but eventually we will also use them to represent the underlying anatomical parts—for example, the bones, muscles, nerves, and blood vessels.

As you study the anatomy, notice that the names of specific anatomical parts reflect the surface region. For example, the femoral region represents the thigh portion of the leg. In that area, the bone is called the femur, the major blood vessels are called the femoral artery and vein, the major nerve is called the femoral nerve, and the two major muscles are called the biceps femoris (one of the hamstrings) and rectus femoris. Making this connection between the names of surface regions and names of organs will help you learn and retain the memory of the many regions, vessels, and organs you will encounter.

Let's look at the front and back surface regions of the human body, which in anatomy are described as the **anterior** and **posterior** surface regions.

Anterior and Posterior Surface Regions

Figure 6.1 illustrates the human figure in SAP, both anterior and posterior. Notice that the surface (or body) regions are labeled and include, in parentheses, the common, everyday term for each region.

With a lab partner, study these terms and quiz each other. As you read the labels, pronounce the words (aloud or silently) as best as you can. You will find that if you can "hear" each word (aloud or silently), your brain will register and remember it much more than if you "look" at the letters of each word. In fact, after "looking" at about five new terms, you will not be able to retain any more. What happens if you mispronounce the word? You will still remember it better than if you do not pronounce it at all. Some brain traces are still better than none! Remember the purpose of this exercise is for you to see, understand, and retain the material, not simply to turn in correct answers.

ACTIVITY 3

Identifying Anterior and Posterior Surface Regions

1. Using Figure 6.4a and b, fill in the blanks (without referring to Figure 6.1). Then check your work against Figure 6.1. Noting those regions you identified incorrectly, discuss ideas with your lab partner(s) for techniques to help remember these regions. Now take a second look at your work. If you were to cover up the labels, could you do it again without referring to the definitions?
2. Draw the anterior and posterior views of any upright organism in SAP. Label the surface regions. Add in your personal touches, such as hair, skin ornamentation, cartoon figures, or alien figures. Make it your work of art; have some fun with this.

3. Identify the following surface regions, indicate if they are anterior or posterior terms, and indicate if there are any terms that apply to both the anterior and posterior surfaces:

Knee

____

Breastbone

____

Arm

____

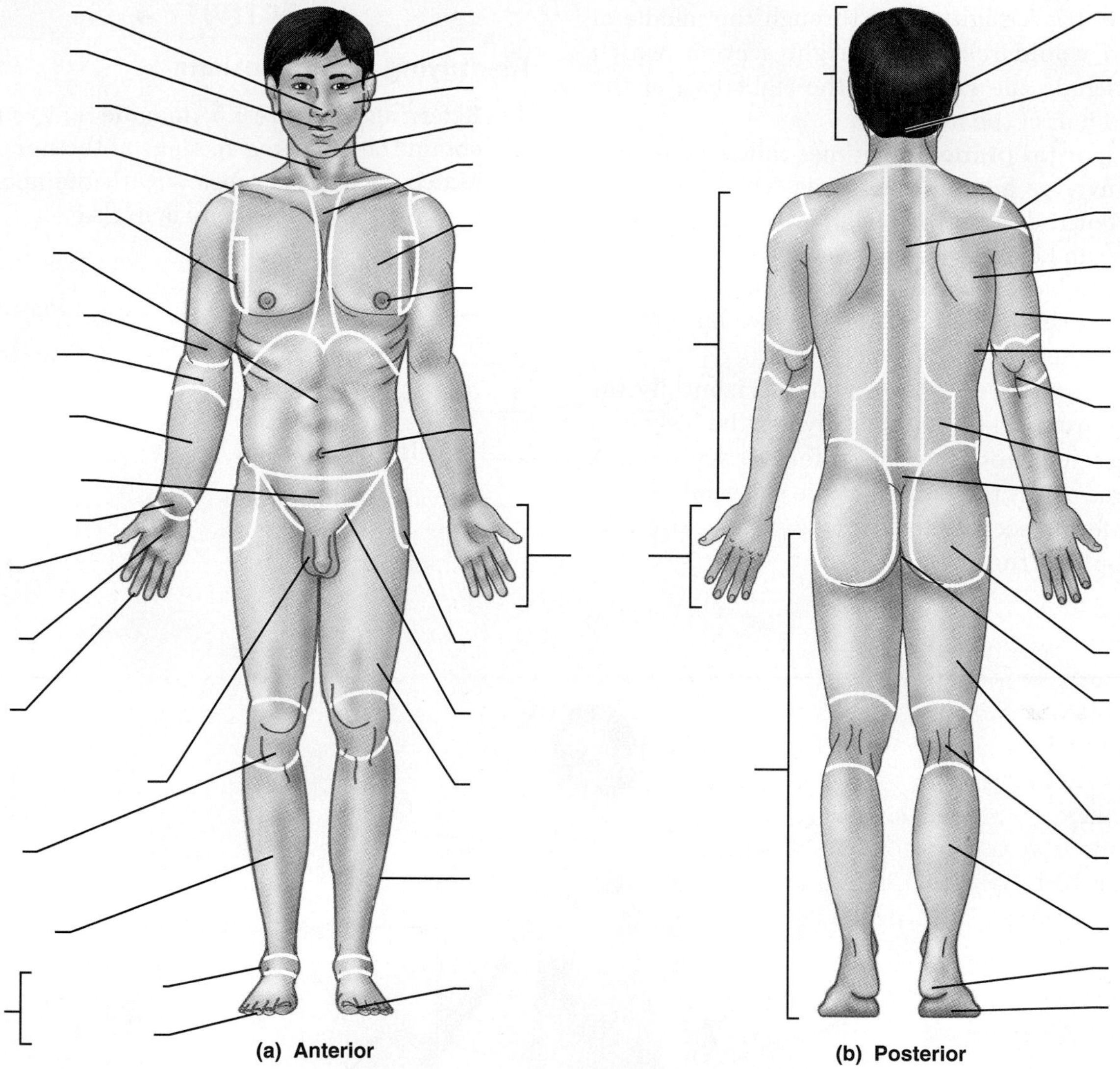

FIGURE 6.4 Surface landmarks for identification. (a) Anterior. (b) Posterior.

Shoulder blade

____

Heel

____

Planes

It is sometimes useful to show and discuss the sections that result from a straight cut of the body or through an organ. A **plane** is a flat surface that conceptually cuts through a tissue, organ, or part of a body. Two traditional terms for sections are **longitudinal section** (ls), indicating a lengthwise cut, and **cross section** (cs) for a crosswise cut. But how was the cut performed? How does lengthwise indicate if the cut was made from the front to the back or from the top to the bottom? In anatomy, those questions do not arise because three basic anatomical planes are used to describe various cuts and sections made to a body.

A **sagittal plane** (not shown in Figure 6.5) runs longitudinally (top to bottom), dividing the body into right and left sections. When it divides the body or an organ, top to bottom, into *exactly equal* sections, it is called a **median (midsagittal) plane,** as shown in Figure 6.5. A sagittal cut through the middle of the head would result in a right section, which would include the right eye, the right half of the nose, and half of the mouth.

The **frontal plane,** sometimes called a **coronal plane,** involves another lengthwise cut from top to bottom, but it divides the body (or organ) into front (anterior) and back (posterior) sections. A frontal cut through the back of the eye would result in an anterior section that would include the face but not the back of the head.

The **transverse plane** runs horizontally (or crosswise from side to side). It divides the body into top (superior) and bottom (inferior) sections. A transverse cut at the bottom of the eye would result in an inferior section that included the nose and mouth, but not the eyes.

ACTIVITY 4

Identifying Body Sections

1. Referring to Figure 6.5, imagine that you need to obtain the following sections of the human body. Match each section below with the appropriate cut (a–c). *Hint:* The body is in SAP.

_______ Right section	a.	Sagittal cut
_______ Left section	b.	Frontal cut
_______ Anterior section	c.	Transverse cut
_______ Posterior section		
_______ Inferior section		
_______ Superior section		

■

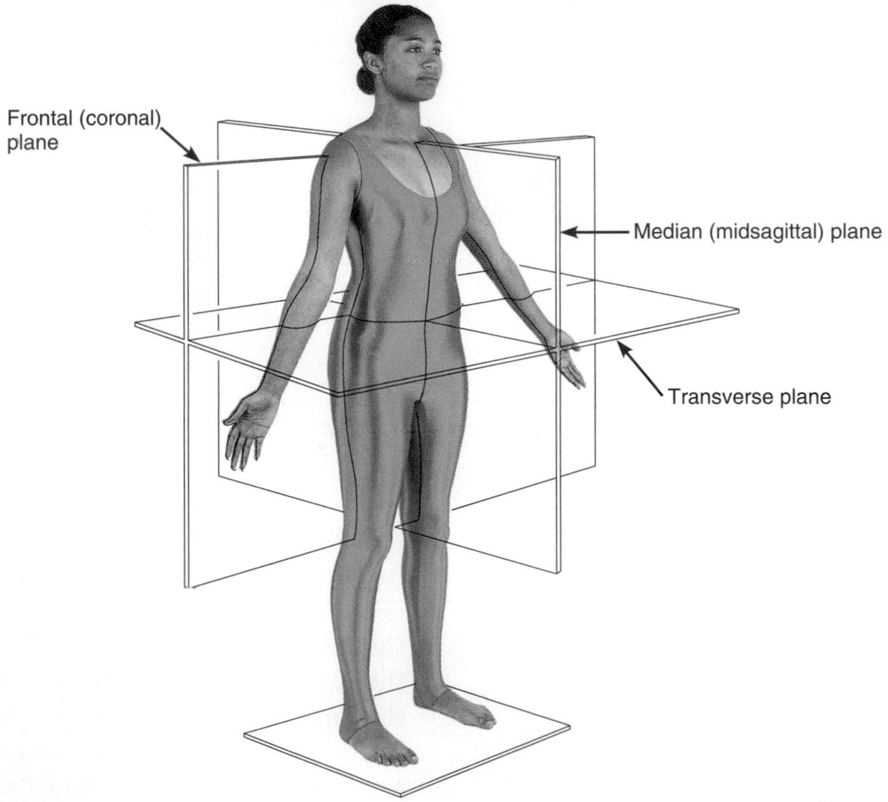

FIGURE 6.5 Anatomical planes.

Section ____________________ Name ____________________

Date ____________________ Instructor ____________________

EXERCISE

6

ORIENTATIONS TO THE HUMAN BODY

Review Questions

Directional Terms

Fill in the directional term for the questions below. More than one term may be acceptable.

1. Where are the shoulders relative to these parts?
   a. Throat ____________________
   b. Hip ____________________
   c. Ears ____________________
   d. Hands ____________________
   e. Eyes ____________________

2. Where is the nose relative to the same parts?
   a. Throat ____________________
   b. Hip ____________________
   c. Ears ____________________
   d. Hands ____________________
   e. Eyes ____________________

Surface Regions

1. Fill in the anatomical terms for these anterior surface regions:
   a. Shoulder ____________________
   b. Armpits ____________________
   c. Wrist ____________________
   d. Thigh ____________________
   e. Forearm ____________________
   f. Side of leg ____________________
   g. Foot ____________________
   h. Chest ____________________
   i. Navel ____________________
   j. Hip ____________________

2. Fill in the anatomical terms for these posterior surface regions:
   a. Ear ______________________
   b. Thigh ______________________
   c. Spine ______________________
   d. Heel ______________________
   e. Elbow ______________________
   f. Calf ______________________
   g. Head ______________________
   h. Sole ______________________
   i. Arm ______________________
   j. Back ______________________

3. Imagine that you are rock climbing with your best friend. Your friend slips down a rock about 50 feet, and lands awkwardly on his right side. The figure below shows the location and types of your friend's injuries. Using your knowledge of surface regions and directional terms you learned in this exercise, you call for help and need to describe the injuries as accurately as possible. How would you describe each injury to the emergency dispatcher assisting you?

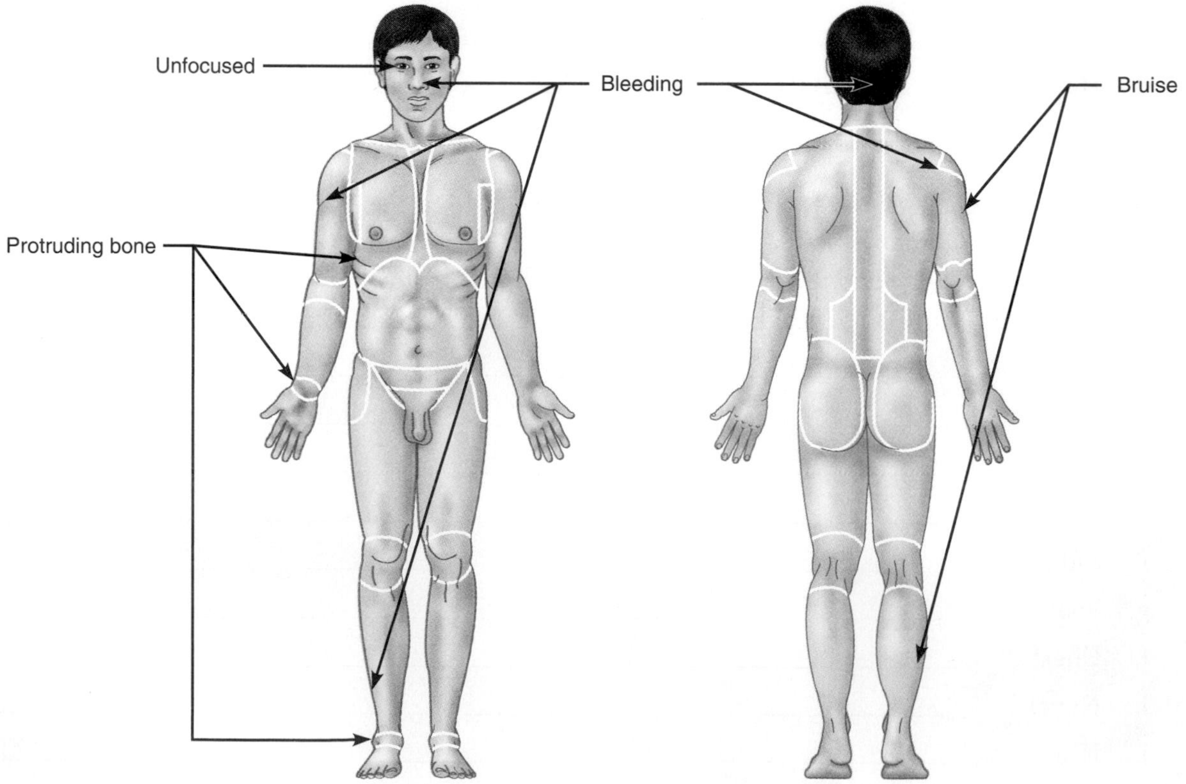

Planes

Draw and label a picture of a banana cut in the following ways:

a. Transverse plane

b. Frontal plane

c. Sagittal plane

d. All three planes

EXERCISE

7

THE INTEGUMENTARY SYSTEM

Objectives

After completing this exercise, you should be able to

1. Describe the general functions of the skin, or the integumentary system.
2. Explain the basic structure and organization of the skin.
3. Name the layers of the epidermis, and describe the characteristics of each layer.
4. Describe the dermis and its primary features and functions.
5. Name each skin derivative found in the dermis, and describe the function of each.
6. Identify under the microscope the epidermis, dermis, and hypodermis.

Materials for Lab Preparation

Equipment and Supplies

- ❑ Compound light microscope

Prepared Slides

- ❑ Human skin (with and without hair follicles)

Human Skin Models (three-dimensional if possible)

- ❑ Regular skin models (scalp and palm omitted)
- ❑ Skin models of scalp, body skin, and palm or sole
- ❑ Skin model showing epidermal layers
- ❑ Human skin charts

Introduction

If we conducted an informal street survey and asked people to name the largest organ in the body, few would say skin; some might say the lungs, while others might choose the heart. Recall, however, what you learned about how the body is organized: cells form tissues; tissues form organs; two or more organs that work together to carry out a general function important to the survival of the whole organism are the basis of an organ system. Now think again about your body, both externally and internally, and you quickly will realize that the **skin** (or **integument,** from the Latin *integere,* meaning "to cover") is the organ with the largest surface area. The skin and its derivatives, such as hair, nails, sweat and oil glands, form the **integumentary system.**

Functions of the Skin

What functions are performed by the skin? The most obvious function is to provide protection from injury, such as abrasion, but there are other equally important ones. They include the following:

- Protection from dehydration
- Defense against bacteria and viruses
- Regulation of body temperature
- Synthesis of vitamin D
- Receipt of sensation—touch, vibration, pain, and temperature
- Provision of a blood reserve

Structure of the Skin

Skin is a tissue membrane, and it contains two basic regions, the epidermis and dermis, which are organized as shown in Figure 7.1.

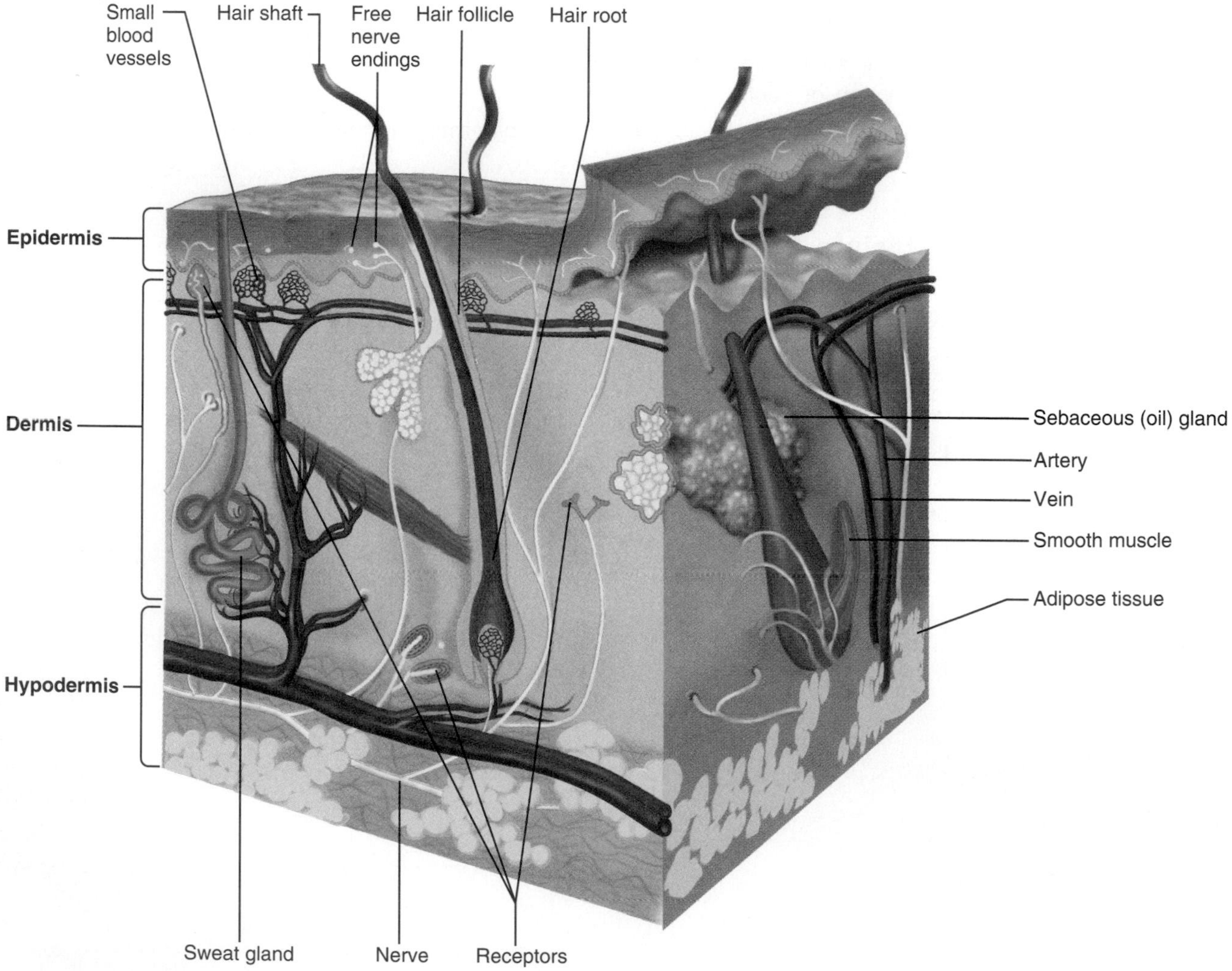

FIGURE 7.1 A three-dimensional view of human skin, with the underlying hypodermis.

The outer, "veneerlike" region of the skin is called the **epidermis,** which is composed mainly of stratified squamous epithelial tissue, which you were introduced to in Exercise 5, Tissues. There are five layers of cells in the epidermis, which we will examine in more detail later in this exercise. For now, keep in mind that the outermost cells of the epidermis are dead and continually slough off. They are replaced by cells from the deeper epidermal layers. The principal function of the epidermis is to provide protection from everyday wear and tear on underlying body tissues.

The **dermis** is the thicker region of the skin, for the most part, and it is composed primarily of dense connective tissue and a variety of fibers: reticular, collagen, and elastic. This combination of tissue and fibers gives the dermis its strength and flexibility. There are abundant capillaries in the superficial dermal region, which provide nutrients to the epidermal layers above.

The uneven, wavelike protuberances on the superior surface of the dermis attach it to the epidermis above. This "fit" is similar to the "tongue and groove" found in finished lumber, which is created by shaping pieces of wood so they can fit together without the use of nails. These protuberances force the inferior surface of the epidermis to fold, and this arrangement is the source of the unique patterns we call fingerprints.

A supportive layer called the **hypodermis** (*hypo-* means "under"), or superficial fascia, is composed primarily of loose connective tissue containing fat cells and a large number of blood vessels. This layer is thickest on the palms and soles, areas that are subject to a great deal of wear and tear, and it is thinner on the eyelid, which suffers considerably less wear.

The hypodermis anchors the dermis to muscle, while being flexible enough to allow the skin to move and bend. Its fat cells provide insulation to protect the body from excessive heat loss and absorb shock.

ACTIVITY 1

Identifying Skin Layers and Organization

Materials for this Activity

Skin models
Skin charts

1. Using the skin models or charts available in your lab, examine the structure of each layer and compare their differences and similarities. Notice that there are no blood vessels in the epidermis. How does the lack of blood vessels affect this layer?

   ____

   ____

   In a first-degree burn, only the epidermis is damaged. This is not a serious burn, and it tends to heal in a few days without special treatment. Can you think of a common example of a first-degree burn?

   ____

2. Look at the thickness of the dermis. The dermis appears to be a network of thick-to-thin fibers and connective tissue holding together many "skin derivatives," such as hair cells, sweat glands, and oil glands. The vertical blood vessels of the dermis connect two horizontal blood vessel supplies, one at the top of the dermis and one at the top of the hypodermis. If the dermis were damaged, would it affect the epidermis at all? If so, why?

   ____

   ____

   Notice the smooth muscle attached to the base of the hair follicle. What is its effect? (Think about frightened dogs or cats and what happens to their hair.)

   ____

   ____

3. Look at the main features of the hypodermis: adipose tissue (or fat cells), nerves, and blood vessels. The hypodermis, like the dermis, is a supportive structure. A network of thin fibers and connective tissue is also found here and anchors the skin to the underlying skeletal muscles. Notice the adipose tissue is located in this layer. What is the function of the adipose tissue? Of the three layers, what are the advantages for the location of adipose tissue in this layer?

   ____

In a second-degree burn, the epidermis and part of the dermis are damaged. This condition is more serious than a first-degree burn. With proper care, skin regeneration may take place in 3 to 4 weeks, and scarring may be avoided. What are the effects on skin functions if the blood vessels, sweat glands, and oil glands are seriously damaged?

___

___

A third-degree burn penetrates past the dermis; a first-degree burn affects only the epidermis. What additional skin functions are lost from a third-degree burn as compared to a first-degree burn?

___

___

Structure of the Epidermis

The epidermis is composed of two major cell types. **Keratinocytes,** which are cells that produce keratin, provide a waterproofing material that prevents our skin from absorbing too much water and getting soaked. **Melanocytes** are cells that produce melanin, a brown pigment that protects the nucleus of the skin cells from the ultraviolet rays of the sun.

The epidermis consists of four to five layers, called strata, that vary in thickness. There are five strata in **thick skin,** which covers the palms, the fingertips, and the soles of the feet. **Thin skin,** which covers the remainder of the body, consists of only four strata, which are thinner than those found in thick skin. The five strata of epidermis, from the outermost stratum to the deepest, are as follows:

1. **Stratum corneum:** most superficial portion, 25–30 layers
   - This is also called the corn layer.
   - It consists of dead, dried-out keratinocytes.
   - No nucleus is evident.
2. **Stratum lucidum:** 3–5 layers
   - This is also called the clear layer.
   - It consists of dying keratinocytes.
   - No nucleus is evident.
   - This layer, which is missing from thin skin, is mainly found on palms and soles and provides additional protection for these heavy-traffic skin areas.
3. **Stratum granulosum:** 3–5 layers
   - This is also called the grainy layer.
   - It consists of living keratinocytes, which are becoming flatter.
   - The nucleus is barely evident.
4. **Stratum spinosum:** 8–15 layers
   - This is also called the spiny layer.
   - It consists of living keratinocytes, which are still cuboidal.
   - The nuclei are evident.
   - The cells are protected by melanin granules produced by the underlying melanocytes.
5. **Stratum basale:** deepest portion, 1 layer only
   - This is also called the base layer.
   - It consists of living keratinocytes and melanocytes.
   - The nuclei are evident, and cell division is in progress.
   - A basement membrane supports this germinal layer.

ACTIVITY 2

Identifying the Layers of the Epidermis

Materials for this Activity

Skin models with epidermal layers

Obtain a skin model and, in conjunction with Figure 7.2, look at the five strata of the epidermis. Observe the differences between the strata in terms of the number of layers, whether the nucleus is evident, the shape of the cells, and the type of cells.

1. What are your observations?

___

___

___

2. How does the structure of the epidermis compare to the bark of a tree?

___

___

___

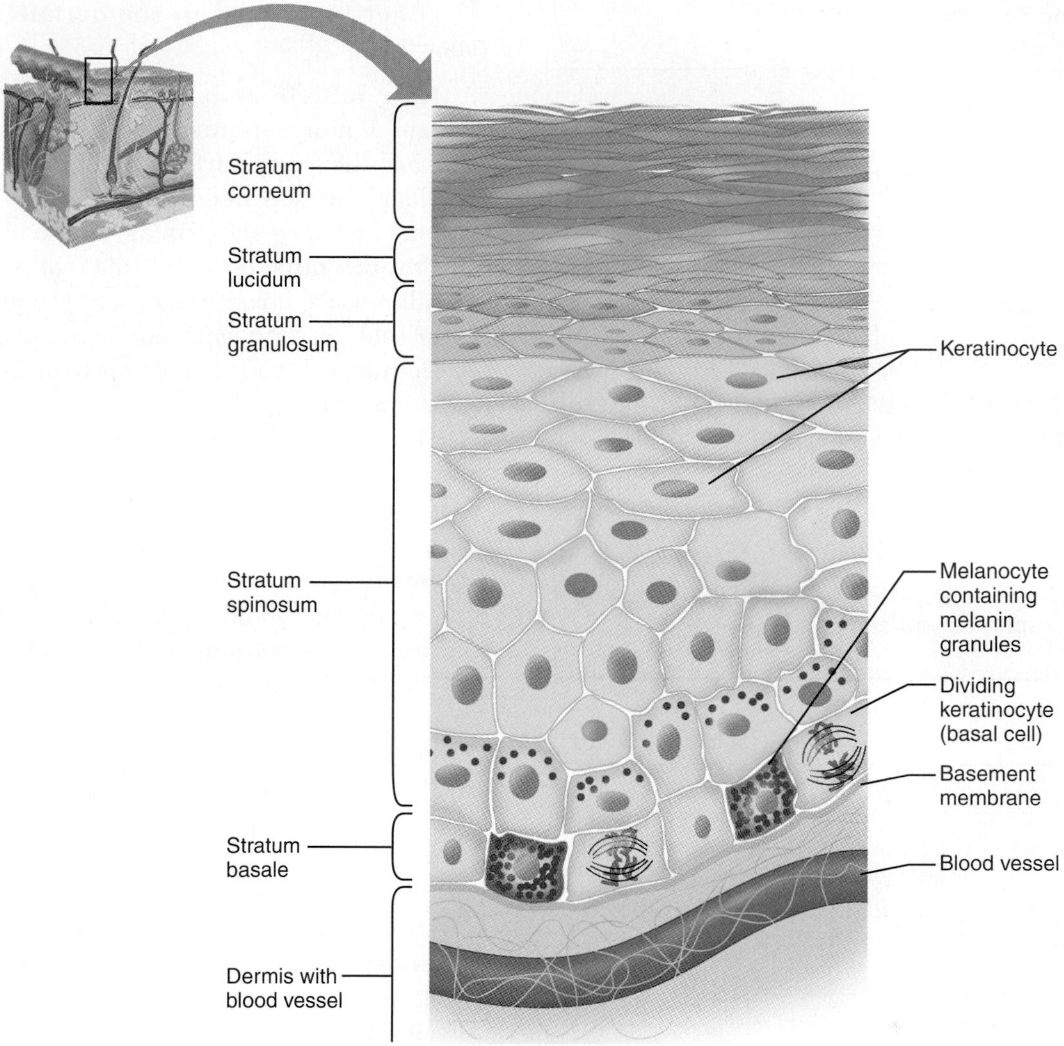

FIGURE 7.2 The epidermis, showing five epidermal strata.

ACTIVITY 3

Observing Slides of Human Skin

Materials for this Activity

Prepared slides of human skin (both with and without hair follicles)

Compound light microscope

1. Under the light microscope, scan a human skin slide without hair follicles, using the scanning-power lens. Look for the surface of the skin. You may see what appears to be thin peeling flakes. After you have focused on the epidermis, change to the low- and then high-power lens. The thin flakes may now appear to be a bundle of thin lines. Remember to look toward the surface of the skin. This is the epidermis, the most superficial portion of the skin. The five strata illustrated in the skin models and charts were discovered with the help of powerful microscopes called electron microscopes, which are capable of up to 10,000-fold magnifications. The light microscope used in the laboratory will not be capable of discriminating this level of detail, but certain features may be observed.

   What do the peeling flakes look like? Just thin lines?

   ____

   ____

Can you observe any structures?

____

____

How many layers of the epidermis can you observe?

____

____

2. Using the scanning-power lens of a light microscope, focus on a prepared human skin slide with hair follicles. Find the epidermis, and focus below it into the dermis. Change to the high-power lens.

   Compare your tissue slide to Figure 7.3a–d. Identify as many structures shown in the photomicrographs as you can.

   What features can you clearly observe in addition to the hair follicles?

____

____

Are you able to distinguish between the dermis, epidermis, and hypodermis? What features make each tissue layer distinct?

____

____

Recall your use of stain in Exercise 2 when preparing a wet mount slide. If you're now examining a particularly poor slide, how would staining improve it?

____

____

______________________________________ ■

Structure of the Dermis

The primary feature of the dermis is the fibers: **collagen, elastic,** and **reticular fibers** embedded in a ground substance. The fibers allow the skin to stretch when we move, and they give it strength to resist abrasion and tearing. As we age, our skin becomes less flexible and more wrinkled because the number of fibers in the dermis decreases.

Other structures in the dermis include the following:

1. **Hair follicle cells:** Our hair is formed at the root of hair cells, and a shaft pushes upward toward the skin's surface. The longer the hair, the older and less maintained it is, so the ends of long hair may show up as "split ends."
2. **Smooth muscle:** This is also called the arrector pili muscle. It contracts when you are frightened or cold, causing your hair to become more erect. Because our body hair tends to be fine, the result is "goose bumps."
3. **Oil glands:** These are also called the sebaceous glands. The oily fluid serves to moisten and soften the hair and skin. If the oil is trapped and accumulates, it causes "pimples."
4. **Sweat glands:** These were formerly called the odoriferous glands. The salty, watery fluid contains dissolved ions and metabolic wastes. The evaporation of sweat on the outer body surfaces serves to remove body heat and help lower our body temperature.
5. **Blood vessels:** These supply both the epidermis and dermis with nutrients and remove their wastes. They also regulate body temperature through dilation when we are too hot and constriction when we are too cool.
6. **Sensory nerve endings:** Specialized receptors detect heat, cold, light touch, deep pressure, and vibration. These sensations enable us to feel the wind, the corners of a table, the coolness of a cold drink, the heat of a cup of tea, or the onset of an earthquake.

ACTIVITY 4

Examining Different Skin Types

Materials for this Activity

Models of different skin types: scalp, body skin, and palm or sole

1. Obtain a skin model with different types of skin. First, identify the structures of the dermis discussed previously. Your model may contain more features than were indicated previously. In a later class in human anatomy, you will learn about these additional features. For now, focus on the features listed here.

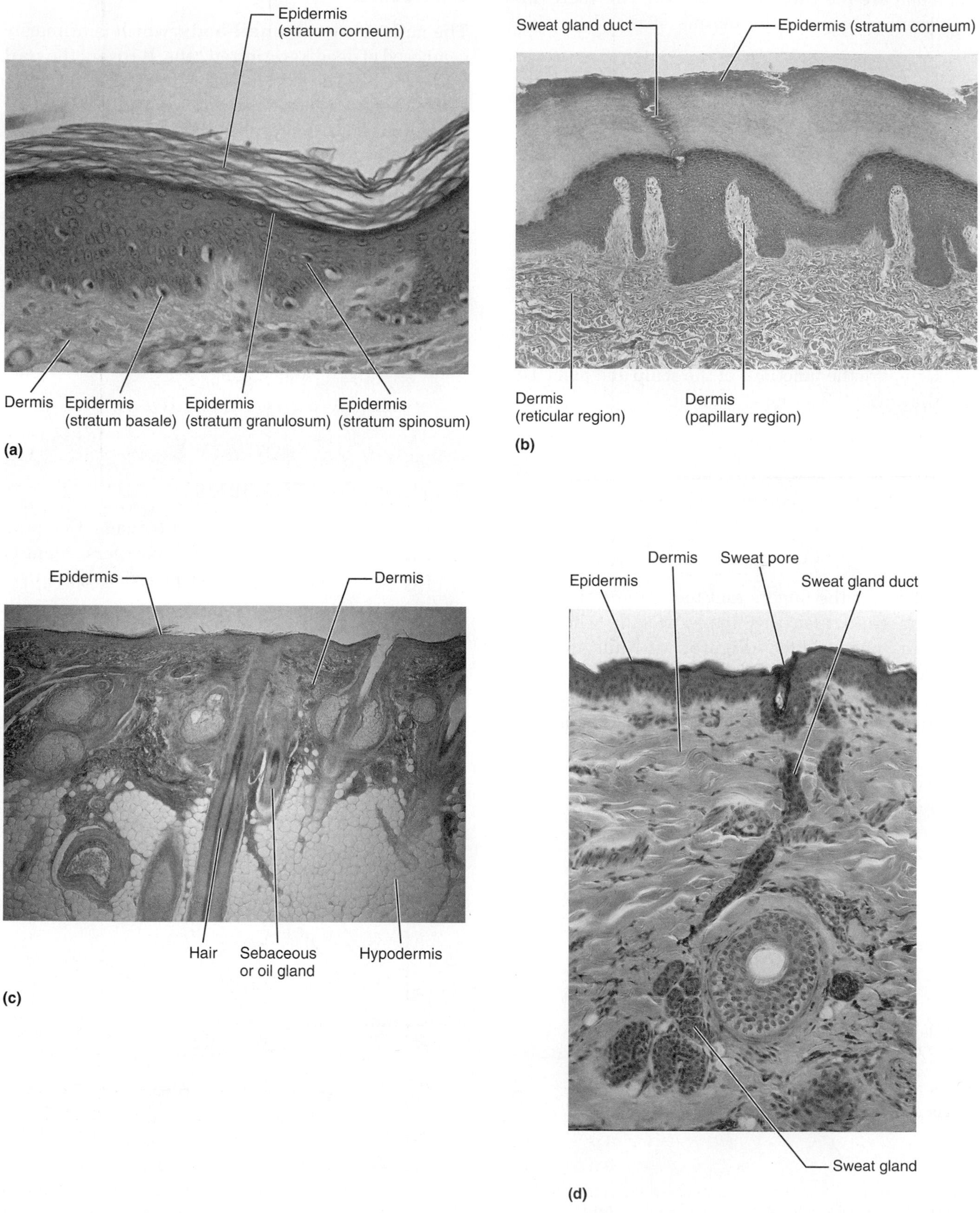

FIGURE 7.3 Photomicrographs of human skin. (a) Thin skin. (b) Palm (thick skin). (c) Scalp. (d) Sweat gland.

2. Compare the three types of skin. Are there differences between the dermis of the three skin types?

 ____
 ____
 ____
 ____

3. Why does the scalp have the most hair and the palm have the least hair?

 ____
 ____

4. Compare the functions of the scalp and palm. Do they differ? If so, how?

 ____
 ______________________________ ■

Nails

Nails form at the fingers and toes, where they protect and strengthen the tips, especially when you use your hands and feet (Figure 7.4). Nails are clear and colorless, but they appear pink because of the blood supply in the underlying skin, except for a white crescent-shaped area called the **lunula** (Figure 7.4a).

Structure

The nails consist of a **nail body,** which is primarily composed of dead keratinized cells. It covers the **nail bed,** which is the skin underneath the nails (Figure 7.4b). There is a **free edge** at the most distal portion of the nail body and a **nail root** at the most proximal portion, which is the site of nail production. An epithelial skin fold **(eponychium** or **cuticle)** projects onto the nail body.

The **nail matrix,** which is a portion of the skin underneath the nail, produces the modified tissue we call nails (Figure 7.4b). As the nail cells grow out and extend from the base of the nails, they become keratinized and die. Thus, like hairs, nails are mostly nonliving material.

ACTIVITY 5

Examining Nail Structures

1. Look at your fingernails and toenails. Compare them with those of your lab partners. Identify the nail structures described previously.
2. What are some general differences between your fingernails and toenails? Compare the size, shape, thickness, and texture of the thumb and big toe.

 ____
 ____
 ____

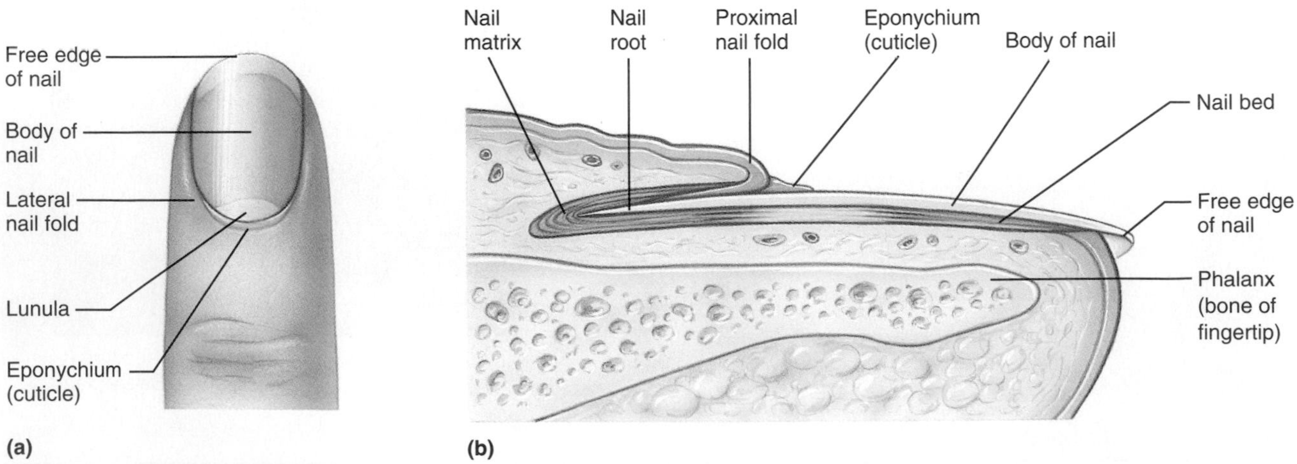

FIGURE 7.4 Structure of a nail. (a) Frontal view of nail. (b) Sagittal view of nail.

3. What are the differences (size, shape, thickness, and texture) between each of your individual fingernails?

4. Compare the thumb and last finger. Why do you think the thumbnail is thicker and larger?

Section Name

Date Instructor

EXERCISE 7

THE INTEGUMENTARY SYSTEM

Review Questions

1. Fill in the following structures of the skin.

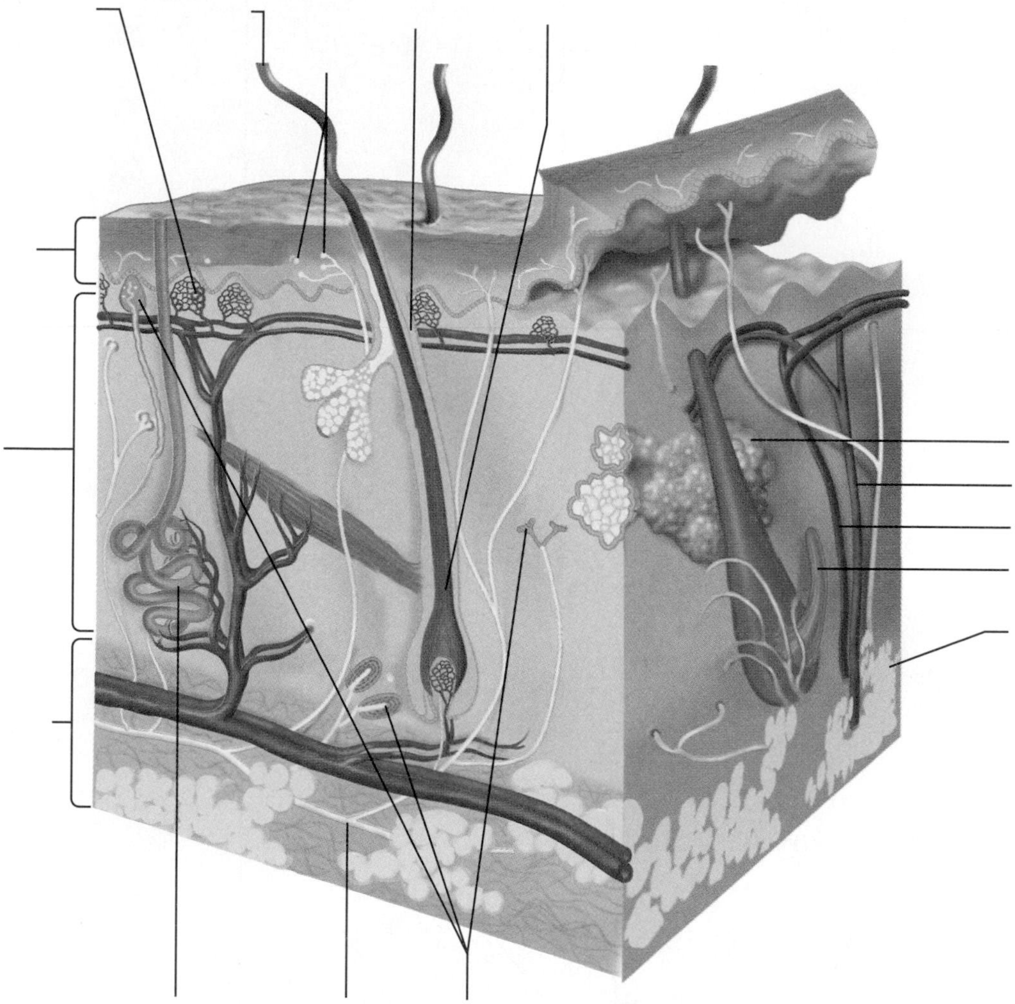

2. Fill in the five strata of the epidermis.

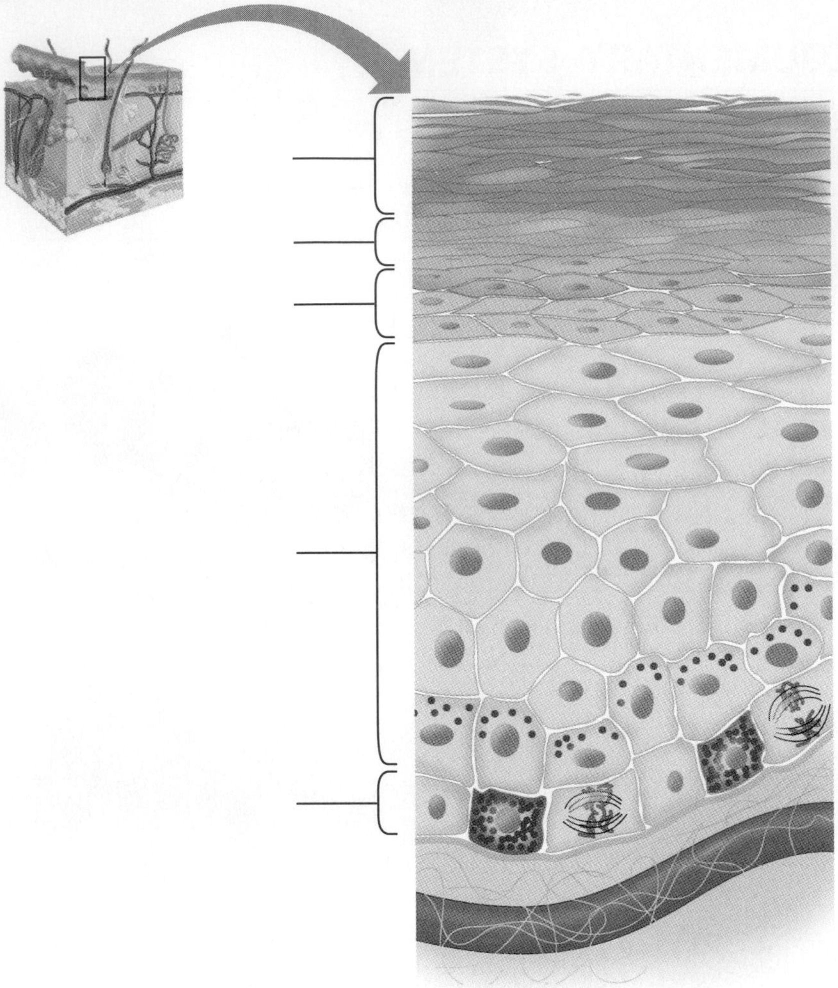

3. Fill in the following structures of a nail.

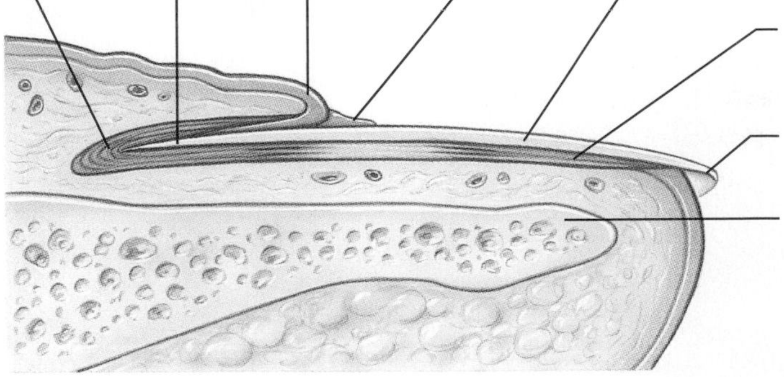

EXERCISE

8

THE SKELETAL SYSTEM

Objectives

After completing this exercise, you should be able to

1. Describe five functions of the skeletal system.
2. Identify the major growth areas of long bones.
3. Locate and describe the major components of the axial skeleton.
4. Locate and describe the major components of the appendicular skeleton.
5. Identify the major components of the knee joint.
6. Illustrate eight major synovial joint movements.

Materials for Lab Preparation

Equipment and Supplies

- ❑ Whole (articulated) human skeletons on individual rolling racks
- ❑ Loose (disarticulated) human bones, stored in individual boxes
- ❑ Fresh (if available) or clean, dry, long bones (cow, pig, or sheep, with most of the meat, ligaments, and tendons removed), both whole and cut longitudinally
- ❑ Dissection equipment
- ❑ Disposable gloves
- ❑ Paper towels
- ❑ Disinfectant in spray bottle

Models

- ❑ Knee joint
- ❑ Leg or knee muscles

Introduction

If you consider the wide scope of activities that you may perform in a single day—from an act as simple as crawling out of bed, to feeding your cat, stepping into the shower, writing a term paper, riding your bike to school, or cooking a meal—and then contemplate the body movements that each of these activities require, you will have some notion of the remarkable maneuverability and complexity of the human skeletal system.

However, the skeletal system has several important functions in addition to **locomotion.** Think, for a moment, about the bones in your body and about the role they might play in all of the systems of the whole body. Discuss your ideas with a lab partner, and write down five functions of the skeletal system.

1. ______________________
2. ______________________
3. ______________________
4. ______________________
5. ______________________

Support of the organs and the entire body and perhaps **protection** of the delicate tissues and soft organs were probably on your list. The following functions may require a bit more background: **storage** of minerals (calcium is one) and lipids, **production** of blood cells, and **leverage** to change the amount and direction of muscle force during movement.

As you compare your original ideas about the role of bones to these additional functions, you will begin to appreciate the fact that each part of your body is multifunctional.

Bone Structure

The human skeleton is composed of two connective tissues: cartilage and bone. In this exercise we will focus primarily on bone.

Bones are the hard elements of the human skeleton. Muscles attach to bones to facilitate movement. Bones also provide support and protection for internal organs. Bones consist of living bone cells, nerves, and blood vessels. There are four categories of bone based on shape: **long, short, flat,** and **irregular.** Long bones include the bones of the limbs and the fingers; short bones, which are short and wide, are the bones of the wrist; flat bones are thin and flattened, such as the cranial bones and the sternum; and irregular bones, such as the coxal (or hip) bones and the vertebrae, have shapes that do not fit into the other categories. Though the bones of the human skeleton vary in size, shape, and composition, we can learn something about the structure of all of them by examining the long bone.

Long Bone Anatomy

Long bones, for example, the femur, consist of a long shaft (called the **diaphysis**) with two rounded heads at each end (Figure 8.1). Each head is called an **epiphysis.** Let's examine a long bone.

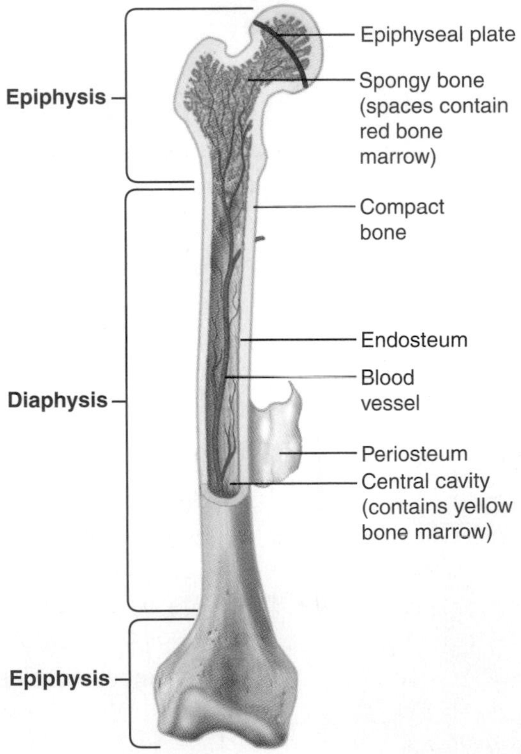

FIGURE 8.1 A long bone with a partial cut to reveal bone structure.

ACTIVITY 1

Examining Long Bone Anatomy

Materials for this Activity

Fresh or prepared long bones from a large animal (e.g., pig, cow, or sheep), whole bones and bones cut longitudinally
Disposable gloves
Dissection tool kit
Paper towels
Disinfectant in spray bottle

1. Prepare an area of your counter for the study by placing several paper towels side by side. Obtain the bones from the supply area. *If you are examining fresh bones, use disposable gloves.* Ask your instructor to identify the source of the bones (cow, pig, or sheep) and the area from which they came (usually the femur [leg] bone). Why would you expect a human leg bone to be proportionally heavier and longer than a sheep or cow bone?

   It performs the same functions

   Looking first at the whole long bone, notice that if the bone is very fresh, the bits of tendon and muscle will still be pink or reddish and soft, the bone will be slightly sticky to the touch, and it will feel heavier and wetter. If the bone is older, the nonbone material will tend to be brownish and hardened, somewhat like beef jerky; the bone will have a rough, nonsticky touch, and it will feel lighter and drier.

2. Check the ends of the whole long bone for a thin translucent line, made of hyaline cartilage, called the **epiphyseal (growth) plate.** As indicated earlier, during puberty, the osteoblasts in the epiphyseal plate become active and form new bone material, which contributes to the length of long bones. If the bone came from an older animal, this line would be barely discernible.

3. With forceps or tweezers, carefully pull away the fibrous, transparent membrane, called the **periosteum,** which covers the outside of the bone. This layer contains active bone cells **(osteoblasts)** that contribute to the diameter or girth of long bones. Have you ever wondered why dogs chew bones? They are scraping the delicious periosteum off the bone with their tongues.

   As bones grow longer and thicker, they become stronger but also heavier. How would heavier bones affect your movements?

   It'll let you move more stable

4. Using a long bone cut lengthwise, use a scalpel or knife to tease or scrape off the transparent membrane, called the **endosteum,** that lines the cavity in the shaft of the bone. This **central cavity** is used to store **yellow marrow,** a fat deposit that serves as an energy reserve. During puberty, the **osteoclasts,** located in the endosteum, erode the inner surface of the bone to counteract the additional weight caused by the bone formation on the outer surface of the bone. The effect is to increase the size of the central cavity.

   As the size of the central cavity increases, the weight of the bones decreases. How does this affect bone strength and body movement?

   It lets you walk onstable and can fall.

5. Examine the cut long bone, and notice the pinkish **spongy bone** in the epiphysis. The color is derived from **red marrow,** which produces the blood cells for the circulatory system. Blood cell production is so important for gas transport and defense that spongy bone is found throughout the body, including the thicker portions of skull bones, where it exists sandwiched between two layers of compact bone.

Run your fingers over the cut portion of the spongy bone. Notice the rough, needlelike texture. Spongy bone consists of a lattice of interconnected calcified rods and plates. The open areas inside the lattice provide space for the red marrow.

Imagine that spongy bones were located in the diaphysis instead of the epiphysis. What might be some of the consequences?

The middle would be growing
& the end would be weaker

6. Look at the white or creamy **compact bone** in the diaphysis. Knock on it, and notice that it is rigid and strong. Why are these characteristics important in terms of bone function?

It can protect the marrow.

Run your fingers over the surface of the shaft of the bone. Notice that the texture of compact bone is considerably smoother than spongy bone. The smooth texture reduces the friction as bone surfaces rub against the surrounding muscles, tendons, and ligaments.

7. If you have been examining fresh long bones during this activity, now notice their weight. To what would you attribute this? How do you think their weight would compare to that of dry bones?

The dry bone would be lighter
due to the lack of water, meat, etc.

8. Return the bones to the supply area. Clean your counter by wiping it down with disinfectant and paper towels. Dispose of your gloves, and wash your hands. ■

Organization of the Skeletal System

The adult human skeleton is formed from 206 bones and various other types of connective tissue that hold them together. The skeleton provides general structural support, as well as support for soft organs. It also protects certain organs from injury, such as the brain, which is enclosed within the bones of the skull. Another vital function of the human skeleton is that it allows for flexible movement at the joints. Finally, the human skeleton is organized into the **axial skeleton** and the **appendicular skeleton** (Figure 8.2).

Axial Skeleton

The axial skeleton consists of 80 bones, including the skull, vertebral column, ribs, and sternum (or breastbone). The **skull,** which consists of the cranium and the face, is the most superior feature, and

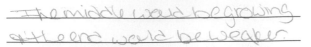

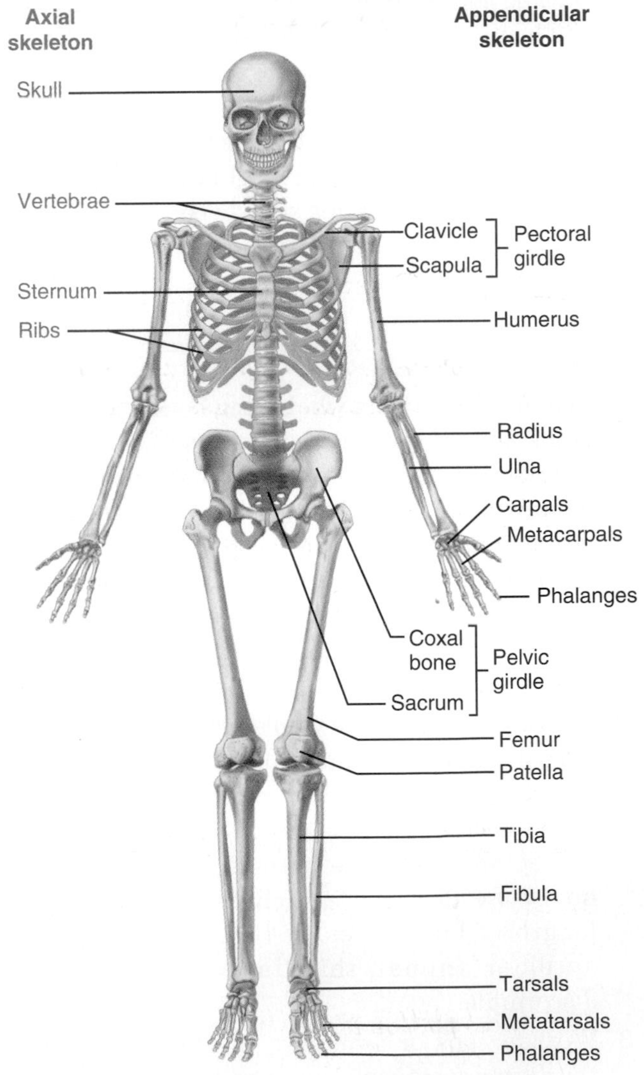

FIGURE 8.2 The human skeleton. Bones of the axial skeleton are indicated by blue labels, and bones of the appendicular skeleton are indicated by black labels.

it rests securely on top of the vertebrae. The **vertebral column** provides a central axis that supports the head, protects the spinal cord, and anchors the body trunk.

The **rib cage** is formed by the ribs, sternum, and vertebral column. It surrounds and shields the chest organs, such as the heart and lungs. Speculate on the reason for the cartilage connecting the ribs to the sternum. Would elongation of the rib bones to replace the cartilage be a better design?

____

____

ACTIVITY 2

Examining the Skull

Materials for this Activity

Whole (articulated) human skeleton on rolling rack

or

Loose (disarticulated) individual human bones in boxes

1. Obtain a whole human skeleton on a rack and roll it to your work area, or use the individual bones. Be careful not to jar or catch the limbs if you are using the articulated skeleton, especially the hands and legs, as you navigate through the lab. *We encourage you to touch and handle the individual bones, but do so gently.* Firmly and gently run your fingers along the surfaces, feeling the different textures, bumps, holes, and ridges.
2. Examine the skull first. The skull is sometimes called a helmet. Why would this be an appropriate description?

   It protects the brain.
3. The **facial bones** make up the front of the skull.

   The **maxilla** is the largest facial bone. The upper portion forms part of each eye socket, and the bottom portion forms the upper jaw and contains the sockets for the upper row of teeth.

   The **mandible,** or lower jaw, contains the sockets for the lower row of teeth. This is the only bone that is not tightly joined with the rest of the skull bones. The mandible is attached by the jaw joint to the temporal bone, which is discussed later. This separation from the skull allows us a full range of motions, for example, the ability to chew and speak. Notice that the mandible is quite thick and strong. Can you offer an explanation for this?

   ____

   ____

   ____
4. Compare the lateral and inferior views of the skull to Figure 8.3a–b. Identify on the skull each of the bones labeled in the illustration by starting with the lateral view at the top (Figure 8.3a).

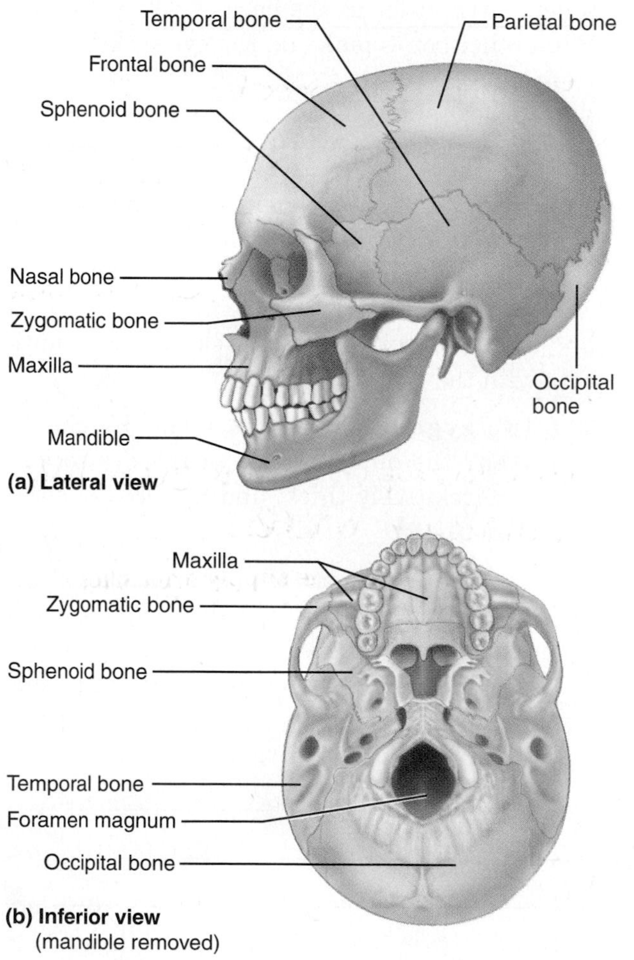

FIGURE 8.3 The skull. (a) Left lateral view. (b) Inferior view with mandible removed.

The **frontal** bones make up the forehead and the upper portions of the eye sockets. They serve to protect the part of the brain involved in thinking, memory, learning, emotions, and personality. The **parietal and temporal bones** are located on the left and right of the skull. Together, they make up the majority of the back half of the skull. They serve to protect the part of the brain involved in coordinating body movements and awareness of sensations, such as touch, hearing, smelling, and so on.

Examine the texture, ridges, and depressions of these four bones. Notice the smooth surfaces, interior and exterior, as opposed to the rough material inside the skull bones. The rough material is called spongy bone, and serves to produce blood.

5. A small portion of the **sphenoid** is visible in the temple area. It forms the back of the eye socket. What other bones make up the eye socket?

   Ethnoid, temperal

6. The back and base of the skull is formed by the **occipital bone,** which contains a large opening called the **foramen magnum.** This opening is where the vertebral column connects to the skull and the spinal cord enters the skull to communicate with the brain.
7. The two **zygomatic bones** form the cheekbones, and the outer portion of the eye socket. They are relatively thick, and protect the sides of our face.
8. The **nasal bones** form the bony bridge of the nose; the rest of the nose is made of cartilage and other connective tissues. Notice these bones are not that thick, and can easily be broken. If your nose is broken, why does it never heal properly without reconstructive surgery?

   Your nose is mostly made out of cartilage so without surgery the cartilage will heal in the wrong direction.

ACTIVITY 3

Examining the Vertebral Column

Materials for this Activity

Whole (articulated) human skeleton on rolling rack

1. Looking at the whole articulated skeleton, you can see that the **vertebral column** is composed of 33 vertebrae—24 single bones and 9 fused bones.
2. The vertebral column has an "S," or sinusoid, pattern. Speculate on the advantages of a curved, versus a straight, backbone.

   It can support you body + brain (head) better. Good for the nerves + comfort.

3. Compare the skeleton to Figure 8.4, and identify each region starting at the top. Notice the **cervical curve** in the neck is formed from the first seven bones, the **cervical vertebrae.** Why are the cervical vertebrae the smallest and most delicate of all vertebrae in the human skeleton?

   It connects to the cerebellum.

   The thoracic curve in the upper back is formed from the next 12 bones, the **thoracic vertebrae.** The **lumbar curve** in the lower back follows, and it is formed from the next five bones, the **lumbar vertebrae.** Notice that the lumbar vertebrae appear larger and stronger than the vertebrae above them. Can you provide an explanation for this phenomenon?

   The lower back has more weight bearing it.

   The **sacral curve** is last, and it is formed from the **sacrum,** which is a single bone composed of five fused bones. The **coccyx,** or tailbone, continues the sacral curve. It is a single bone

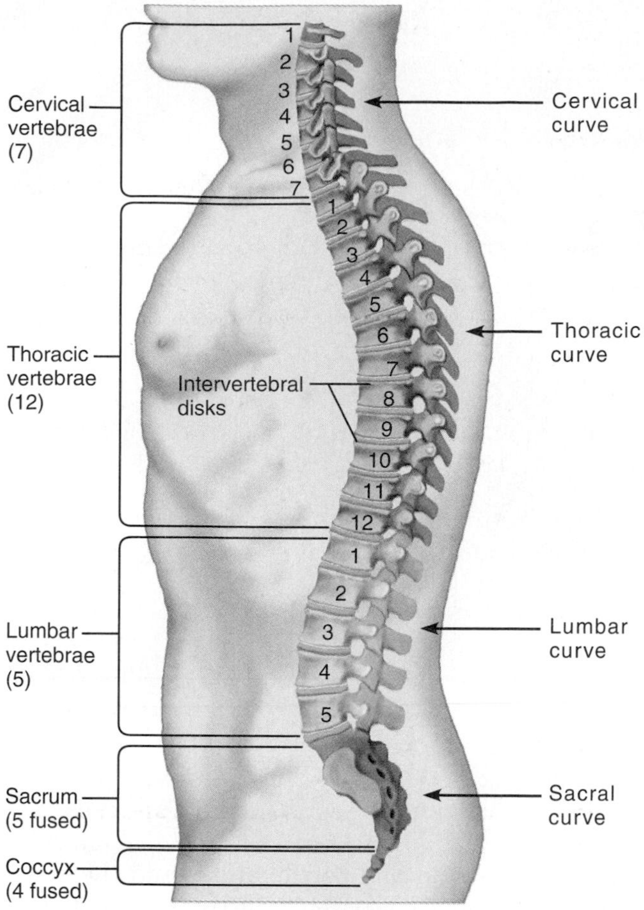

FIGURE 8.4 The vertebral column.

formed from four fused bones. Which bones of the vertebral column might be considered vestigial (from *vestiges,* or remains, meaning something that no longer serves a purpose)?

Coccyx

4. Notice that the individual vertebrae are stacked, but they are cushioned from each other by the **intervertebral disks,** which are made of a strong connective tissue called fibrocartilage. Speculate on the function of these disks (think about your various physical activities during the day).

They are used to seperate the vertebrae so they won't rub on eachother decrease their capacity.

Sudden impact or movement can compress an intervertebral disk, forcing the softer center to balloon outward and cause intense back pain. This condition is called a "herniated disc." What would the effects be, if any, on mobility if the damaged disk was surgically removed?

Spine would get smaller + the back will contract

5. Protection of the spinal cord is a crucial function of the vertebral column. When injured on the playing field or during an automobile accident, injured parties are always required to lie absolutely still until they have been examined by a medical professional. ■

ACTIVITY 4

Examining the Rib Cage

Materials for this Activity

Whole (articulated) human skeleton on rolling rack

1. Humans have 12 pairs of **ribs,** which are anchored on one end of the vertebral column in the thoracic region (Figure 8.5). Let's begin our examination of the rib cage at the other end though, with the **sternum.** The sternum is more commonly known as the breastbone, and is a flat blade-shaped bone, consisting of three separate bones that have fused. The **manubrium** is at the top of the sternum, followed by the **body**, which is the bulk of the sternum, and then the **xiphoid process.** This process serves as an attachment for abdominal muscles.
2. Now look at the ribs. The topmost seven ribs are called **true ribs** because they attach *directly* to the sternum. The second set of five ribs are called **false ribs** because they attach *indirectly* to the sternum, in conjunction with the cartilage of rib 7. The last set of two ribs, which are also grouped with the false ribs, are called **floating ribs** because they do not attach at all to the sternum.

Which ribs are the longest? What might be the reason for this?

True bones

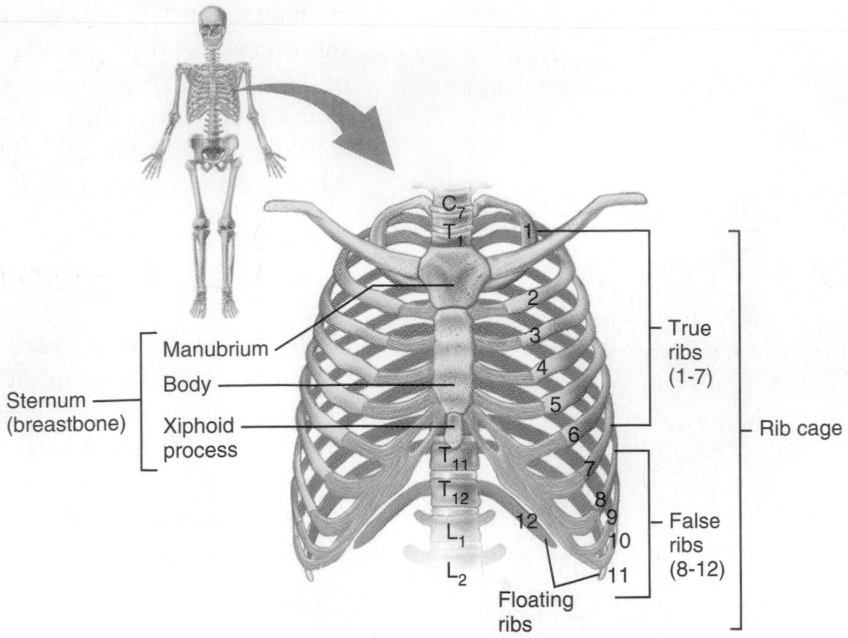

FIGURE 8.5 The rib cage.

Why is the rib cage narrowest on the top and widest at the bottom?

The lungs need room to expand.

Why would it not make sense for the rib cage to be formed of two solid sheets of bones instead of individual rib bones?

When they are individual, they can cover more.

Notice that the rib cage does not extend into the abdominal area. Speculate on the reasons for this.

You need room for the intestines.

Appendicular Skeleton

The appendicular skeleton consists of those parts of the body that attach, or are *appended* to, the axial skeleton, which forms the longitudinal axis of the body. The pectoral and pelvic girdles and the bones of the upper and lower limbs of the body make up the appendicular skeleton. The **pectoral** (shoulder) **girdle** attaches the arms to the axial skeleton and allows for a lot of mobility of the arms. In contrast, the **pelvic** (hip) **girdle,** which attaches the legs to the axial skeleton, must support the body weight of the upper body against gravity, and allows a lot less mobility of the legs.

Try swinging your arms and legs in as big an arc as possible. Which one swings more easily? Try to explain the reason for differences in mobility between arms and legs.

The Arms because it contains less weight than than the leg

Compare the function of the two girdles, using the words "flexibility" and "stability."

Pectoral girdle function: Stability

Pelvic girdle function: Flexibility

ACTIVITY 5

Examining the Pectoral and Pelvic Girdle

Materials for this Activity

Whole (articulated) human skeleton on rolling rack

Pectoral Girdle, Arm, and Hand

1. The pectoral girdle (Figure 8.6) consists of the **clavicle** or collarbone, which extends across the top of the chest, and connects to the sternum or breastbone. The clavicle anchors many muscles and serves to brace the shoulders away from the chest. When the clavicle is broken, the shoulder region will collapse toward the center.

   Look at the slender clavicle. Do they look very strong? If you were to fall on your upper body, what might happen to your clavicles? If they don't heal, what might happen to your shoulder?

   It breaks. The shoulder ______

   can collapse. ______

2. The second part of the pectoral girdle is the **scapula** or shoulder blade. It connects to the clavicles and provides a socket for the arms. Notice the scapula is triangular.

   In a dislocated shoulder, the arm is no longer anchored to the scapula. How does this affect the effective use of the arm?

   It won't be able to move up. ______

   ______

3. Each arm consists of the **humerus, ulna,** and **radius.** What joint connects these three bones?

   ______

4. Each hand is made of two regions: 8 **carpals** and 5 **metacarpals** make up the palm, and 14 **phalanges** make up the fingers.

   What joint connects the hand to the arm? ______ How many bones make up the two hands? ______ Which finger has only two bones? ______

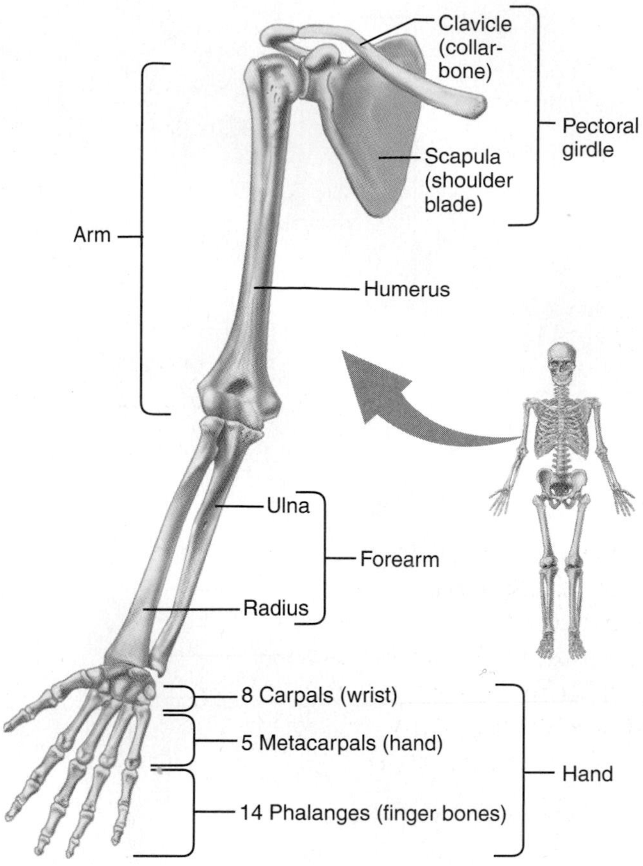

FIGURE 8.6 The pectoral girdle, arm, and hand.

5. Examine the arrangement of the "opposable thumb." What are the benefits of this hand design? *Hint:* pick up an object with just the fingers.

   ______

   ______

   ______

Pelvic Girdle, Legs, and Feet

1. Examine the **pelvic girdle,** and compare it to the representation in Figure 8.7. The pelvic girdle supports the weight of the upper body, protects the pelvic organs, and attaches the leg to the vertebral column.

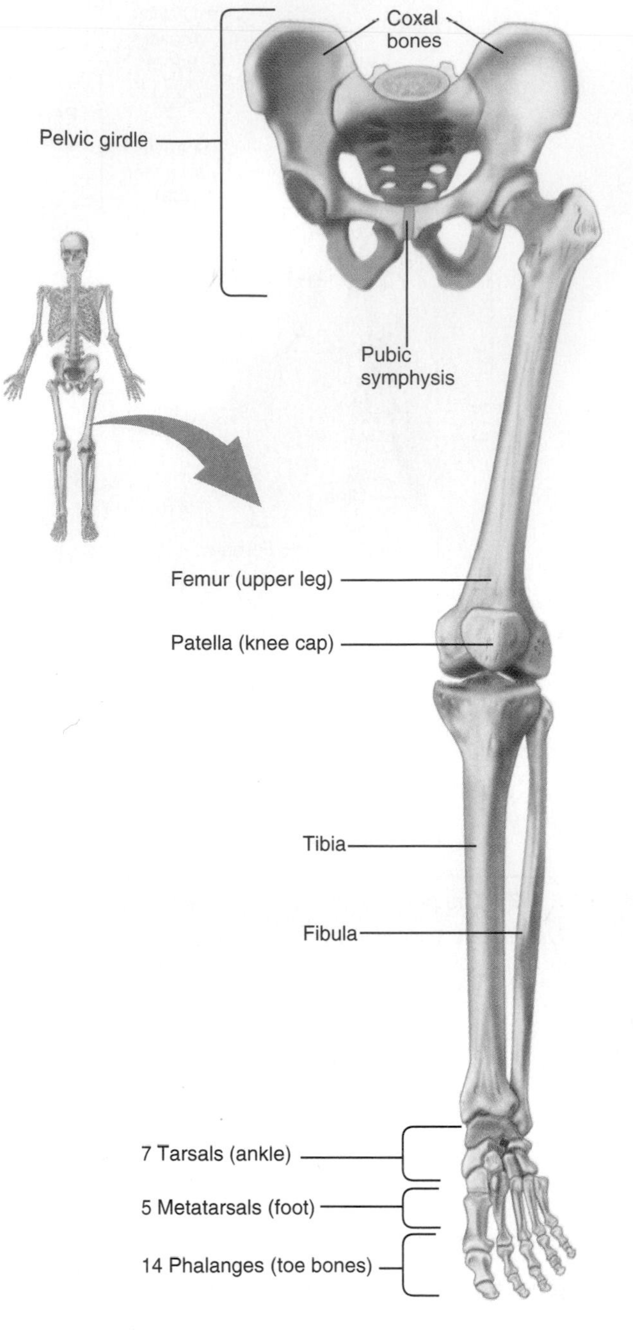

FIGURE 8.7 The pelvic girdle, leg, and foot.

2. The left and right **coxal bones** make up the pelvic girdle. Notice, these bones connect the lower extremities to the axial skeleton via the sacrum and coccyx. In adult women, this girdle is broader and shallower than in men, and the opening is wider. Speculate on the reasons.

____

____

3. Examine the thigh now, which is formed by the **femur,** the longest and strongest bone in the body. The lower leg is formed by the **tibia** and **fibula.** From a superior view, the tibia connects with the femur and **patella,** or kneecap, to form what structure?

____

From an inferior view, they connect with the tarsals to form what structure?

____

4. Each foot is formed by three regions: 7 **tarsals** make up the ankle and heel, 5 **metatarsals** make up the sole of the foot, and 14 **phalanges** form the toes. Like the thumb, the big toe only has two bones, yet it is thicker than the other phalanges. Speculate on the reason.

____

____

Joints

Joints, also called articulations, are the points of contact between bones. They are attached to the bones and stabilized through ligaments and tendons. Joints vary considerably in how much freedom of movement is allowed. There are three types of joints. Fibrous joints are immovable, and they are found in the sutures, which firmly connect and stabilize the skull bones. Cartilaginous joints are slightly movable, and they are found in the intervertebral discs that cushion and connect the vertebrae in the backbone and allow a small degree of flexibility. Synovial joints are the most freely movable. They are found in the knee, elbow, hip, and many other places.

ACTIVITY 6

Exploring Joint Movements

1. Let's perform some simple joint movements. Loosen your clothing, especially the collar and sleeves. Step to a clear area of the lab, and take turns demonstrating the movements depicted in Figure 8.8 with your lab partner. Notice that joint movements usually occur in pairs. A movement away is complemented with a movement that returns the body part to the original point.

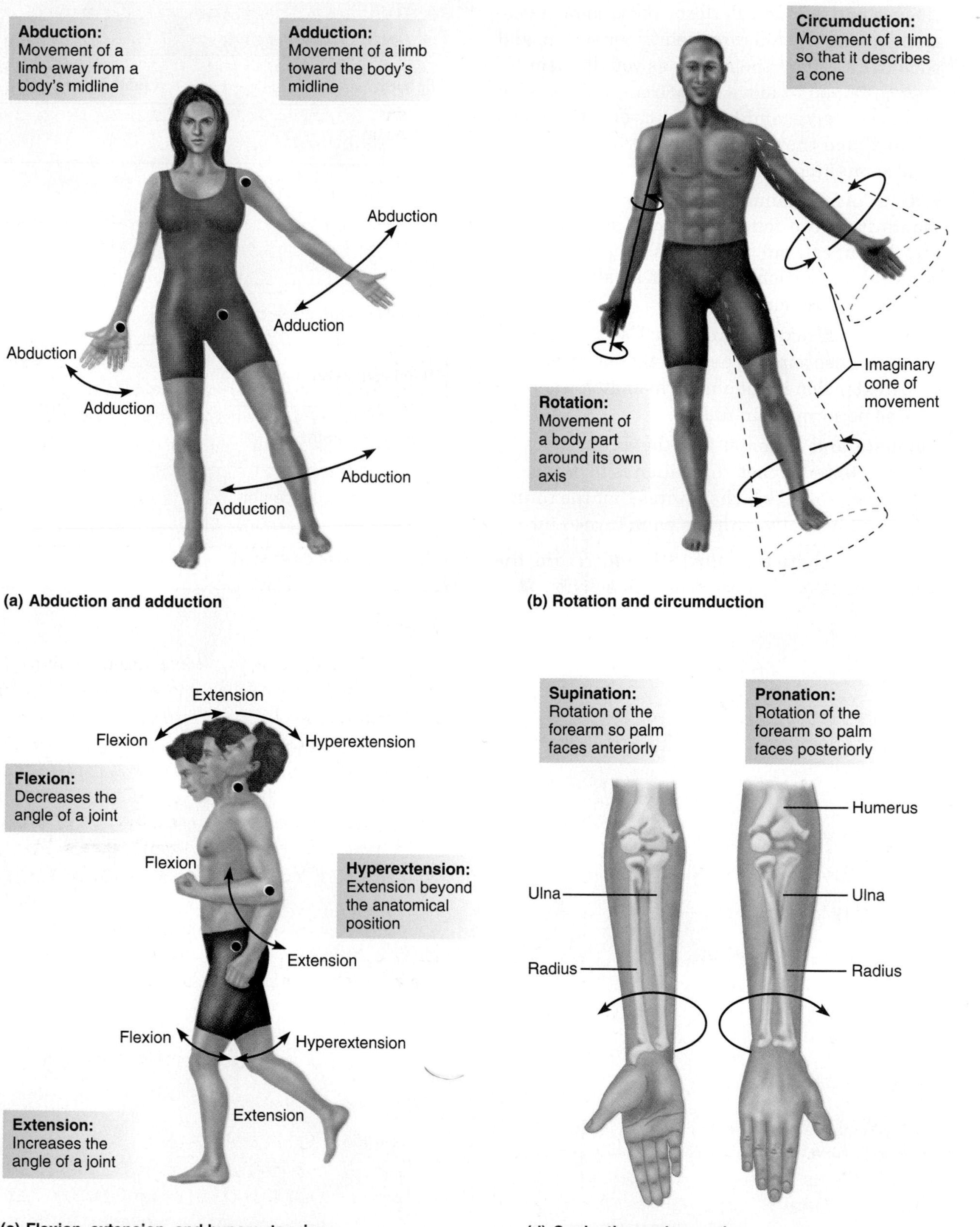

FIGURE 8.8 Various movements made possible by the synovial joint structure. (a) Abduction and adduction. (b) Rotation and circumduction. (c) Flexion, extension, and hyperextension. (d) Supination and pronation.

2. Abduct and Adduct: Perform these joint movements using the following joints: wrist, hip, and shoulder. Call out the names as you do them. To *abduct* means to move away from midline. Can you see the resemblance between the word "abduct" and the word "kidnap," which means to steal something or someone away from home?
3. Rotate and Circumduct: Perform these movements from the following joints: shoulder and hip. Again, call out the names as you do them. Notice that rotation is a tighter, smaller circular move than circumduction.
4. Flex, Extend, and Hyperextend: Use the neck, elbow, and knee joints to perform these movements. Hyperextension is not a natural anatomical position, so perform it carefully.
5. Supinate and Pronate: Use the wrist joint. An old trick to remember supination is to connect it to "holding soup," which requires that the thumb be turned laterally with the palm turned up.

Perform the activities in Box 8.1, and record the joint movements used to perform each activity. ■

Box 8.1 Joint Movements

Activity	Joint Movement
Lift a cup	
Put down a cup	
Twist your arm (turn a doorknob)	
Turn your palm upward	
Bend your head way back	
Move a leg to the side	
Move a leg to the center	
Turn your palm downward	

EXERCISE

9

THE MUSCULAR SYSTEM

Objectives

After completing this exercise, you should be able to

1. Explain the role played by muscle tissue in the human body.
2. Describe the microstructure of skeletal muscle cells.
3. Describe how skeletal muscle is stimulated.
4. Identify the major skeletal muscles, and describe their functions.
5. Explain the basic mechanism of muscle contraction.
6. Explain the basic physiology involved in muscle contraction.

Materials for Lab Preparation

Equipment and Supplies

- ❑ Compound light microscope
- ❑ Gripper or other dynamometer capable of measuring the force of contraction
- ❑ Assorted dumbbells of various weights
- ❑ Cloth tape measure

Prepared Slides

- ❑ Skeletal muscle (teased)
- ❑ Neuromuscular junction
- ❑ Muscle (cross section)

Models

- ❑ Muscle cell showing parts of a myofibril
- ❑ Muscle cell and motor neuron showing the neuromuscular junction
- ❑ A whole muscle and/or skinned chicken leg with muscle attached to bone
- ❑ The major human muscles

Introduction

As you learned in Exercise 5, Tissues, muscle tissue is unique in its ability to shorten or contract. Contraction of muscle cells is used by the human body to do work in a variety of ways. **Smooth muscle** is used to push things through many internal organs, and it is also used to regulate the pressure and flow of blood throughout the body. **Cardiac muscle** is formed into a hollow structure that is used to pump blood. The heart is a major topic, which is covered later in this lab manual. Finally, **skeletal muscle** is the voluntary muscle tissue that is typically attached to the skeleton for movement. This exercise focuses on skeletal muscle tissue and the organs it forms. However, it may be useful for you to review the muscle tissue section of Exercise 5 to help put this exercise in proper perspective.

Skeletal Muscle Cell Microstructure

In Exercise 5, you learned that skeletal muscle cells are **striated,** that is, they have a cross-striped appearance (Figure 9.1a). The reason for these alternating bands of light and dark areas is a precisely arranged pattern of thin and thick contractile proteins, or **myofilaments.** The thin myofilaments are mostly made of a protein called **actin,** and the thick myofilaments are mostly made of a protein called **myosin.** These myofilaments are bundled together within the muscle cell into structures called **myofibrils** (Figure 9.1b).

Figure 9.2 depicts how these proteins are arranged in a myofibril. Note that the area where the

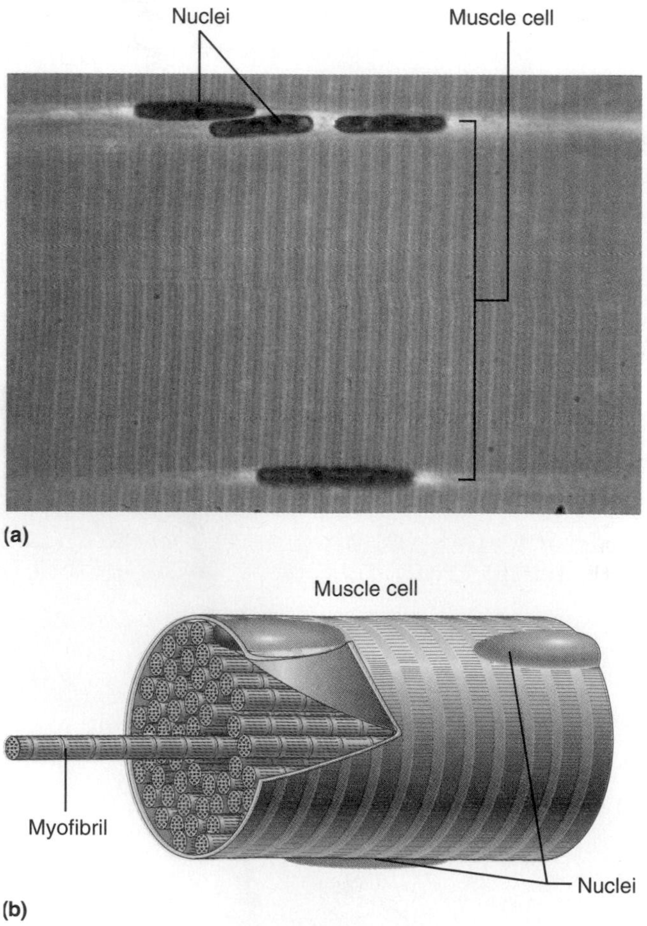

FIGURE 9.1 Muscle cells. (a) Photomicrograph of several muscle cells (2000×). (b) A single muscle cell containing myofibrils.

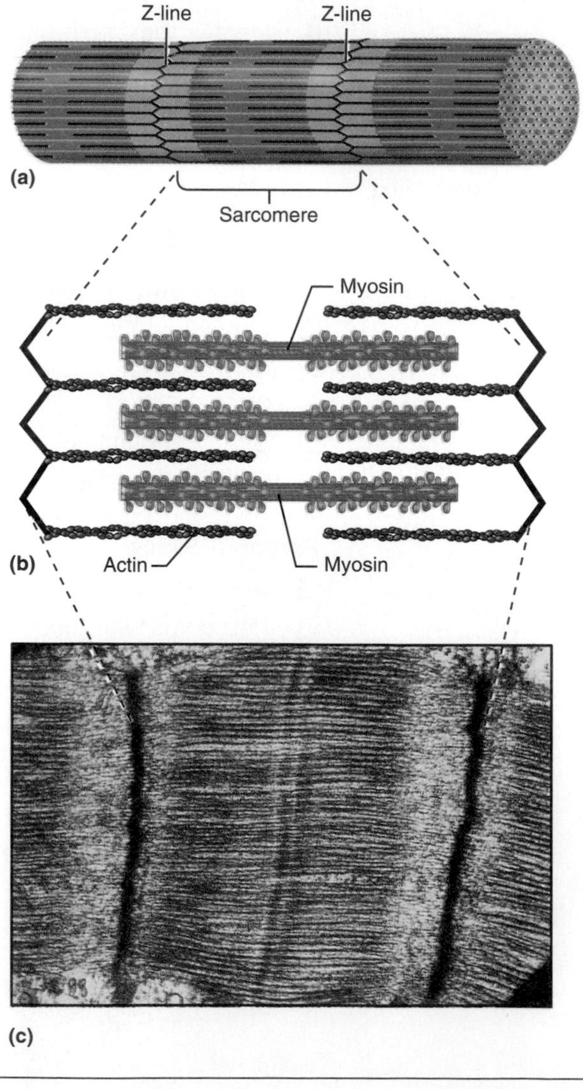

FIGURE 9.2 (a) Structure of a myofibril. (b) Structure of a sarcomere. (c) Photomicrograph of a sarcomere.

thick proteins are located has a darker appearance, while the area containing the thin proteins has a lighter appearance. Also note the jagged actin backbone, which is called the **Z-line.** The area between any two Z-lines is the basic contractile unit of skeletal muscle, called the **sarcomere.**

✱ ACTIVITY 1

Learning Skeletal Muscle Microstructure

Materials for this Activity

Compound light microscope

Prepared slide of skeletal muscle (teased)

Model of muscle cell showing parts of a myofibril

View a slide of skeletal muscle, and carefully focus and adjust the light so that you can see the striations, or alternating light and dark areas. The light areas are where we mainly have the thin actin proteins. The darker areas are where we mainly have the thick myosin proteins. Now view a model of a muscle cell showing the details of a myofibril. Identify the thick and thin myofilaments, Z-lines, and sarcomeres. Sketch a sarcomere in the space provided below, labeling the previously mentioned structures. ■

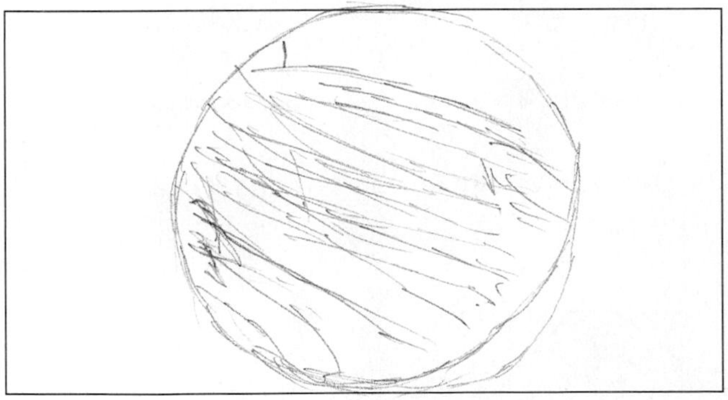

Stimulation of Skeletal Muscle Cells

Nearly all skeletal muscle is under the direct control of our conscious brain, which is why it is also referred to as voluntary muscle. Although the diaphragm and some of the muscles in the neck are sometimes exceptions, we usually consciously direct movement of skeletal muscles. The **motor neuron** is the nerve cell, which carries the message from the brain to the muscle cells we want to stimulate. The specialized area where the motor neuron interacts with a muscle cell is called the **neuromuscular junction.**

Figure 9.3 illustrates a neuromuscular junction. The neuron's long, slender process, the **axon,** may branch and lead to more than one muscle cell. A motor neuron and all of the muscle cells that it will stimulate is called a **motor unit.** Also see Figure 9.4, which shows this branching of the motor neuron.

Also note that the motor neuron and muscle cell do not quite touch, leaving a space in-between called the **synaptic cleft.** Unlike electrical wires and the devices they supply, the stimulus from a motor neuron is brought to the muscle cell by using chemicals called **neurotransmitters.** In the case of most muscle cells, the neurotransmitter is acetylcholine.

ACTIVITY 2

Learning About the Neuromuscular Junction

Materials for this Activity

Compound light microscope

Prepared slide of a neuromuscular junction

Model of a muscle cell and motor neuron showing the neuromuscular junction

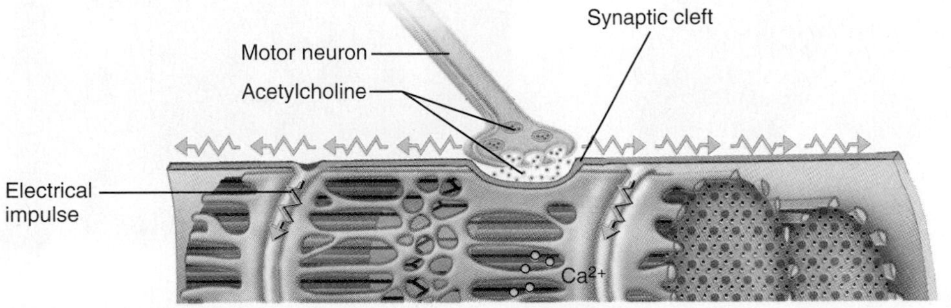

FIGURE 9.3 The neuromuscular junction.

View a slide and/or a model of the neuromuscular junction. Note the branching of the motor neuron to several muscle cells (Figure 9.4a,b) and the small gap between the neuron and the plasma membrane of the muscle cell. Make a sketch of the neuromuscular junction in the space below. ■

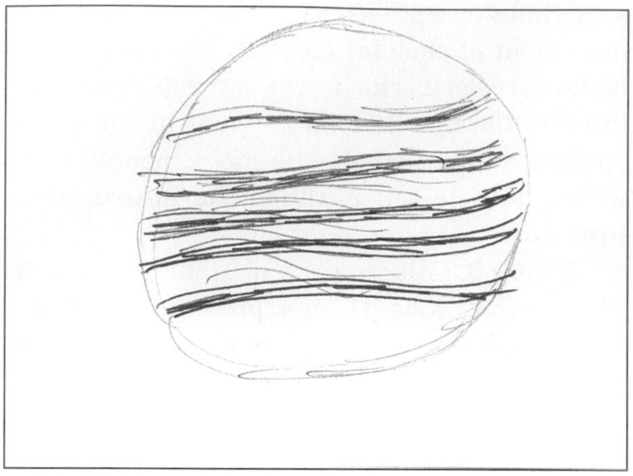

Whole Skeletal Muscle Organization

Motor neurons may stimulate only a small percentage of skeletal muscle cells when lifting or moving something that is lightweight. Nonetheless, the muscle must contract as a whole. To accomplish this, the muscle cells must be fastened together with connective tissue.

Figure 9.5 illustrates how a whole muscle is organized. Note the individual muscle cells, each one surrounded by a thin sheath of connective tissue (called the endomysium). Muscle cells are bundled into **fascicles** and are also surrounded by connective tissue (called the perimysium). Finally, all of the fascicles are wrapped together by connective tissue (called the epimysium). These connective tissues typically merge to form a **tendon,** which attaches the muscle to a bone or some other structure.

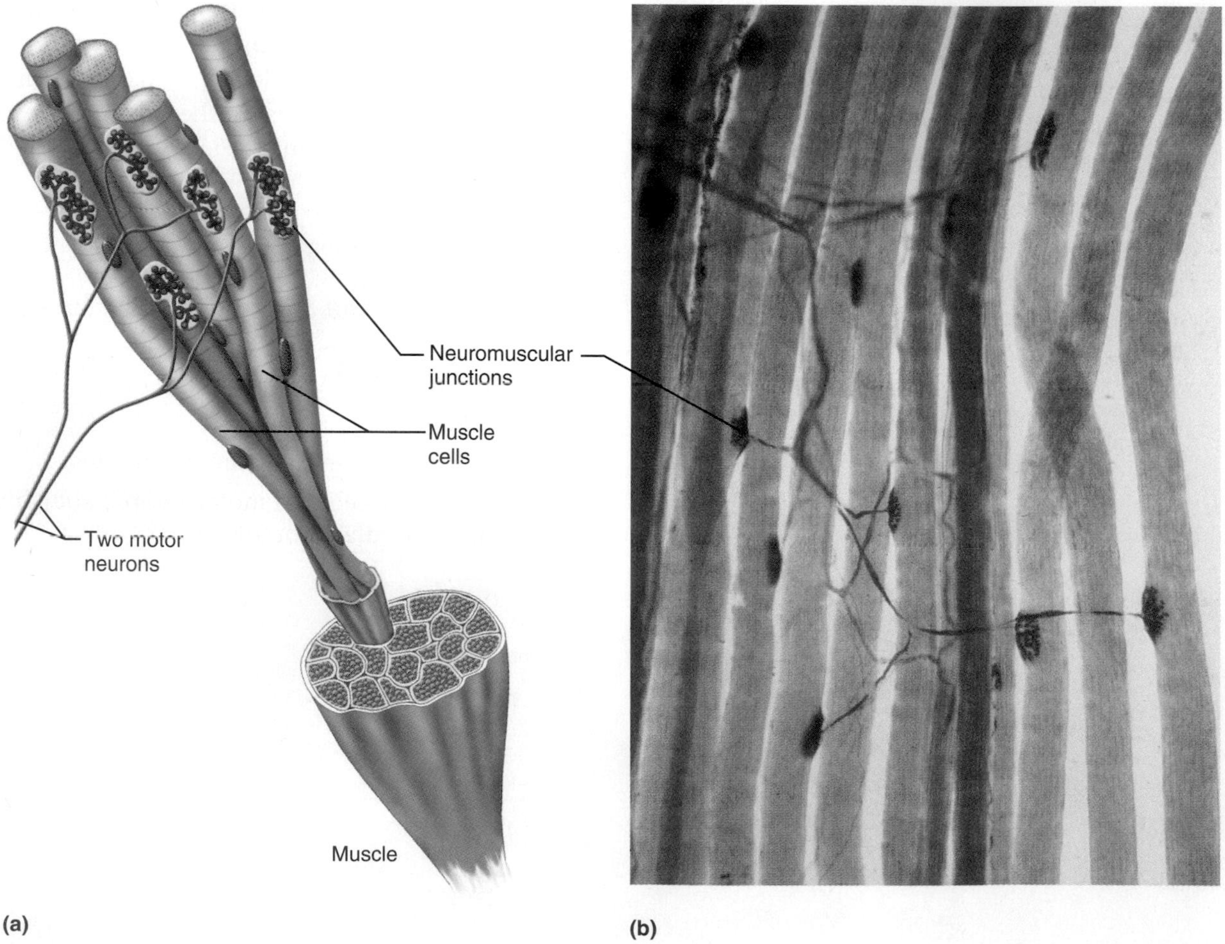

FIGURE 9.4 Photomicrograph and drawing of the neuromuscular junction.

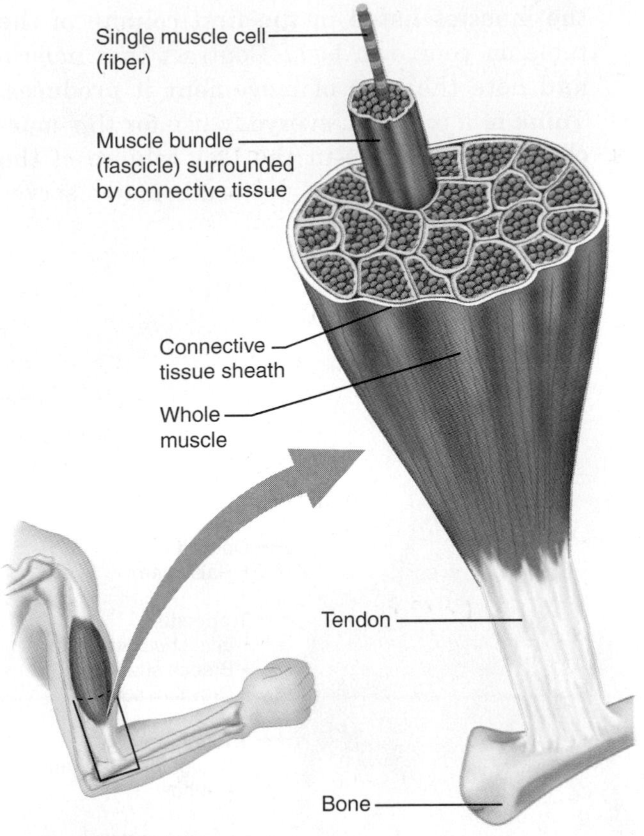

FIGURE 9.5 Organization of whole muscle.

ACTIVITY 3

Observing Whole Muscle Organization

Materials for this Activity

Compound light microscope

Model of a whole muscle and/or skinned chicken leg with muscle attached to bone

Prepared slide of muscle (cross section)

1. Observe the muscle model and/or a skinned chicken leg. Observe the connective tissue, which surrounds the muscle and tapers down to the tendon. If using a chicken leg for examination, feel the muscle, fascia, and tendons. *Note:* Wash your hands after handling the chicken leg to prevent the spread of bacteria.
2. View a prepared slide of a muscle cross section. Note that the muscle cells are grouped into bundles (fascicles) and are surrounded by connective tissue. Sketch the muscle cross section in the space provided below, labeling the fascicles, perimysium, and epimysium. ■

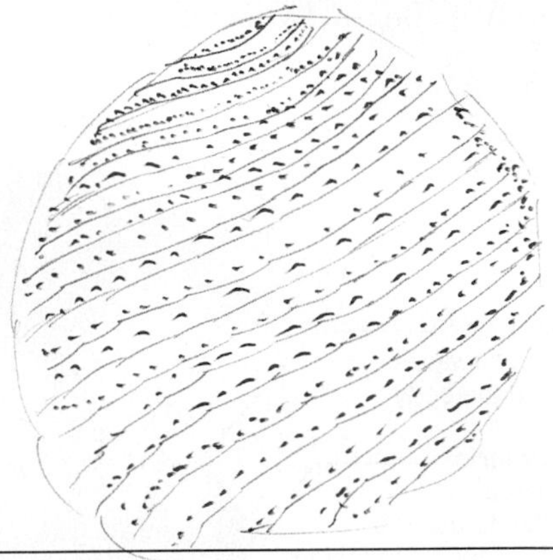

Gross Anatomy of Skeletal Muscles

Muscles are most commonly studied by learning their **origin** (usually the stationary or anchoring point of attachment), **insertion** (usually the more movable point of attachment), and **action** (what movement they cause when they contract). This movement is described using the terminology covered in the joints section of Exercise 8. Review Figure 8.8 to ensure that you are familiar with the terms that describe movement at a joint.

ACTIVITY 4

Observing the Gross Anatomy of Skeletal Muscles

Materials for this Activity

Models displaying the gross anatomy of the major human muscles

1. Observe the models of human muscles available for study in your laboratory, and consult Figure 9.6 to aid you in identifying the selected muscles.
2. Complete the final column (Example of Use) in Table 9.1 using the knowledge you have obtained so far in this lab. Figure 9.6 is also a good reference. Begin by visualizing the location of the muscles listed in the first column of the table on your own body. Contract that muscle and note the type of movement it produces. Think of a common, everyday use for this muscle. Write this use in the last column of the table. The first muscle, pectoralis major, serves as an example. ■

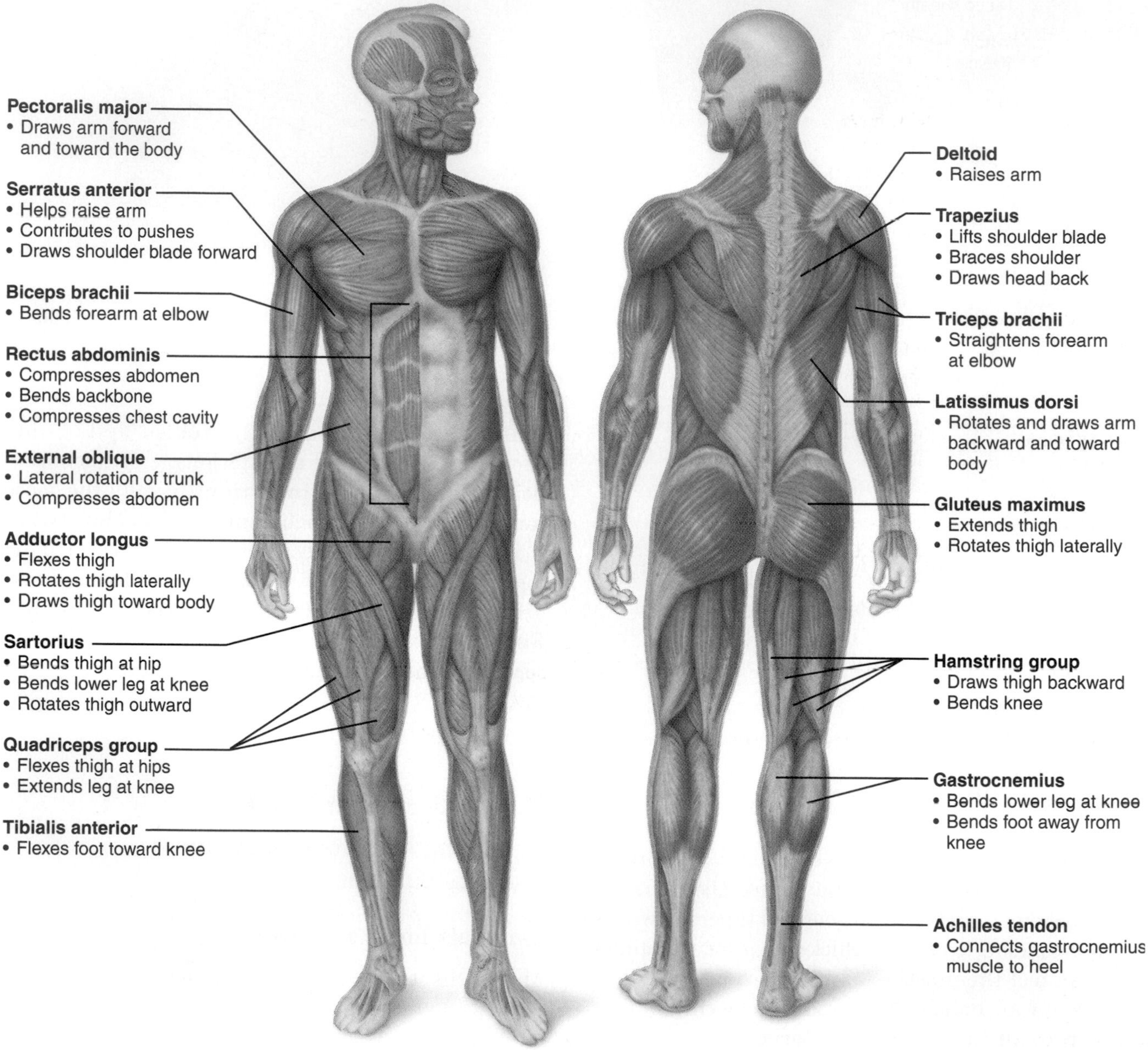

FIGURE 9.6 Gross anatomy of selected muscles.

Table 9.1 Selected Skeletal Muscles

Muscle	*Location*	*Function*	*Example of Use*
Pectoralis major	Mostly under the breast area; spans from sternum to humerus	Pulls the arm forward and down	Hugging someone around the waist
Serratus anterior	On the side of the body under the armpit (axillary) area	Stabilizes the shoulder and contributes to arm movements such as pushing	
Biceps brachii	On the anterior side of the upper arm (humerus)	Bends (flexes) the arm at the elbow joint	
Rectus abdominis	Runs from the ribs to the pelvis on the anterior side of the body	Bends (flexes) the backbone; compresses the abdomen	
External oblique	Lateral sides of the abdomen; angles down toward the center of the abdomen	Acts the same as the rectus abdominis when both are contracted together; rotates the trunk laterally	
Adductor longus	Inner thigh near the groin	Flexes the thigh; rotates thigh laterally; draws the thigh toward the body	
Sartorius	Long, belt-shaped muscle that runs diagonally across the thigh from hip to inside of the knee area	Bends the thigh at the hip; bends the lower leg at the knee; rotates the thigh outward	
Quadriceps (femoris) group	Group of four large muscles on the anterior side of the thigh	Raises (flexes) the thigh; extends the lower leg at the knee	
Tibialis anterior	In the front and slightly to the side of the tibia	Raises (flexes) the foot	
Deltoid	Covers most of the shoulder	Raises (abducts) the arm	
Trapezius	Large, diamond-shaped muscle of the upper back	Pulls the head back (extension); moves and braces the shoulder	
Triceps brachii	On the posterior side of the upper arm (humerus)	Extends the arm at the elbow joint	
Latissimus dorsi	Large muscle of the middle back	Pulls the arm back and down (adducts the arm)	
Gluteus maximus	Covers most of the buttocks area	Kicks and rotates the leg backward (extends the leg)	
Hamstring group	Group of three large muscles on the posterior side of the thigh	Flexes the leg (bends the knee)	
Gastrocnemius	Posterior calf muscle	Bends the foot downward (extension)	

Contraction of Skeletal Muscles

As explained earlier, skeletal muscles contract when stimulated by neurons. But what exactly makes the cells shorten or contract? The answer lies in our previous explanation of the microstructure of skeletal muscle cells. The thick myosin proteins have small projections called **myosin heads** that grab the actin fibers and pull them toward the center of the sarcomere (Figure 9.7). Thus, the myofilaments (actin and myosin) appear to slide along one another. The myosin heads have enzymes that split ATP molecules. It is the energy from the splitting and release of ATP that allows the myosin heads to cock back and then pull the actin fibers. Consult your textbook for a more thorough description of the **sliding filament mechanism** of muscle contraction.

The ATP-splitting enzymes seem to work more efficiently at elevated temperatures, which is one of several reasons why a muscle that has been properly warmed up by light exercise performs better and is less likely to suffer injury.

ACTIVITY 5

Measuring the Force of Skeletal Muscle Contraction

Materials for this Activity

Gripper or other dynamometer capable of measuring the force of contraction

Work in pairs. Obtain a gripper or other dynamometer, and read the directions for use.

1. Squeeze the gripper five times, and record the average strength of contraction.

   50 60 60 65 65 120 170 130 5)300 60

2. Rest for 1 minute, squeeze the gripper five times, and record the average strength of contraction.

   5)310 62

3. Rest for 1 minute, squeeze the gripper five times, and record the average strength of contraction.

   5)320 64

Which average was the lowest?

60

Propose a reason for your results.

The more you rest, the stronger you can be ■

Energy Use by Skeletal Muscles

As explained earlier, myosin heads require a molecule of ATP in order to cock back and then move the actin fibers. Considering that there are many millions of myosin heads in the smallest of muscle

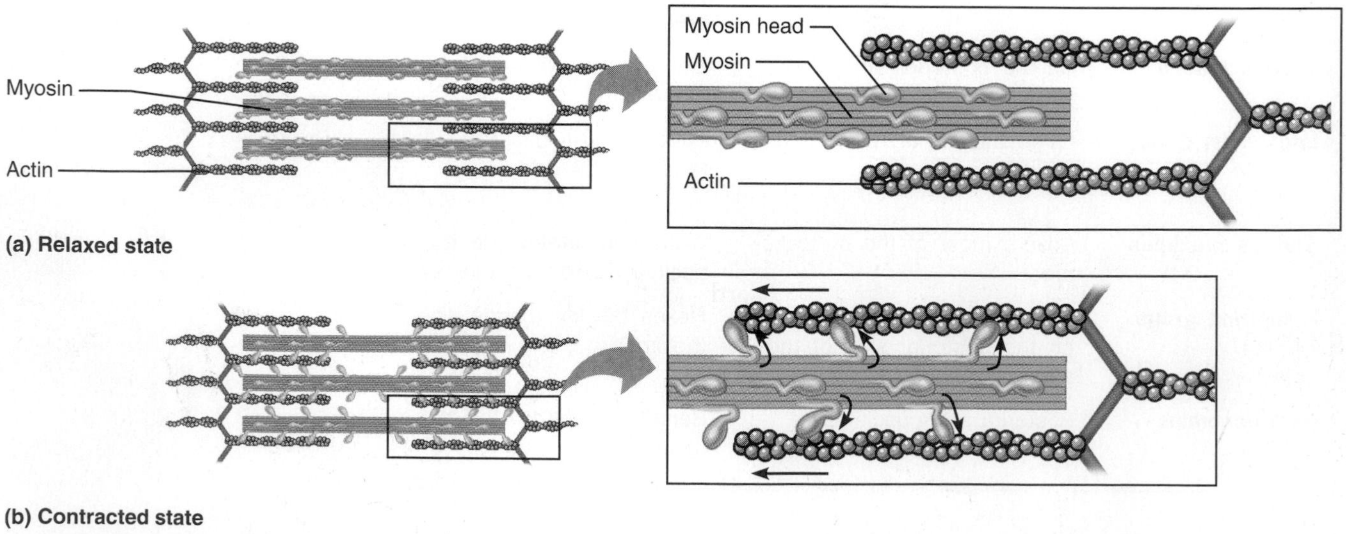

(a) Relaxed state

(b) Contracted state

FIGURE 9.7 The sliding filament mechanism.

cells, contraction clearly requires an enormous expenditure of energy. In addition, there are finite limits to the amount of ATP that a cell can store. These two facts taken together suggest that muscle cells will cease to operate if they cannot rapidly replace the ATP they use. Creatine phosphate and an inefficient mechanism of sugar metabolism called anaerobic respiration provide the fastest ways to regenerate ATP. But these energy sources are also quickly depleted, and anaerobic respiration produces **lactic acid.**

This is not all bad. By triggering pain receptors, lactic acid tells us that we should slow down, if possible, and allow the stressed muscles to recover. Lactic acid also dilates blood vessels that supply skeletal muscles, bringing in the additional nutrients and oxygen needed by the overtaxed muscle cells. However, once ATP levels fall too low in a muscle cell, its strength of contraction may weaken to the point where it cannot perform the task it is being called upon to do. This is referred to as **muscle fatigue.**

ACTIVITY 6

Measuring the Energy Use of Contracting Skeletal Muscles

Materials for this Activity

Assorted dumbbells of various weights
Cloth tape measure

Part 1

Work in pairs. Obtain a tape measure and dumbbell of sufficient weight that it will be challenging for the designated lifter.

1. Measure the arm of the designated lifter around the thickest portion of the upper arm. Do this during both relaxation (extended) and forced contraction (flexion) of the biceps brachii. Record your measurements.

   Relaxed/extended 12¼
   Contracted/flexed 12½

2. Perform 10 curls (flexing the arm by contracting the biceps brachii) with the arm that was measured in the previous step. If you cannot do 10 repetitions (reps), switch to a lighter weight; if 10 reps did not noticeably tire your arm, switch to a heavier weight until 10 reps leaves your arm feeling at least slightly tired. Measure the arm as you did in the previous step, and record your measurements.

   Relaxed/extended 12½
   Contracted/flexed 13

   Explain any differences you note in the measurements made in Steps 1 and 2. *Note:* If you did not obtain any difference, repeat Step 2 until you do.

   Relaxed .25 difference
   Flexed 1/2 inch difference

Part 2

Continue to work in pairs, and use the same dumbbell used in the previous part.

1. Using the same arm as in the previous experiment, perform as many curl repetitions as you can until discomfort or a mild burning sensation is felt in the biceps brachii muscle. Have your partner record the number of reps, and rest for no more than 10 seconds. *Note:* If you can do more than 20 reps, you should switch to a more challenging weight.

   Number of reps 10

2. After no more than 10 seconds of rest, perform as many curl repetitions as you can until the same level of discomfort or mild burning sensation is felt in the biceps brachii muscle. Have your partner record the number of reps.

   Number of reps: 20
   Which number was lower? The top

   Propose a reason for your results.

   More you try, more you can accomplish ■

EXERCISE

10

THE NERVOUS SYSTEM I: ORGANIZATION, NEURONS, NERVOUS TISSUE, AND SPINAL REFLEXES

Objectives

After completing this exercise, you should be able to

1. Explain the organization of the nervous system.
2. Describe the supporting cells, and explain basically what they do.
3. Describe the different classes of neurons, and explain basically what they do.
4. Explain how neurons work together to transmit information.
5. Explain how a simple spinal reflex works.

Materials for Lab Preparation

Equipment and Supplies

- ❑ Compound light microscope
- ❑ Reflex hammer

Prepared Slides

- ❑ Giant multipolar neurons
- ❑ Peripheral nerve (cross section)

Models

- ❑ Vertebral column with nerves and plexuses
- ❑ Sensory and motor neurons

Introduction

As you learned in Exercise 5, **neurons** are used to rapidly transmit information throughout the body. However, there is more to nervous tissue than that. First, not all cells found in nervous tissue transmit impulses. There are also supporting cells called the **neuroglia**. Second, there are different kinds of neurons, each modified for the job it must do. Finally, neurons are only useful when connected to other neurons, so we must first understand how the nervous system is organized and its neurons work together.

Organization of the Nervous System

The nervous system is divided into two major segments: the **central nervous system (CNS)** and the **peripheral nervous system (PNS)**. The CNS contains the **brain** and **spinal cord**, which are also referred to as **integration centers** because they process incoming information and effect a response. The incoming information and response must be carried to and from the appropriate body parts by the PNS. **Nerves** are the typically long structures that carry the extensions of neurons to and from often-distant body parts.

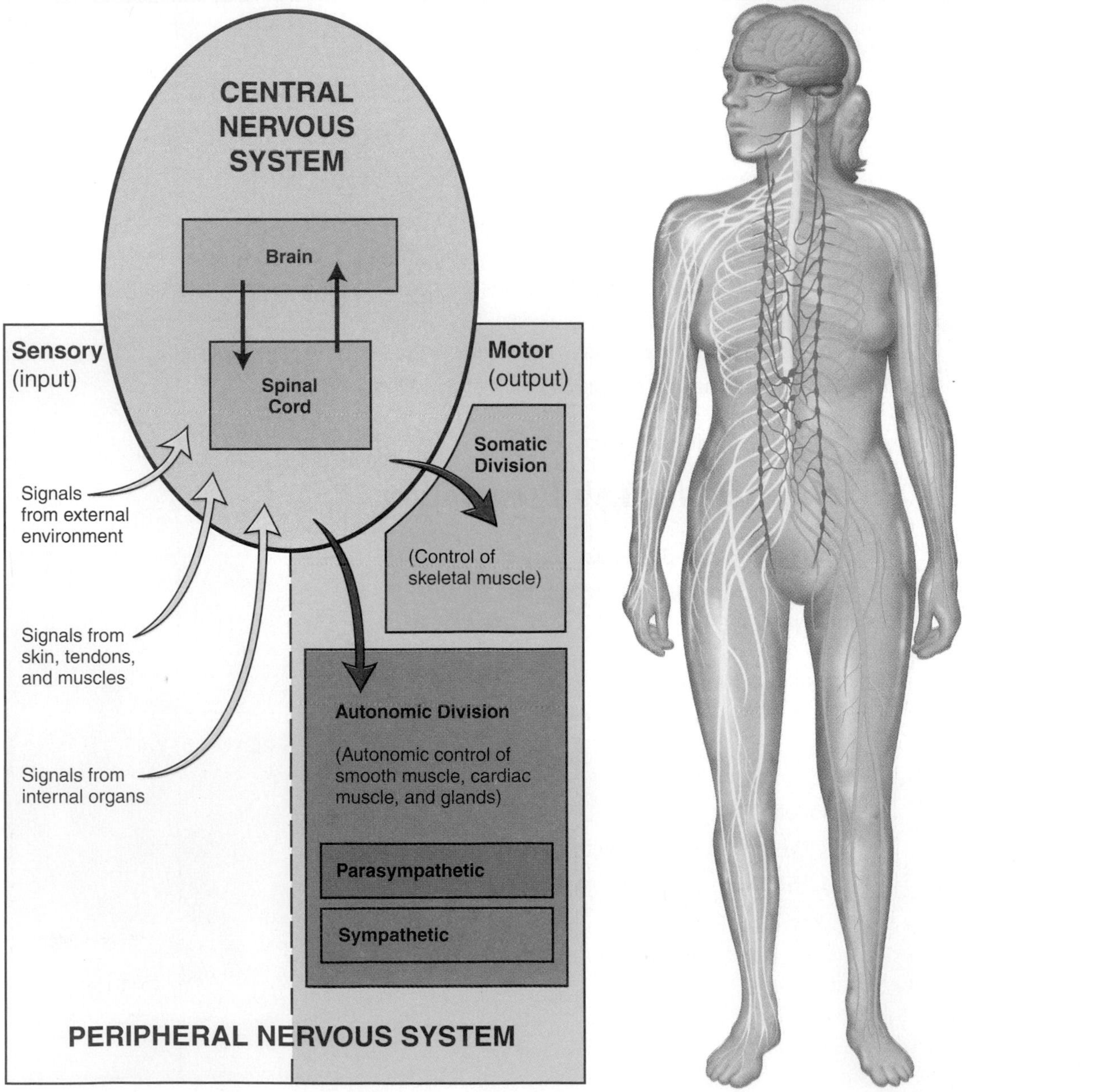

FIGURE 10.1 Organization of the nervous system.

The PNS is divided into **sensory** (input) and **motor** (output) divisions (Figure 10.1). The sensory division brings impulses, or **action potentials,** from receptors to the CNS. The motor division brings action potentials from the CNS to **effectors** such as muscles and glands. This motor division is further divided into **somatic** and **autonomic** divisions. The somatic division's nerves primarily bring information to skeletal muscles, so it is therefore often described as the *voluntary* motor division. As the name implies, the autonomic division mainly stimulates *involuntary* structures such as smooth muscle and glands.

ACTIVITY 1

Identifying the Organization of the Nervous System

Consult Figure 10.1, and match the following items to their places in the nervous system:

_____ Spinal cord	a. sensory division
_____ Nerves to the stomach	b. CNS
_____ Nerves to the deltoid	c. somatic motor division
_____ Nerves from the skin	d. autonomic motor division ■
_____ Brain	

Nervous Tissue

The **neuroglia**, or supporting cells of the nervous system, perform a variety of functions. Some, such as astrocytes and microglial cells, attach to neurons to support and connect them to each other and to other structures. Others, such as oligodendrocytes and Schwann cells, wrap an insulation-like material called **myelin** around neurons. They fulfill important accessory roles, but they do not conduct impulses. Myelin protects neurons and helps to speed impulse conduction by causing the impulse to skip, rather than slowly flow, down the axon. Consult your textbook for a more detailed description of the role played by myelin in the nervous system.

Neurons come in several different varieties, but each has the same basic components (Figure 10.2). They have a large **cell body** that holds the nucleus, several short **dendrites** that pick up stimuli, and a long, slender **axon**, which can transmit action potentials over great distances.

Sensory neurons typically have a cell body attached to an axon with a single, short process (Figure 10.2). For this reason, sensory neurons that look like this are called *unipolar* neurons, that is, they have *one* process attached to the cell body. Some sensory neurons are attached to a short dendrite at one end and a short axon at the other end of the cell

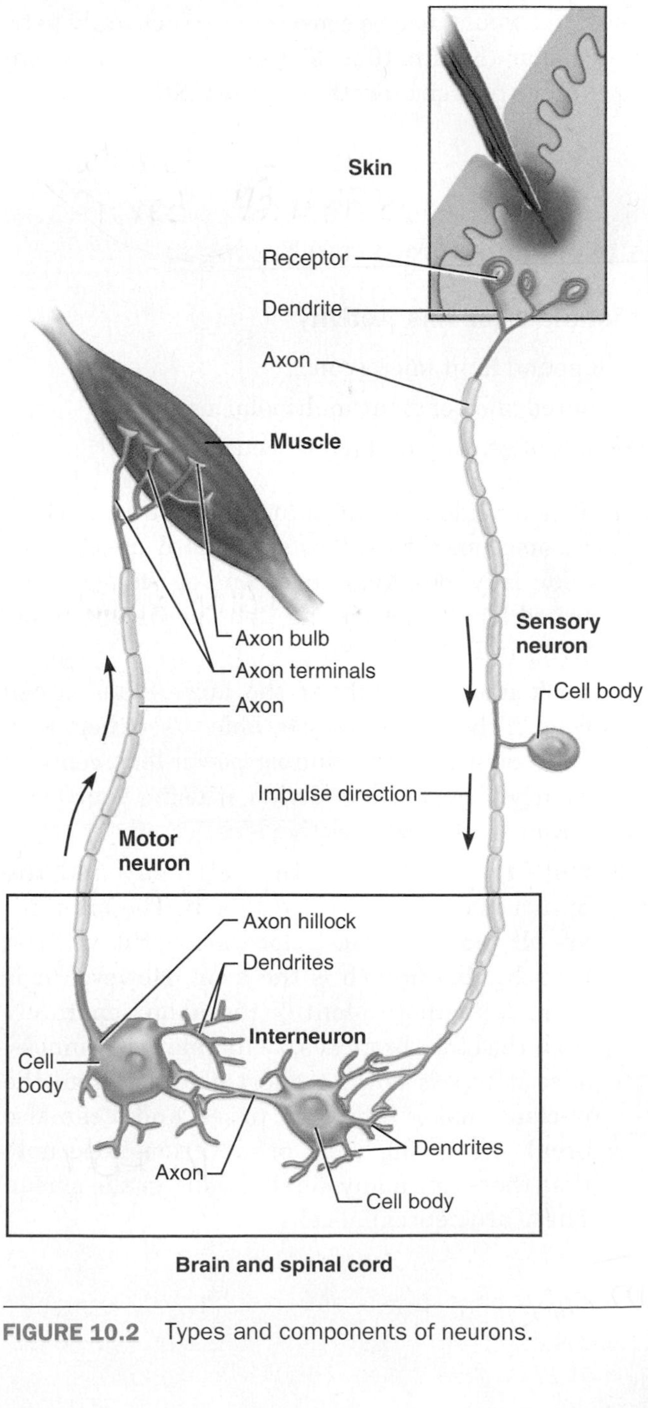

FIGURE 10.2 Types and components of neurons.

body. However, these *bipolar* neurons are rare and found only in certain specialized nervous tissues such as the retina of the eyes. These neurons carry information toward the CNS.

Motor neurons have several branched dendrites attached directly to the cell body and one axon running from the tapered end of the cell body (Figure 10.2). For this reason, motor neurons are also called *multipolar* neurons, that is, they have *many* processes attached to the cell body. These neurons carry information from the CNS to muscles and glands.

Interneurons resemble very short motor neurons and would also be considered structurally to be multipolar (Figure 10.2). These short neurons connect other neurons together in the CNS.

ACTIVITY 2

Observing Neurons and Neuroglia

Materials for this Activity

Compound light microscope
Prepared slide of giant multipolar neurons
Models of sensory and motor neurons

1. View a slide of giant multipolar neurons, which is sometimes labeled "ox spinal cord smear." This slide may look familiar as one of the slides selected for the "Observing Cellular Diversity" activity in Exercise 3.
2. Look more carefully at the large blue-stained cells. Although you may be able to see them with your low-power or scanning-power lens, center a clearly visible cell, and view it using your high-power (high-dry) objective lens.
3. Note the nucleus in the cell body and the branches coming off the cell body. The branches are all dendrites, except for the longest, thickest branch. This branch is the axon. However, it is often difficult to identify the axon separately from the other branches, as its relative thinness, once it tapers down a short distance from the nucleus, makes it difficult to see; and it can also break off during slide preparation. Also note that there are many smaller cells in the smear. These are neuroglial cells.
4. Sketch a small portion of this slide in the space provided below, labeling the cell bodies, dendrites, and axons where you can discern them from the dendrites.

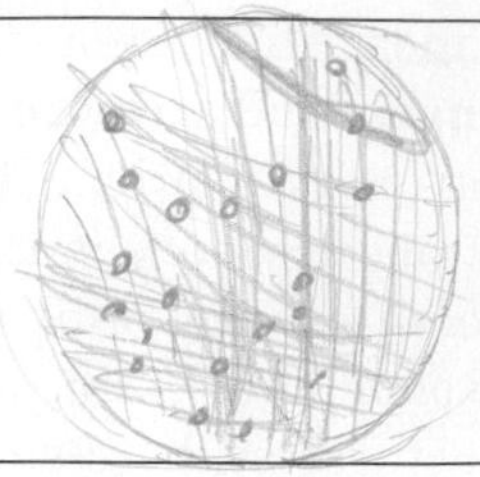

5. Now view models of sensory and motor neurons. Identify the cell body, axon, dendrites, and myelin sheath. Note the small gaps or "cracks" between the myelinated areas. These **nodes of Ranvier** allow for the fast impulse skipping mentioned earlier. Make a general sketch of these models in the space provided below, labeling the previously mentioned structures. ■

The PNS and Nerves

Nerves are simply bundles of axons wrapped by connective tissue into structures that resemble strings and function a little like telephone wires. The layout of a nerve is remarkably similar to that of a muscle. Individual axons are wrapped by a thin layer of connective tissue, a group of axons are wrapped into **fascicles**, and the fascicles are wrapped into the whole nerve.

There are two categories of nerves. They are the **cranial nerves**, which are attached to the brain, and the **spinal nerves**, which are attached to the spinal cord. More details about these nerves and the way they are connected to the brain and spinal cord will follow in the next exercise. For now, we will focus on the spinal nerves specifically where they have exited the spinal cord (Figure 10.3).

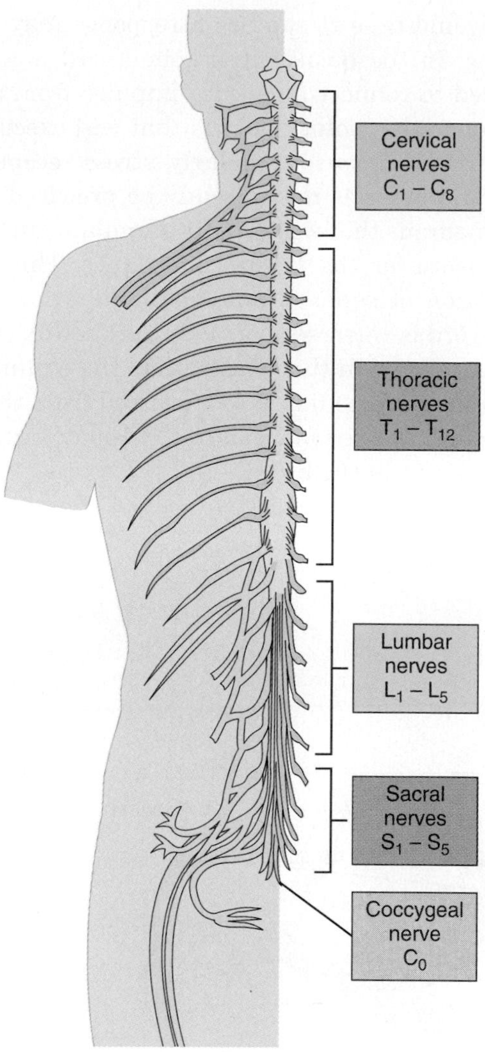

FIGURE 10.3 Spinal nerves.

Note that the nerves are initially named for the region of the vertebral column from which they exit. That is, they are named **cervical nerves** in the neck, **thoracic nerves** in the chest region, **lumbar nerves** in the lower back, and **sacral nerves** in the area of the sacrum.

ACTIVITY 3

Learning About the Spinal Nerves

Materials for this Activity

Compound light microscope
Model of the vertebral column with nerves and plexuses
Prepared slide of a peripheral nerve (cross section)

1. Observe the model of the vertebral column with nerves. Find the cervical, thoracic, lumbar, and sacral nerves.
2. Look carefully at Figure 10.4, and then view the peripheral nerve cross section slide.
3. Use the lower-power magnification, and note how the nerve is arranged in fascicles similar to muscle as described earlier. Then center the lens area on one fascicle so that you may switch to your high-power (high-dry) objective lens. Now observe these individual axons. The axons will appear as dark dots surrounded by a small area of "empty" space. This space is not empty at all but is where the myelin sheath is located. Because myelin does not easily pick up the stain used to color the slide, the area containing the myelin appears empty.
4. In the space provided below, sketch the low-power view of the nerve and the high-power view of the axons. Compare your observations with Figure 10.4. ■

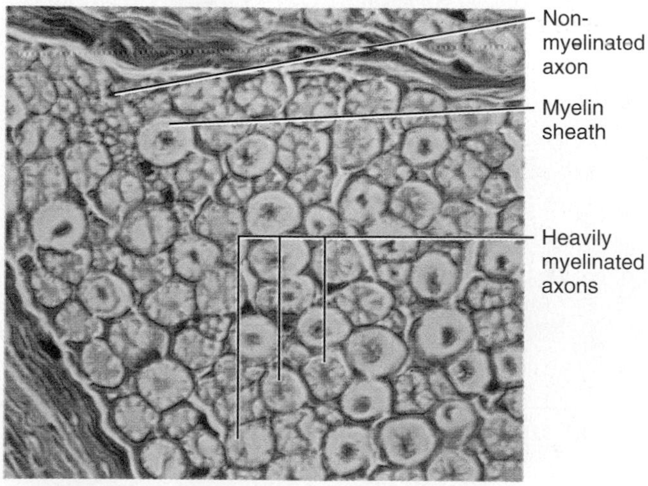

FIGURE 10.4 A photomicrograph of a cross section of a peripheral nerve (560×).

Spinal Reflexes

Although the spinal cord is not covered until Exercise 11, it seems important and appropriate to finish this exercise with an example of a way that neurons are connected and used in the human body. A spinal reflex is the simplest example of this use.

We usually think of the CNS as the place where sensory information is processed so that a decision can be made and action taken in response to an event. That is not always the case, however. There are situations where no thinking or decision making is necessary, only a rapid response. When we contact something harmful (e.g., touching a sharp object or a hot burner on a stove) or when we need to balance when walking, the extra second or more that the brain would take to process a response may be far too long. In these cases, the spinal cord is already prewired to connect a sensory impulse from a pain receptor to the motor neurons that will execute the appropriate response. Similarly, stress receptors in joints and muscles may already be prewired to the motor neurons that will rapidly stimulate muscles to compensate for the failure of balance. This direct connection of sensory and motor neurons in the spinal cord is referred to as a **spinal reflex** (Figure 10.5). As shown in this illustration, the stimulus information is immediately transferred from the sensory neuron to the motor neuron without waiting for a response from the brain.

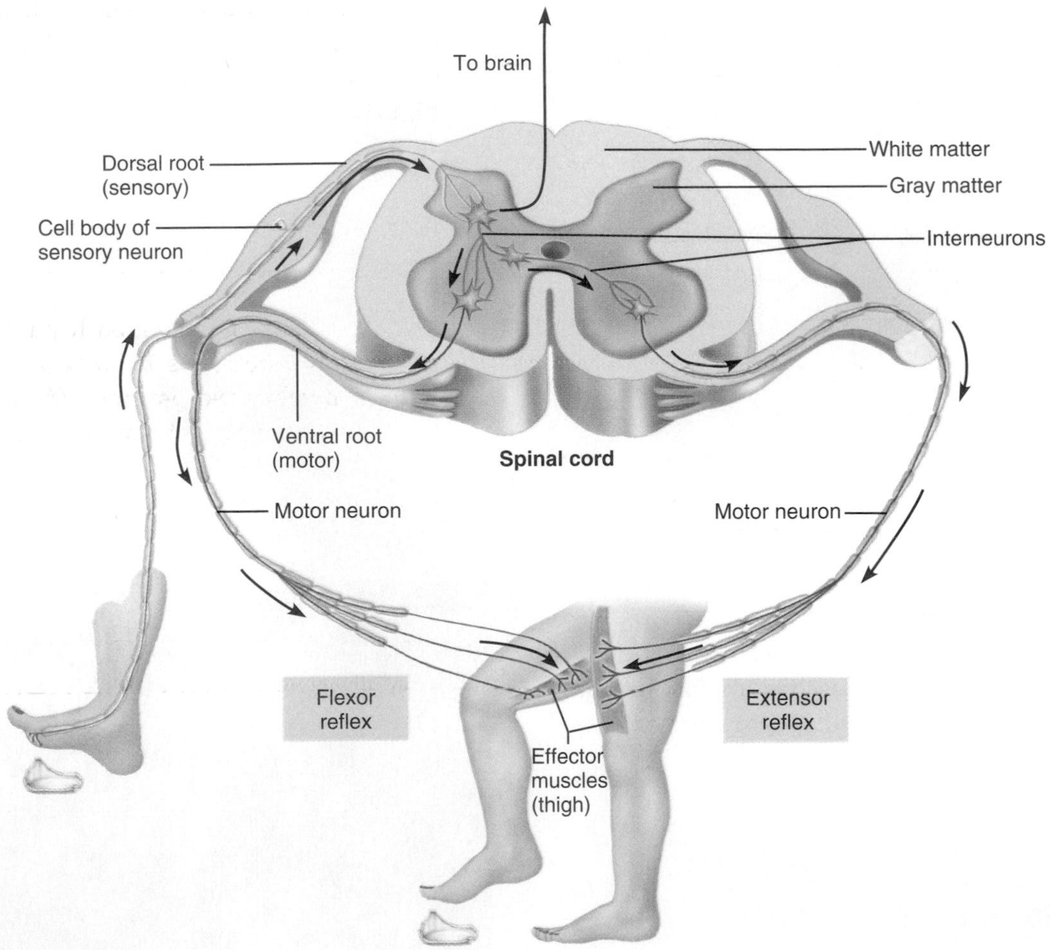

FIGURE 10.5 A spinal reflex: pain withdrawl reflex.

ACTIVITY 4

Testing Spinal Reflexes

Materials for this Activity

Reflex hammer

Part 1: Knee Jerk Reflex

1. Work in pairs, and obtain a rubber reflex hammer. Have one person sit on a chair or table so that the lower legs dangle freely, with a few inches of space below and behind the foot. First, find the softer area directly below the patella (kneecap). Then, tap this softer area with the reflex hammer. This may take several attempts. Describe your observations.

___

___

___

2. Consider what was happening to the nervous system of the subject during this reflex, and compare it to the reflex in Figure 10.5. How is this reflex similar?

___

___

How is this reflex different?

___

___

Part 2: Ankle Jerk Reflex

1. Similar to Part 1 above, the foot should be freely suspended. Use the wider edge of the reflex hammer to strike the Achilles tendon about one inch above the heel bone. Repeat several times and describe your observations below.

___

___

___

2. Compare this reflex to the one in Part 1 above and the reflex illustrated in Figure 10.5. How is this reflex similar to each of the above mentioned reflexes?

___

___

___

From which *one* is it different? How so?

___

___

___

Why is it important for both pain withdrawal and balance reflexes to have pathways using heavily myelinated neurons? (*Note:* it would be unethical for us to ask you to actually perform a pain withdrawal reflex, so answer this based upon your memories of experience with pain stimuli.)

___

___

___ ■

EXERCISE

11

THE NERVOUS SYSTEM II: THE SPINAL CORD, BRAIN, AND AUTONOMIC NERVOUS SYSTEM

Objectives

After completing this exercise, you should be able to

1. Describe the meninges and cerebrospinal fluid.
2. Describe the structure of the spinal cord, and explain its basic functions.
3. Identify the selected parts of the brain.
4. Describe the major parts of the brain, and explain the basic function of each.
5. Explain the basic functions of the two parts of the autonomic nervous system.

Materials for Lab Preparation

Equipment and Supplies

- ❑ Compound light microscope
- ❑ Sheep brain with meninges (if possible)
- ❑ Large, nonserrated knife
- ❑ Dissection microscope

Prepared Slide

- ❑ Spinal cord (cross section)

Models

- ❑ Spinal cord (cross section)
- ❑ Human brain

Introduction

In the previous exercise, you learned that the nervous system is made of many neurons connected together to communicate information. Integration centers are the complex areas of the nervous system where neurons interact to process and store information. The brain and spinal cord are these integration centers in humans. Before we specify how the brain and spinal cord work, we must first study how these delicate organs are protected.

The Meninges and Cerebrospinal Fluid

Three layers of connective tissue surround and protect the brain and spinal cord. These layers of connective tissue are called the **meninges** (Figure 11.1). The **dura mater** is the tough outer layer of connective tissue, the loose **arachnoid** is in the middle, and the soft, thin **pia mater** lies directly on the brain and spinal cord.

The meninges alone would not be completely effective in protecting delicate organs such as the brain and spinal cord, thus these organs are also surrounded by a thin fluid called **cerebrospinal fluid** (CSF). The CSF is produced by specialized capillaries in the brain, and it resembles the fluid found around most of the tissues in the body (Figure 11.2). Consult your textbook for a more detailed description of CSF production.

The CSF fills hollow spaces called **ventricles** in the brain, and it also surrounds the brain and spinal cord in the space between the arachnoid and pia mater (Figure 11.2), where it acts as a shock absorber.

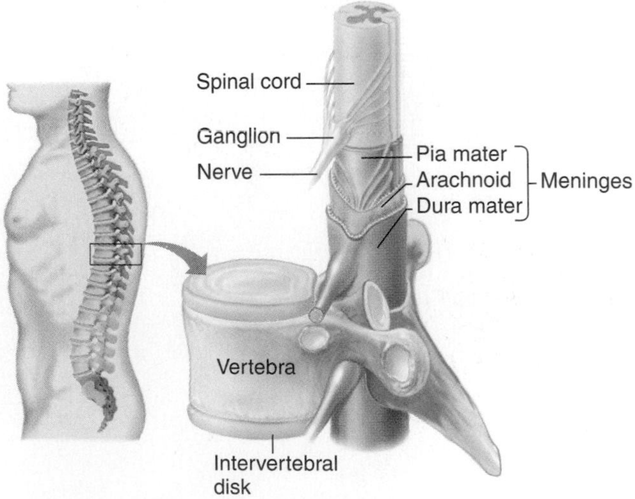

FIGURE 11.1 The spinal cord and meninges.

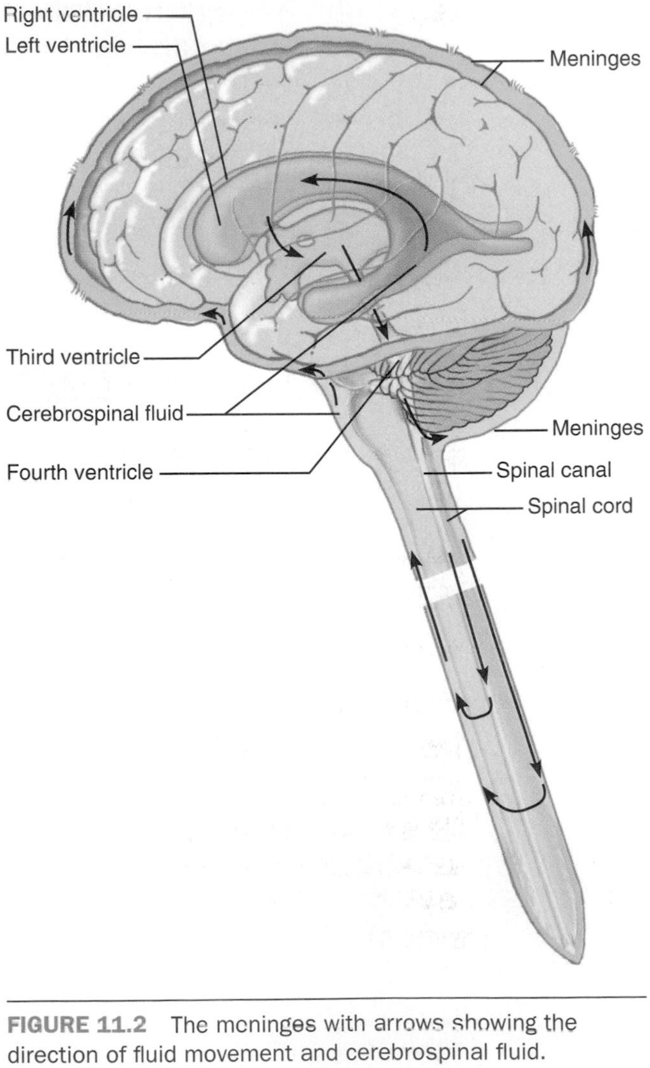

FIGURE 11.2 The meninges with arrows showing the direction of fluid movement and cerebrospinal fluid.

ACTIVITY 1

Observing the Meninges and CSF

Materials for this Activity

Sheep brain with meninges
Spinal cord model

1. Observe a model of the spinal cord. You will examine this model in more detail later, but for now, notice the relationship of the bone to the dura mater. Observe the arachnoid and pia mater if they are present on the model. Is the dura mater easy or difficult to tear?

   ____

   What would be the advantage of having this tissue between the brain and the bones of the skull?

   ____
   ____
   ____
   ____
   ____
   ____
   ____

2. Observe a sheep brain with the dura mater still intact. Notice its toughness. ■

The Spinal Cord

The spinal cord is, in part, comparable to a huge nerve that carries information to and from the brain. But as you learned in Exercise 10, it also makes connections and handles reflexes. This suggests that we may find two very different functional components in the spinal cord, and we do. The **white matter** is packed with myelinated axons that rapidly transmit information to the brain, in ascending tracts, and down from the brain, in descending tracts. The **gray matter** is where we find the cell bodies of motor neurons, as well as interneurons, which make connections between neurons for effects such as reflexes (Figure 11.3).

The spinal cord also serves as the "junction box" for the spinal nerves, which split into dorsal and ventral roots just before attaching to the spinal cord (Figure 11.3). The **dorsal root** veers toward the back of the spinal cord and contains the sensory neurons. Most sensory neurons do not have their cell bodies positioned near one end of the neuron as do motor neurons. Because of this, the cell bodies of sensory neurons are gathered together just outside the spinal cord in a bulging area of the dorsal root called the **dorsal root ganglion.** The **ventral root** veers toward the front of the spinal cord and contains the axons of motor neurons. Cell bodies of motor neurons are positioned near one end, and these cell bodies are found in the gray matter of the spinal cord (Figure 11.3). Thus there is no ventral root ganglion.

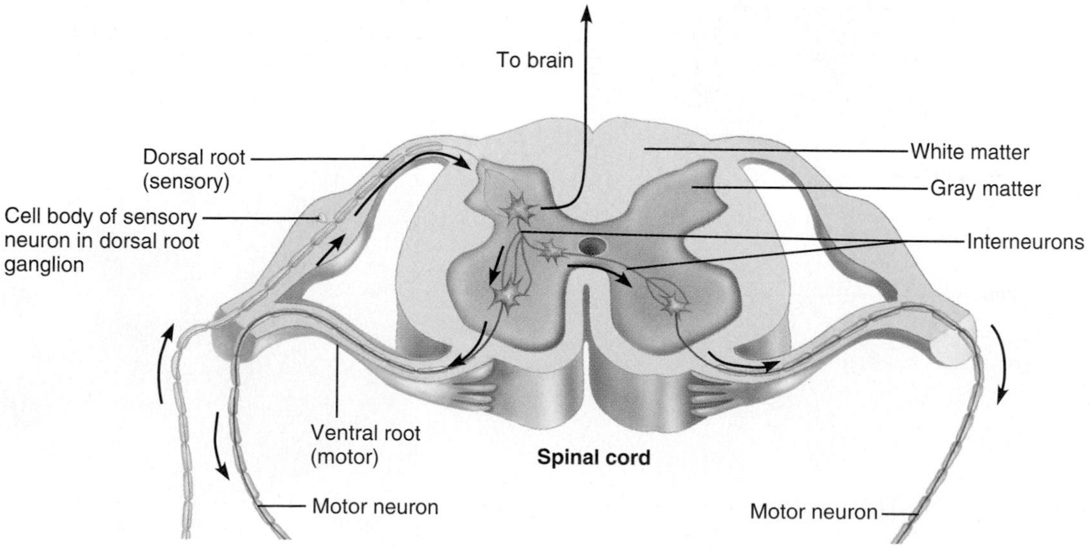

FIGURE 11.3 A spinal cord cross section showing dorsal and ventral roots.

Because it functions as a junction device for spinal nerves, the spinal cord is not uniform in thickness throughout its length. Instead, it is thicker where more neurons enter and exit and thinner where there is less traffic. These thicker areas, which must handle the extra neuron traffic from the arms and legs, are respectively called the **cervical enlargement** and **lumbar enlargement** (Figure 11.4).

Another interesting feature of the spinal cord is that it ends in the lumbar region of the vertebral column instead of extending all the way down into the sacral region. The spinal nerves that must exit from the sacrum and lower lumbar region are found projecting down in this area (Figure 11.4). This cluster of spinal nerves is called the **cauda equina** due to its resemblance to a "horse's tail."

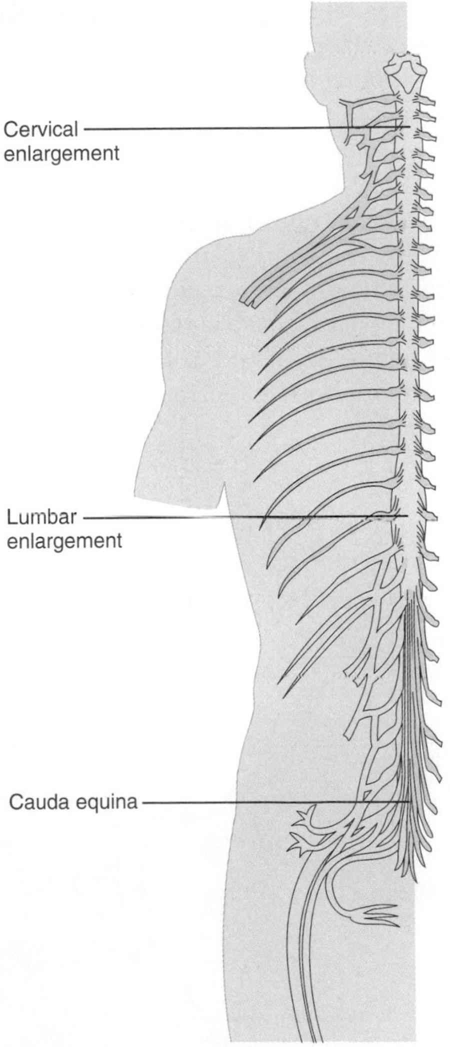

FIGURE 11.4 A longitudinal view of the spinal cord, showing the cervical and lumbar enlargements.

Suggest a reason why the spinal cord appears to "come up short" in the vertebral column.

____

____

ACTIVITY 2

Examining the Spinal Cord

Materials for this Activity

Compound light microscope
Dissection microscope
~~Prepared~~ slide of the spinal cord (cross section)
Model of a spinal cord cross section

1. Observe the model of a spinal cord cross section. Note the "butterfly-shaped" area of gray matter surrounded by white matter. Also note the dorsal and ventral roots of the spinal nerve and the dorsal root ganglion.
2. Using the dissection microscope, view a prepared slide of a spinal cord cross section. You should again be able to see the "butterfly-shaped" area of gray matter surrounded by white matter. Compare your slide to Figure 11.5.
3. Use your compound light microscope to view the same prepared slide. You may not be able to see the entire spinal cord in your field of view, even under the low power, so scan around the slide to become familiar with this view. Then, center the slide on the thickest area of gray matter, and switch to a higher magnification. You should now be able to see the large cell bodies of motor neurons in this area of the gray matter. ■

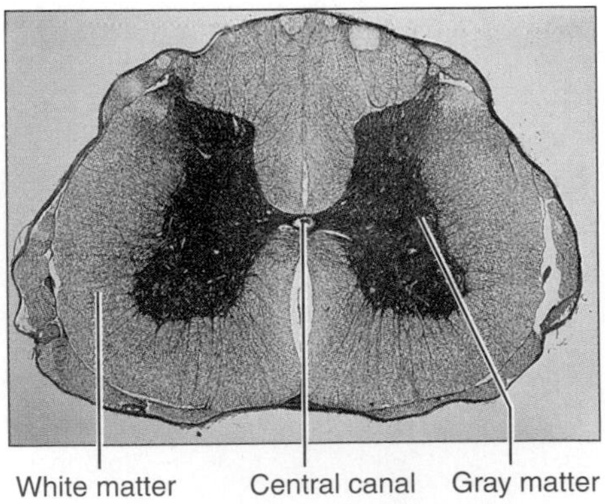

FIGURE 11.5 A photomicrograph of a spinal cord cross section.

The Brain

The brain is the complex integration center that processes, stores, and makes decisions about the information we obtain from the millions of sensory neurons in the body. There are several specialized regions of the brain, each representing a level of brain evolution and/or embryonic development. Although valid schools of thought differ on how to divide the regions of the brain for study, we will use the functional divisions of the **hindbrain, midbrain,** and **forebrain** found in your textbook.

The Hindbrain

The hindbrain, which makes up most of what is referred to by some as the **brain stem,** consists of the **medulla oblongata,** the **pons,** and the **cerebellum** (Figure 11.6). The medulla oblongata controls various autonomic life-support functions such as breathing (respiratory center), blood pressure (vasomotor center), and heart rate (cardiac center). The pons is a bridge connecting neurons from the spinal cord with higher brain centers. The cerebellum coordinates basic and complex movements.

The Midbrain

Sometimes referred to as part of the brain stem, along with the pons and medulla oblongata, the midbrain relays information (mainly vision and hearing) to higher brain centers (Figure 11.6). It is also the control center for some movement and for regulating our degree of alertness or wakefulness. The midbrain and hindbrain are considered to be the most primitive parts of the brain.

The Forebrain

The forebrain contains the parts of the brain that are more highly evolved. The **thalamus, hypothalamus, cerebrum,** and various diffuse systems coordinate a huge array of sensory information, and they produce our complex responses (Figure 11.6).

Together, the thalamus and hypothalamus mostly constitute an area often referred to as the **diencephalon.** The thalamus receives a great deal of sensory information, and it is responsible for the initial processing of the information before relaying it

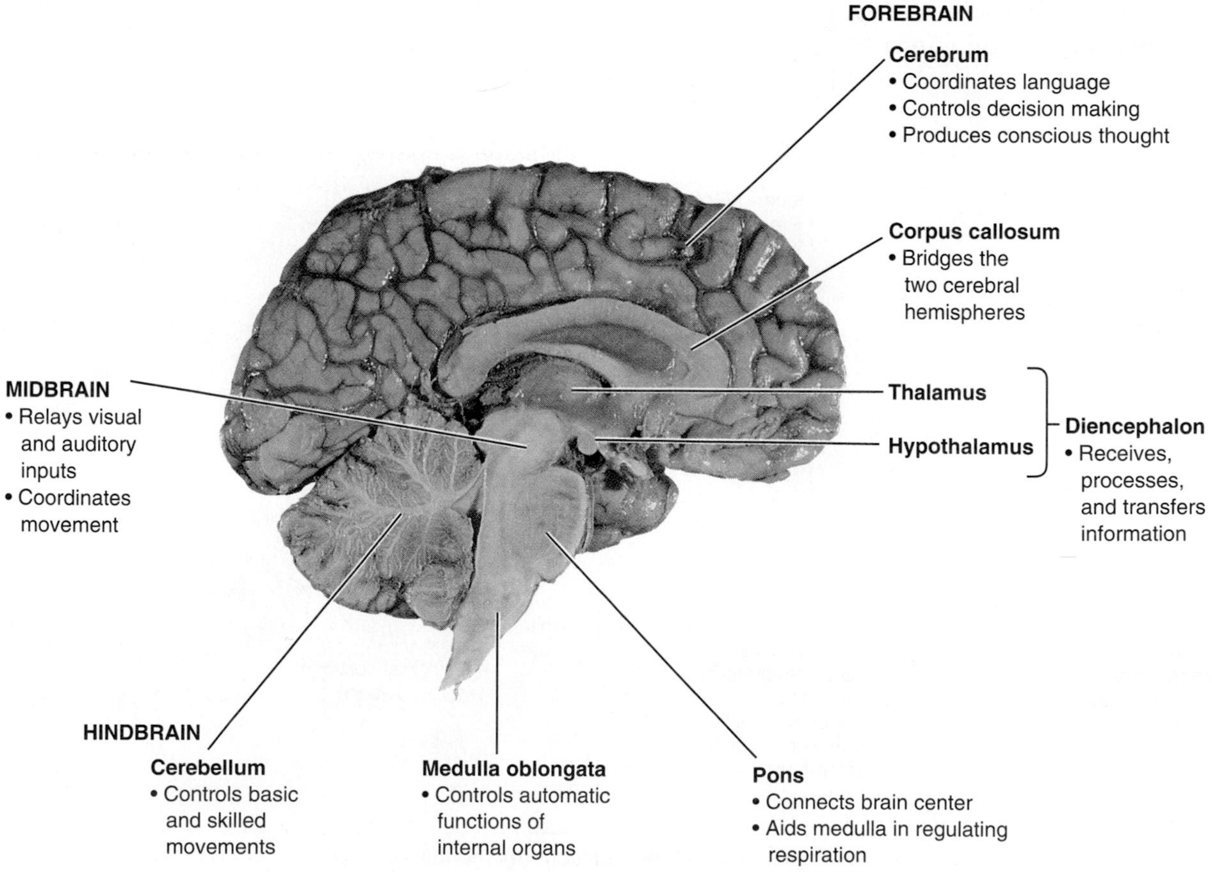

FIGURE 11.6 The functional divisions of the brain.

on to the cerebrum for more precise interpretation. The hypothalamus is the center of highly sophisticated life-support systems. Control over body temperature, hunger, thirst, and other features critical to homeostasis is regulated by the hypothalamus.

The highly folded cerebrum is where sensory information is processed, stored, and sometimes forwarded for decision making (Figure 11.7a). The folding is a neat trick of Mother Nature to increase surface area on the outside (cortex) where the gray matter is found. This way, the amount of gray matter, which is used for storing and processing information, could be increased without expanding the overall size of the brain. Some of these folds separate the cerebrum into several **lobes,** each with a specialized function. These lobes are named for the bone that covers them.

The **occipital lobe** receives and processes visual information. The **temporal lobe** is the center of our hearing interpretation. Because spoken language came about long before written language, the temporal lobe is also the area where language interpretation is centered. The **parietal lobe** is where information from the skin, primarily the sense of "touch," is located. The largest of the lobes, the **frontal lobe,** is involved in nonsensory functions such as initiating movement, speaking, and thinking (Figure 11.7a).

A deep fold separates the right and left halves, or hemispheres, of the cerebrum (Figure 11.7b). Communication between the right and left cerebral hemispheres occurs via a band of white matter called the **corpus callosum** (Figure 11.17b).

ACTIVITY 3

Observing the Brain

Materials for this Activity

Model of the human brain
Sheep brain
Large, nonserrated knife

Part 1: Human Brain Model

1. Observe the model of the human brain, and use Figures 11.6 and 11.7 to assist you in locating the cerebellum and the four lobes of the cerebrum discussed previously.
2. Separate the halves of the brain model, and locate the medulla oblongata, which appears to be a simple enlargement of the spinal cord. Find the somewhat rounded pons, directly above the medulla oblongata. Directly above the pons, find the midbrain.

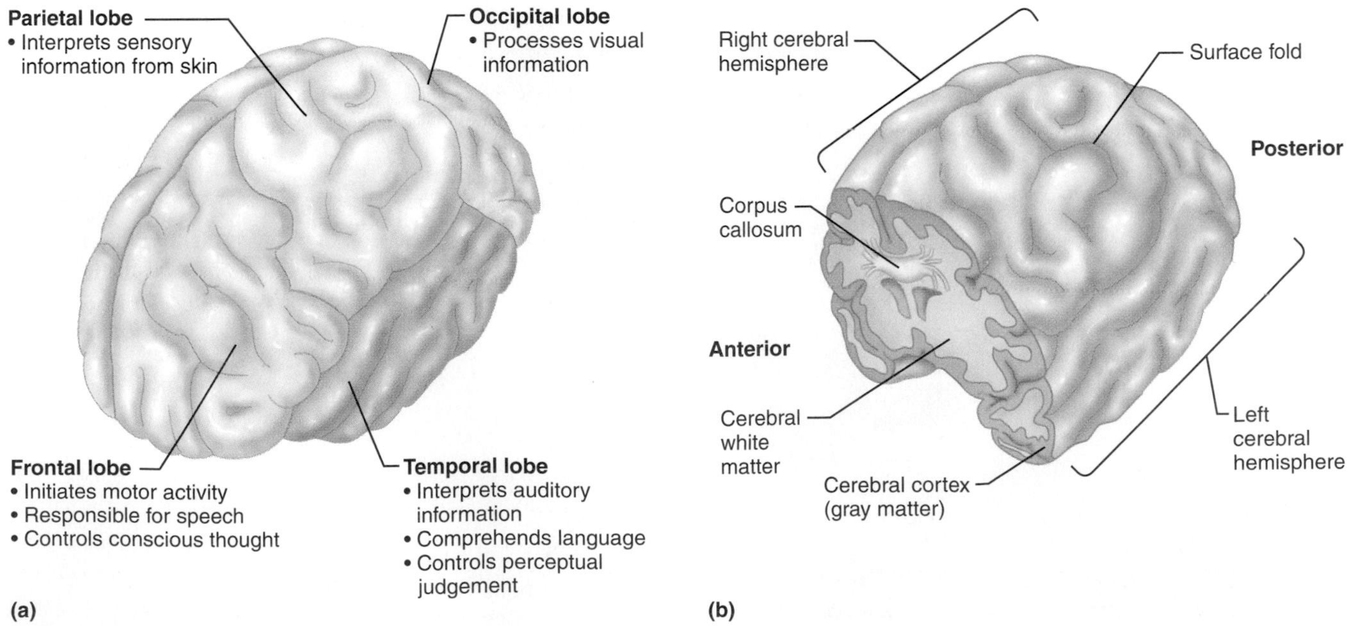

FIGURE 11.7 The cerebrum. (a) Each lobe has a specialized function. (b) The left and right hemispheres are depicted, including the corpus callosum, which allows communication between them.

3. Next, locate the cerebellum directly behind the pons, medulla oblongata, and midbrain. Notice that the cerebellum has a highly branched core of white matter.

   After reviewing the function of the cerebellum, propose a reason why the branched white matter is found there.

   ____

   ____

   ____

   Continue your examination of the midsagittal view of the brain model. Note the somewhat rounded thalamus above the midbrain. Below the thalamus, angling down toward the marble-like pituitary gland, you will find the hypothalamus. Above the thalamus, locate the curved strip of white matter called the corpus callosum. Finally, observe the cerebral hemisphere and its highly folded appearance. Does the corpus callosum seem as if it would be connected to both cerebral hemispheres?

   ____

   What would happen if the corpus callosum was cut in a living human being?

   ____

   ____

   ____

Part 2: Sheep Brain

1. Obtain a preserved sheep brain. If the brain still has the meninges intact, carefully remove the dura mater, noting its toughness. Observe the whole brain from the side. You will note that it looks more straight or linear than the human brain. Suggest a reason for this.

   ____

   ____

   ____

2. View the whole sheep brain from the top (the dorsal view), and note the deep fissure that separates the left and right cerebral hemispheres (Figure 11.8c). Although the separation of the lobes is less distinct than it appeared in the human brain model, approximate the location of each lobe. Observe the cerebellum.
3. View the sheep brain from the bottom (the ventral view), and identify the medulla oblongata, the rounded pons, and the midbrain (Figure 11.8a, b). In the area of the midbrain, you should see two thick, tough trunklike structures crossing over each other. These are the optic nerves, pointed out here as an example of the **cranial nerves,** which are attached directly to the brain rather than the spinal cord. Some of the cranial nerves are too tiny to be easily seen or may have become detached when the meninges were removed. Observe any other cranial nerves that you can find and the flaplike olfactory bulbs near the anterior end.
4. Next, use a nonserrated knife to slowly and carefully cut the sheep brain along the midsagittal line. From the inside of this midsagittal view, identify the medulla oblongata, pons, midbrain, cerebellum, thalamus, hypothalamus, corpus callosum, and cerebral hemisphere (Figure 11.9 on page 127). Which of these structures are considered part of the more primitive hindbrain?

   ____

   ______________________________ ■

The Autonomic Nervous System

As mentioned in Exercise 10, the autonomic branch of the nervous system is composed of **sympathetic** and **parasympathetic** divisions, and it controls a wide variety of involuntary actions such as heart rate and digestion (Figure 11.10, on page 128). Many autonomic functions are controlled by cranial nerves, particularly in the parasympathetic division.

Although the sympathetic and parasympathetic divisions seem at odds, these two systems have a lot in common. They work together to maintain homeostasis, and they share an unusual two-neuron pathway scheme (Figure 11.10).

The sympathetic division is commonly referred to as controlling our "fight or flight" response. That is, it automatically prepares the body for some kind of emergency response. Speeding heart rate, increasing blood pressure, and activating sweat glands are just a few activities stimulated by the sympathetic nerves within seconds of perceiving a potentially dangerous situation. Most sympathetic neurons originate in the thoracic and upper lumbar region of the spinal cord.

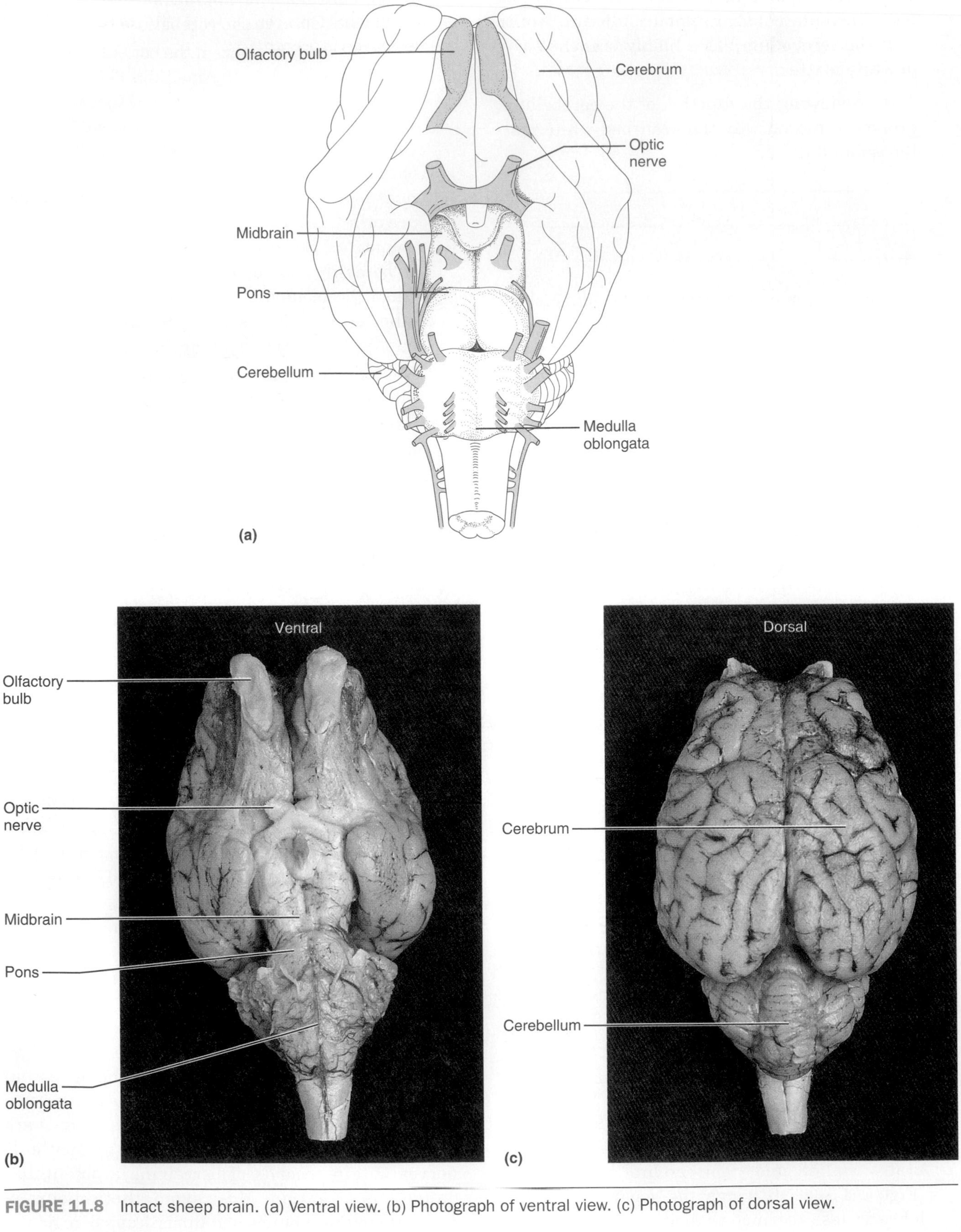

FIGURE 11.8 Intact sheep brain. (a) Ventral view. (b) Photograph of ventral view. (c) Photograph of dorsal view.

The parasympathetic division is the opposite of the sympathetic. It controls maintenance functions such as stimulating the digestive system and slowing the heart rate. Most parasympathetic neurons originate from the cranial nerves, with a small percentage coming from the sacral region of the spinal cord (Figure 11.10).

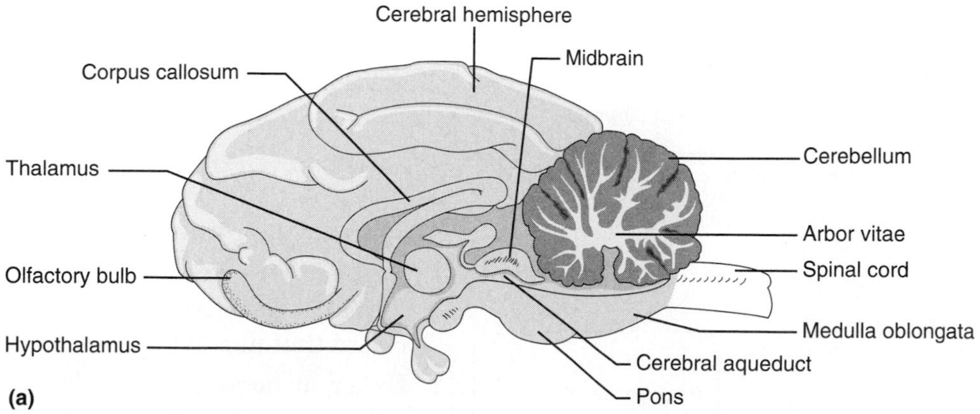

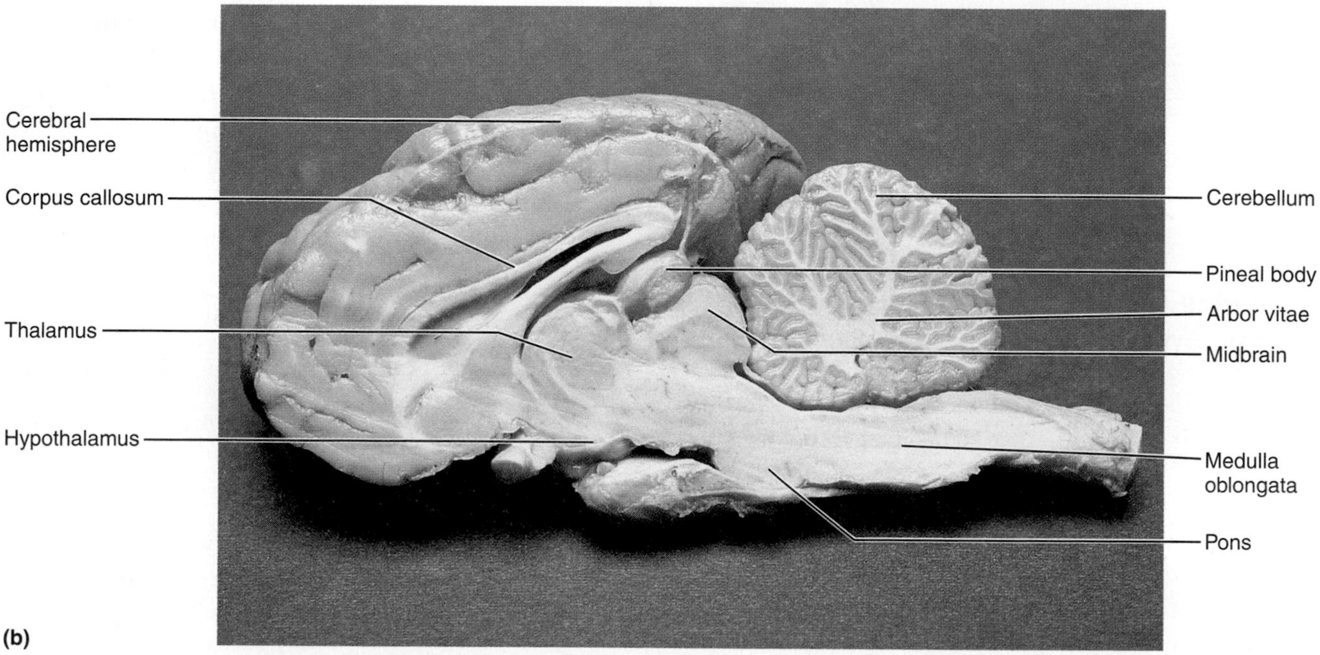

FIGURE 11.9 Midsagittal section of the sheep brain.

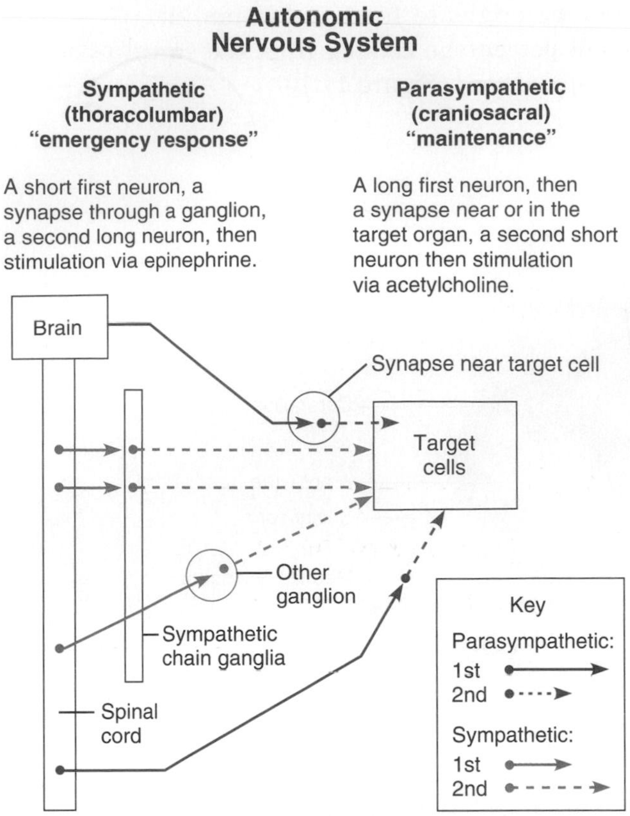

FIGURE 11.10 A schematic drawing of the autonomic nervous system.

ACTIVITY 4

Measuring the Effect of the Autonomic Nervous System on Heart Rate

Although it is possible to perform this activity in the laboratory, it may be better to do it at home for the best results.

1. Sit comfortably with your right or left arm on a table. Place your index finger over the pulse, which is the pulsing area of the wrist below the thumb. Count the number of pulses over a 15-second period, and then multiply by 4. For example, if you count 20 pulses during a 15-second period, then $20 \times 4 = 80$. The heart rate is 80 beats per minute. Take your pulse three times and calculate the beats per mintue for each. Record it as your "baseline" pulse rate in Box 11.1. Calculate the average pulse rate, and record that number as well.
2. Either at home or in the laboratory, take your pulse again after some relaxing activity such as eating a meal, listening to *relaxing* music, or breathing slowly and deeply. Record this pulse rate in the "parasympathetic" row. Repeat these activities twice more, and then calculate and record your average pulse rate.
3. After some stressful activity (e.g., driving home in heavy traffic or watching a suspenseful movie), take your pulse and record it in the "sympathetic" row. Repeat this activity twice more, and then calculate and record your average pulse rate.

Box 11.1 Pulse Rate Data

	Trial 1	Trial 2	Trial 3	Average
Baseline				
Parasympathetic				
Sympathetic				

Which average was the highest?

____

Which average was the lowest?

____

Explain your results below. If you did not obtain the "expected" results, propose a reason why this may have occurred.

____

____

EXERCISE

12

THE SENSES

Objectives

After completing this exercise, you should be able to

1. List and describe the receptors involved in general sensation.
2. List the main structures of the eye, and describe their role in vision.
3. Describe selected vision tests.
4. List the main structures of the ear, and describe their role in hearing and balance.
5. Describe selected hearing and balance tests.

Materials for Lab Preparation

Equipment and Supplies

- ❑ Cotton swabs
- ❑ Powdered or granulated sugar (sucrose)
- ❑ Lemon juice
- ❑ Seven flat-head dissecting pins
- ❑ Three corks or Styrofoam cubes (about 1-inch cubes)
- ❑ Metric ruler
- ❑ Fine-point marker
- ❑ Paper cups (8-ounce size)
- ❑ Tap water
- ❑ Beaker or flask
- ❑ Sheep or cow eye
- ❑ Dissection equipment
- ❑ Dissection tray
- ❑ Snellen eye chart
- ❑ Astigmatism eye chart
- ❑ Medium-size test tube
- ❑ Pencil
- ❑ Tuning fork
- ❑ Chair with wheels
- ❑ Stopwatch or clock
- ❑ Thermometer

Models

- ❑ Human ear
- ❑ Human eye

Introduction

In Exercise 11, you learned that the brain interprets sensory information. But where does this sensory information come from? Specialized portions of sensory neurons called **receptors** are capable of responding to environmental stimuli and eventually converting the stimuli into action potentials. These receptors are usually grouped into two major categories for study: the **special senses** and **general sensation.** Receptors for the special senses include the eye and ear, organs that will be covered as separate topics in this exercise. Taste and smell are also special senses based on chemoreceptors. General sensation includes the category of **somatic sensation,** which is detected by the sensory receptors in skin and muscle that are generally spread out over the entire body (Figure 12.1).

Receptors

Receptors are classified according to the stimuli to which they are sensitive, and their names are often self-explanatory. **Photoreceptors,** which we will study during examination of the eye, are sensitive to light. **Chemoreceptors** are sensitive to chemicals. **Thermoreceptors** respond to temperature, and **pain receptors** are sensitive to pain. Technically, the mechanism for thermoreceptors and pain receptors may also qualify them as chemoreceptors, as it seems to be the chemicals released from damaged or stressed cells and tissues that trigger these receptors.

The largest class of receptors is the **mechanoreceptors,** which include all receptors that respond to mechanical energy. This category includes the following:

- **Touch** or **tactile** receptors, which are usually found in the skin and detect contact and/or vibrations (Figure 12.1).
- **Pressure** receptors, which detect pressure on the skin or joints and even subtle changes in blood pressure (Figure 12.1).
- **Stretch** receptors, which are found in muscles, tendons, joints, and the lungs and detect stretching or stress (Figure 12.2).

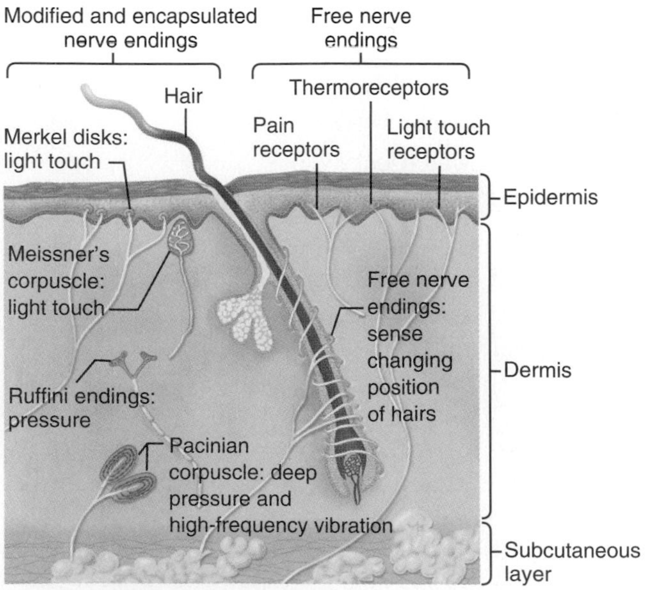

FIGURE 12.1 Sensory receptors in skin for sensing touch, pressure, and vibration.

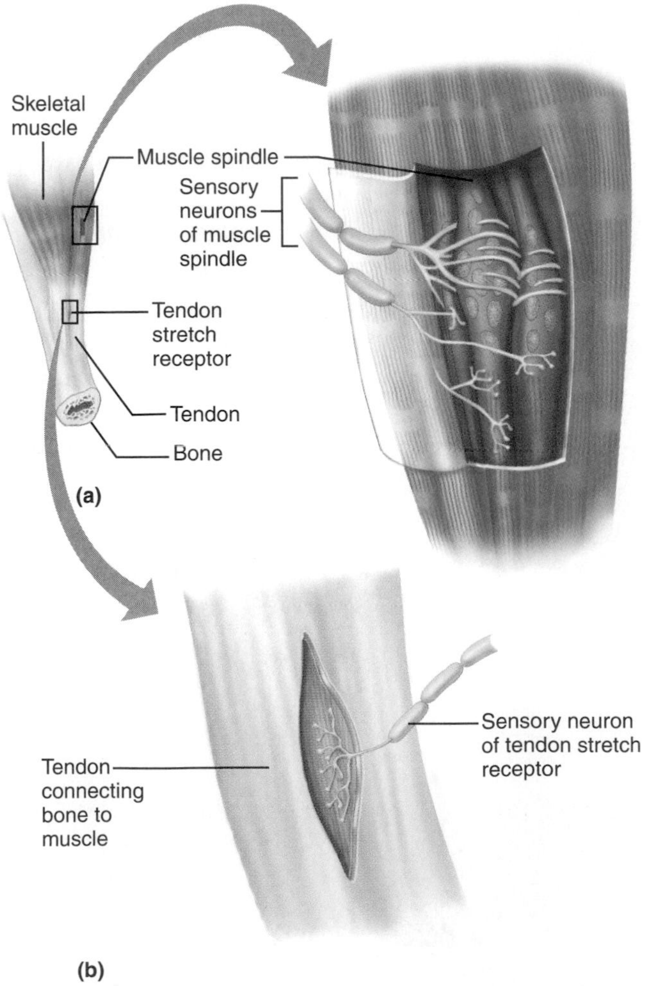

FIGURE 12.2 Muscle spindles and tendon receptors that monitor muscle length. Reflexes initiated by these receptors help us maintain balance.

- **Hearing** and **balance** receptors, which are found in the inner ear and will be covered later in this exercise.

With the notable exception of pain receptors and some receptors found in joints and muscles, most receptors are more sensitive to *changes* in stimuli and will eventually stop responding to the same level of stimulus. This phenomenon is referred to as **receptor adaptation.** You probably are not aware of the clothes you are wearing (unless they are uncomfortable) because the receptors in your skin *adapted* to the stimuli induced by the clothing. Touch receptors, as well as the chemoreceptors in the nose, adapt rapidly. If you have ever encountered someone who has put on far too much cologne, or someone who smokes tobacco regularly, you know they do not seem to notice these odors.

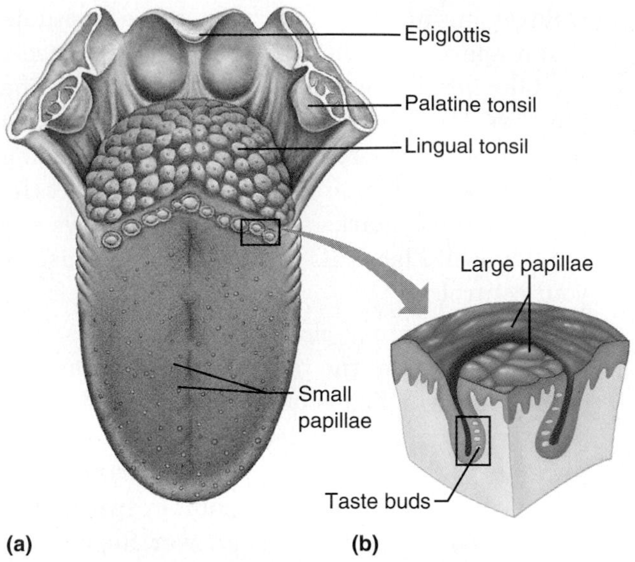

FIGURE 12.3 Structure of taste buds on the tongue. (a) Taste buds are associated with peglike projections (large and small papillae). (b) An illustration of a section of large papillae with the taste buds in its lateral wall.

ACTIVITY 1

Exploring Oral Chemoreceptors

Materials for this Activity

Cotton swabs

Powdered or granulated sugar (sucrose)

Lemon juice

1. Wet a cotton swab, and coat it with sugar. Using a mirror, or the assistance of a lab partner, touch the sugar-coated swab on the surface of the tongue near the tip. *Note:* The tongue should be moist, not dry, for this experiment to work. Do you immediately detect a significant sweet sensation?

   ____

2. Wet the cotton swab with lemon juice until it is saturated. Place the lemon juice-saturated cotton swab on the tongue near the sides. Do you immediately detect a significant sour sensation?

   ____

3. Considering that molecules and chemoreceptors must make a lock-and-key fit, speculate as to whether the same or different chemoreceptors were used during Steps 1 and 2 above.

   ____

   See Figure 12.3 illustrating the tongue, and note that taste buds are located within grooves on the tongue. How is this feature related to the need for the tongue to be moist?

   ____

   ____

ACTIVITY 2

Exploring Touch Receptors

Materials for this Activity

Seven flat-head dissecting pins

Three corks or Styrofoam cubes (about 1-inch cubes)

Metric ruler

Fine-point marker

It would be wasteful and impractical for touch receptors to be placed all over the body in the same density that we find them in our fingertips. Instead, we find a much lower density of touch receptors in areas of the skin where feedback from touch receptors is less essential to our needs. An easy way to determine the relative density of touch receptors is a **two-point discrimination test.** Follow the instructions below to perform a simple two-point discrimination test.

1. Using your marker and ruler, place two dots 5 mm apart near the center of the cork. Insert the pins into these marks, making sure that the pins are straight and level. Then, repeat this procedure for the remaining two corks, placing your marks 10 mm and 25 mm apart. Insert the pins into these marks as well. You will have one pin left over. Place this single pin aside to use as your control.
2. Have your partner close his or her eyes, and place one hand on the table palm up. In random order, touch the heads of the pin sets and single pin to your partner's index finger. After each application of the pin head(s) to your partner's finger, ask if he or she feels one point or two. Do not tell your partner the correct answer. Repeat this experiment until you obtain a consistent trend.
3. Was your partner able to accurately and consistently judge the number of points?

   ____

   Which pin set or sets (if any) was the subject most consistently *incorrect* in identifying as two points.

   ____

   Note: It is important that the subject answer carefully and honestly. The single pin serves as a control so that the subject is not tempted to answer "two" every time.
4. Repeat Step 2, this time using your partner's palm. Was your partner able to accurately and consistently judge the number of points?

   ____

   Which pin set or sets (if any) was the subject most consistently *incorrect* in identifying as two points?

   ____

5. Repeat Step 2, this time using your partner's forearm. Was your partner able to accurately and consistently judge the number of points?

   ____

   Which pin set or sets (if any) was the subject most consistently *incorrect* in identifying as two points?

   ____

   A higher density of receptors makes it easier to discriminate the two different points of contact, as more receptors increase the likelihood that each point will stimulate entirely different receptors. Based upon your results, rank the three tested areas in order of their density of receptors (starting with the highest density).

   ____

   ____

   Speculate as to why this test would be useful or practical.

   ____

   ______________________________ ■

ACTIVITY 3

Exploring Proprioceptor Function

Materials for this Activity

Three paper cups (8-ounce size)
Tap water
Beaker or flask

Mechanoreceptors in joints, tendons, and muscles that help to convey information about the position of our body parts and the stress on joints are categorized as **proprioceptors.** Proprioceptors contribute to our sense of balance by helping the brain rapidly monitor the effect of movement or stress on joints.

To test the function of these receptors, work in pairs, and obtain three paper cups and a beaker or flask.

1. Add water to two of the paper cups so that one is about ¼ full and the second is about ⅔ full. Fill a third cup about ⅓ full, and put it aside for later use.
2. Your partner must close his or her eyes, while you carefully place one cup in each hand. Now ask your partner which cup is heavier or has more water. Take the cups away, and repeat the process (perhaps in a different order of right or left) twice more to reduce the possibility of lucky guessing. Can your partner determine which cup is heavier at least two out of the three times?

   ____

   ____

3. Pour water from the heavier cup into the lighter cup until both are approximately ⅓ full. The exact amount of water is not critical as long as both cups are as close to equal in weight as possible. Once again, after your partner closes his or her eyes, carefully place one cup in each of his or her hands. Pour the contents of the third cup, which you had previously put aside, into one of the cups, and ask your partner to identify which cup received the water. Repeat this procedure twice more to reduce the possibility of lucky guessing. Can your partner determine which cup you added water to at least two out of the three times?

____

____

Explain your results.

____

____________________ ■

ACTIVITY 4

Exploring Receptor Adaptation

Materials for this Activity

Thermometer

Two paper cups (8-ounce size)

Tap water

Stopwatch or clock

1. Prepare two cups of water so that one is approximately room temperature and one is very warm but not uncomfortably hot—approximately 43 to 44°C. The cups should be about ¾ full.
2. Have your partner place the index finger of one hand in the room temperature water. Ask your partner to describe the temperature sensation of the immersed finger, and record the comments.

____

____

3. Next, have your partner remove and immediately place the same finger in the warm water. Ask your partner to describe the temperature sensation of the immersed finger.

____

____

After 1 minute, ask your partner to describe the sensation again, now that the finger has been immersed for a little while.

____

4. Finally, have your partner remove and immediately place the same finger in the room temperature water again. Ask your partner to describe the temperature sensation of the immersed finger.

____

Does this description exactly match the comments you recorded in Step 2? Explain.

____

____________________ ■

The Eye and Vision

The eye is very much like a video camera, with the photoreceptors of the retina acting as the instantly reusable videotape. The three tissue layers of the eye from outside to inside are the **sclera, choroid,** and **retina.** Most of the structures of the sclera and the choroid are simply there to ensure that a focused image is projected on the retina (Figure 12.4).

The Sclera

The sclera is the so-called white of the eye. It is the tough outer layer of connective tissue that is modified into the transparent **cornea** in the front (Figure 12.4). Thus, the delicate structures inside the eye are partly protected by the sclera, but the cornea still allows light to enter the eye.

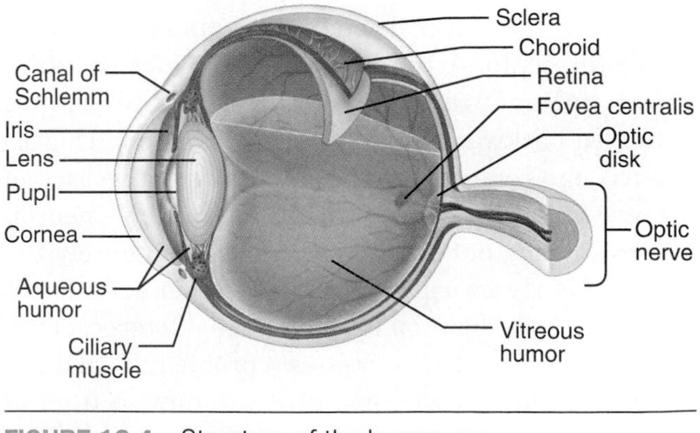

FIGURE 12.4 Structure of the human eye.

The Choroid

The choroid is the dark, pigmented, middle layer of the eye (Figure 12.4). Similar to the dark inside of a camera, this dark pigmentation absorbs light and prevents the bouncing or scattering of light that would distort the image. Most of the blood supply needed to nourish the retina is contained in the choroid.

Several special structures are modified from the choroid coat (Figure 12.4). The **lens** is a transparent, somewhat oval, marble-sized piece of tissue that bends light that passes through it. The thickness of the lens is modified by the **ciliary muscle,** which causes the flexible lens to become thicker or thinner to assist in focusing on near or far objects. Another muscular structure, the **iris,** adjusts the diameter of the **pupil,** so that more light enters the eye when it is darker and less light is permitted to enter the eye under brighter conditions.

A thin liquid called **aqueous humor** is found between the cornea and the lens. The **vitreous humor** is a thicker fluid found in the large chamber behind the lens. These fluids are important for transporting nutrients and maintaining the shape of the eye (Figure 12.4).

The Retina

The retina, as previously mentioned, contains the photoreceptors of the eye (Figure 12.5). The **cones** (for color) are active in brighter light conditions, and the **rods** (for black and white) are active in dimmer light conditions. A high concentration of cones called the **macula lutea** is found in a small area in the back of the eye called the **fovea centralis** *(center of focus).* When we squint to try to make out the details of a small object, we are attempting to center the focus of the object on this area.

There are three layers of specialized neurons in the retina: the rod and cone cells (commonly known as photoreceptors), the **bipolar cells,** and the **ganglion cells.** Interestingly, the three layers are arranged backward from a design viewpoint. That is, the rod and cone cell layer points toward the back of the eye into the choroid, with the other two neuron layers sticking out toward the front. Fortunately the retina is fairly transparent, and the other cells do not really block the light on its way to the photoreceptors. However, this structure creates a problem in that the axons of a million ganglion cells are now in front of the photoreceptor layer, and they must somehow exit the eye in order to connect to the brain. All of these axons are bundled together and exit the eye in an area called the **optic disk,** or blind spot, because the exiting axons displace the photoreceptors (Figure 12.4).

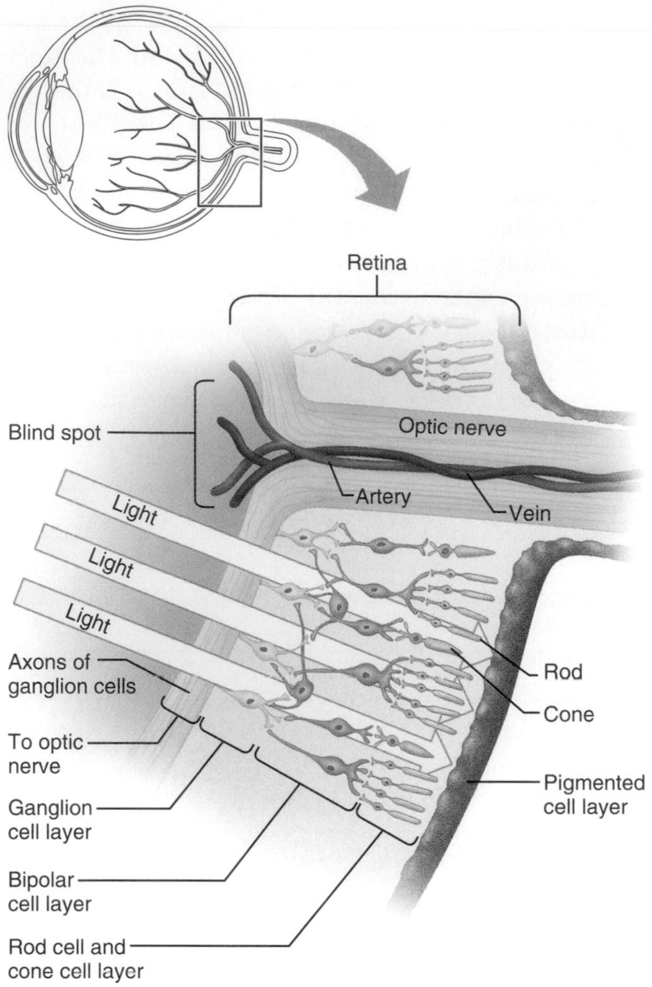

FIGURE 12.5 Structure of the retina.

ACTIVITY 5

Examining the Structure of the Eye

Materials for this Activity

Model of the human eye
Sheep or cow eye
Dissection equipment
Dissection tray

Sheep or Cow Eye

1. Obtain a sheep eye or cow eye, and place it cornea side up on a dissection tray. Cut away any muscles that may still be attached to the eye.
2. Using a sharp scalpel, carefully make a small incision in the sclera about ¼ inch to the outside of the transparent cornea.
3. Insert the point of a scissors blade into this incision, and cut a circle around the cornea, leaving a ¼-inch border of gray-white sclera. Once you have cut all the way around it, remove the front section of the eye.
4. First examine the inside of the smaller front section. You will see a marble-shaped structure that is relatively hard. This is the lens, which would have been more flexible when the specimen was living.
5. Use forceps to remove the lens. Now you will see the pupil and some of the cornea from the inside. Forming the pupil is the thin, dark iris. To the outside of the iris is the dark ciliary body, which is made largely of the ciliary muscle previously described. The ciliary body is easily discernable from the iris because of its radiating grooves, while the iris is smooth in appearance.
6. Now examine the back section of the eye. The cream-colored or light yellow retina may have already become detached from the dark choroid. If not, peel the retina away from the choroid. You will notice that there is one place where the retina will not detach. This is the optic disk, where the retina attaches to the optic nerve. Note the position of the optic disk, flip the eye over, and find the optic nerve to the outside of this point. You may have to cut away some muscle or fat tissue in order to find the optic nerve on the outside. Next, carefully separate the choroid from the sclera in one small area so that you can observe the three tissue layers of the eye and note their relative thickness to each other.

   You may notice that there is a lighter blue area in the back of the choroid coat. This is called the **tapetum lucidum,** a somewhat more reflective area found in the eyes of some animals, which improves night vision. This reflective area is a large part of the reason why a cat's eyes seem to "glow" at night. The eyes do not actually glow; they simply reflect more light than the cat's surroundings.

Eye Models

Using Figure 12.4 as a reference, locate the following structures on the eye models provided for you in the laboratory.

- Cornea
- Lens
- Pupil
- Iris
- Ciliary body/muscle
- Retina
- Choroid
- Sclera
- Optic nerve
- Optic disk ■

ACTIVITY 6

Testing Visual Acuity

Materials for this Activity

Snellen eye chart
Astigmatism eye chart

1. Stand 20 feet away from the Snellen eye chart, and gently cover one eye. Begin reading the letters in each row, starting with the top row, while your lab partner stands next to the chart. Your partner will stop you when you make a mistake, and he or she will read the numbers to the left of the last row that you read correctly. Repeat this procedure for the other eye. Record the numbers for each eye.

   Right ____________

   Left ____________

   A pair of numbers are found to the left of each line on the Snellen chart; the first represents the distance you are standing from the chart (20 feet), and the second is the distance a person with "normal" vision would stand to read the same line as his or her last correct line. Thus, someone who can read the large "E" at the top of the chart as his or her only correct line would be reading the line that a person with normal vision would see when standing much farther away. This person's vision would be rated as 20/200. That is, when standing at 20 feet (the first number), this person sees what a person with normal vision would be able to see at 200 feet (the second number).

   Vision rated at 20/20 suggests normal vision—that the subject's eye is correctly reading a line at 20 feet that a person with normal vision would also be reading at that same distance. Second numbers larger than 20 may suggest nearsightedness (myopia), while second numbers smaller than 20 (e.g., 20/15) may either suggest farsightedness (hyperopia) or simply better-than-normal vision.

2. Next, stand 10 feet away from the astigmatism eye chart, and gently cover one eye. Note if any of the lines look *substantially* darker than the others, and record your results.

   Right ________________

   Left ________________

   The lines in the astigmatism eye chart are all equal in thickness. Astigmatism is the result of irregularities in the curvature of the lens or cornea. If some lines look substantially darker than the others, it may indicate an astigmatism. As a result, light is scattered somewhat irregularly, rather than focusing evenly on the retina. Like nearsightedness or farsightedness, an astigmatism also can be easily corrected with eyeglasses. ■

ACTIVITY 7

Testing Binocular Vision

Materials for this Activity

Medium-size test tube
Pencil

The brain is capable of judging depth perception partly because of learned behavior and partly because of the fact that we receive slightly different visual information from each eye. For example, we learned long ago that if a person looks larger than a tree, he or she is probably standing quite a bit closer to us compared to the tree. But for items where size and point of reference are less helpful, we rely on comparing data from both eyes. This is the concept of binocular vision.

1. Work in pairs, and have one partner hold a small test tube as steady as possible. The other partner should stand a full arm's length away from the test tube and hold a pencil with the *eraser end down.* Using a steady and gentle arc-like motion, the second partner should attempt to place the eraser end of the pencil in the test tube. The movement should not be excessively slow, considering this experiment is testing your ability to quickly evaluate the three-dimensional positions of the pencil and test tube. This action is supposed to be somewhat difficult to do, and it would be unrealistic if you are able to get the pencil in the test tube every time. Record the number of times out of 10 that you successfully inserted the pencil into the tube.

   ____

2. Repeat Step 1, but this time, cover one eye. Record the number of times out of 10 that you successfully inserted the pencil into the tube.

   ____

   Propose a reason for your results.

   ____

   ____

   ____

   ________________________________ ■

The Ear and Hearing

The ear is both an organ of hearing and balance. The structures of the ear are divided into three sections: the **outer ear, middle ear,** and **inner ear** (Figure 12.6). We will cover the balance responsibilities of the inner ear in a separate section in this exercise. Most of the structures associated with the ear are involved in our hearing mechanism.

The visible portion of the outer ear, or **pinna,** collects and channels sound waves into the **auditory canal.** These sound waves strike the **ear drum,** or **tympanic membrane** (Figure 12.6).

The middle ear begins on the inside of the eardrum and is an air-filled cavity (Figure 12.7 on page 140). If this cavity were not connected to the outside in some way, changes in air pressure would create painful stress on the eardrum. The **Eustachian tube,** or auditory tube, connects the middle ear cavity with the throat—which explains why swallowing may cause our ears to "pop" and relieve the pressure we feel in our ears when going up or down in elevation (Figures 12.6 and 12.7). In the middle ear, we find three little bones attached to the eardrum. These little bones, the **malleus** (hammer), **incus** (anvil), and **stapes** (stirrup), amplify the vibrations of the eardrum by concentrating the larger

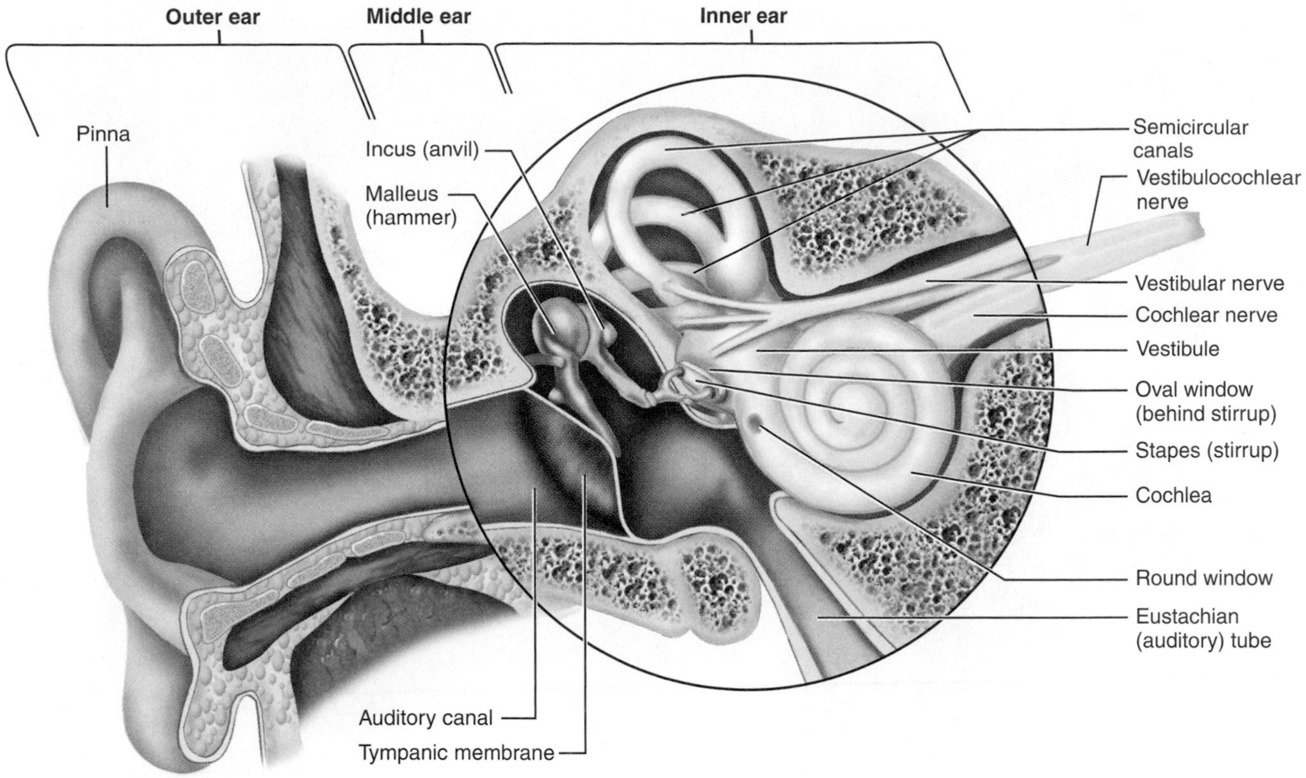

FIGURE 12.6 Structure of the human ear.

eardrum's vibrations down to a single, smaller point. This smaller area at the end of the stapes is called the **oval window,** and it brings us to the inner ear (Figure 12.6).

The inner ear consists of three main parts (Figure 12.6). The **semicircular canals,** as the name implies, are three small tubes bent into a curved, semicircular shape. The **vestibule** is the thick, middle portion of the inner ear. These two structures are mainly involved with the balance function of the inner ear and will be explained in more detail later. The coiled **cochlea** is the hearing portion of the ear, where vibrations are eventually converted into nerve impulses that the brain interprets as sound (Figures 12.6 and 12.7).

Fluid within the long, coiled ducts of the cochlea vibrates when the stapes vibrates against the oval window. Mechanoreceptors called **hair cells** form a long line, running the length of one of these ducts (Figure 12.7c). Bending of the hairlike projections of the hair cells triggers an action potential that travels to the brain for interpretation. Sounds of different frequencies vibrate the cochlea in different areas, thus allowing the brain to interpret the frequency of the sound based upon which part of the cochlea is receiving impulses (Figure 12.7a).

Localizing Sounds

A sound that arrives at your right ear from 100 feet away does not lose much of its volume as it travels the few inches to the other side of your head. Yet, with a fairly high degree of reliability, we can determine the direction of incoming sound. It does seem that the sound entering one's ear from the side closer to the origin of the sound is louder. This differentiation is yet another interesting feature of the human nervous system. Although there is only a small fraction-of-a-second delay in the sound that is farther from the source of the sound reaching the ear, the delay is enough to cause the nervous system to inhibit the impulses from the "second" ear. Thus, even though the volume of the sound reaching both ears is nearly identical in most cases, our *perception* is of a louder sound from the closer ear. We learn to interpret this difference as a way of judging the direction from which a sound originated.

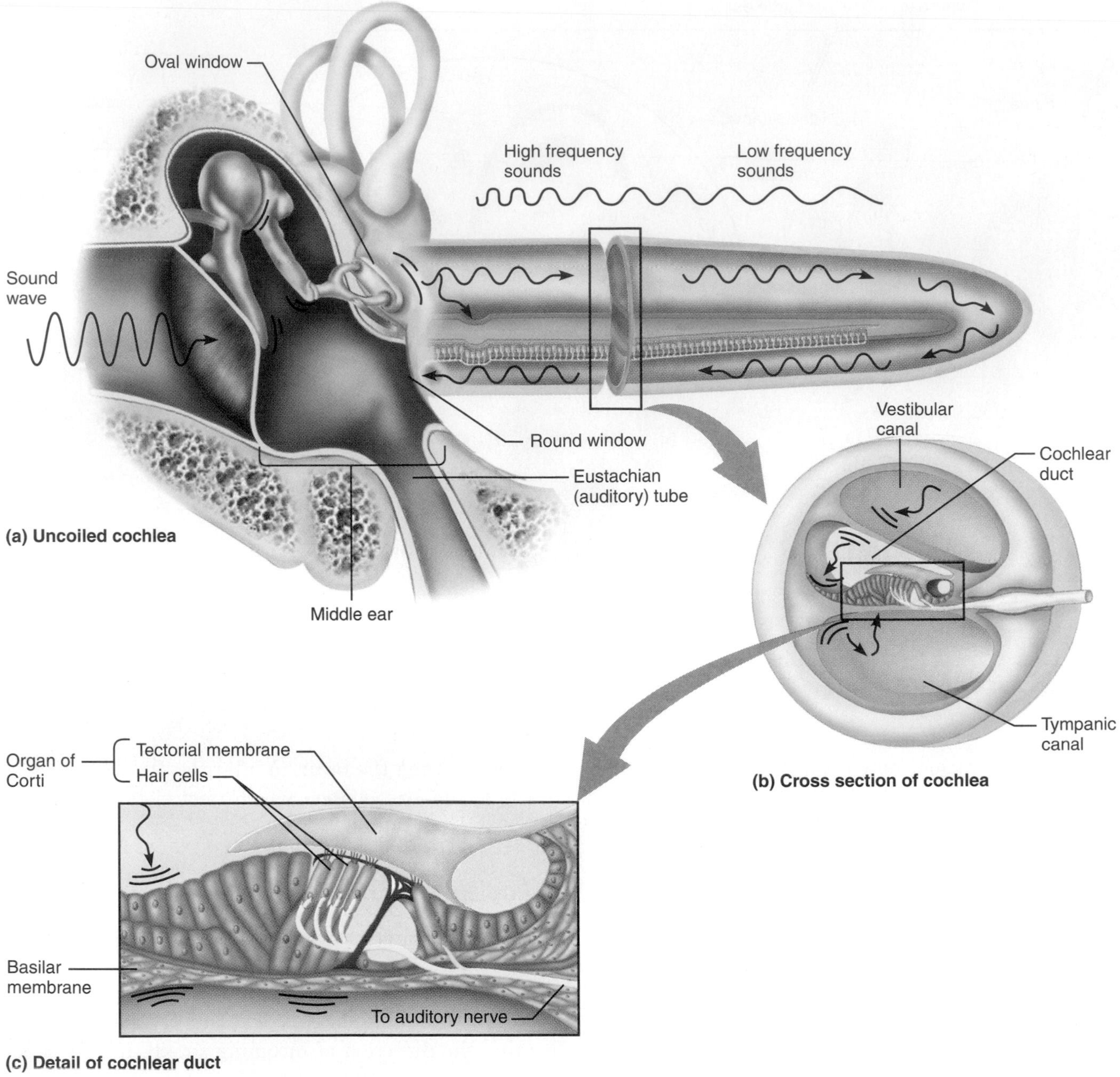

FIGURE 12.7 The middle and inner ear: converting vibrations into nervous impulses.

ACTIVITY 8

Examining the Structure of the Ear

Materials for this Activity

Model of a human ear

Using Figure 12.6 on page 139 as a reference, locate the following structures on the ear models provided for you in the laboratory.

- Pinna
- Auditory canal
- Tympanic membrane
- Eustachian tube
- Malleus
- Incus
- Stapes
- Oval window
- Semicircular canals
- Vestibule
- Cochlea ■

ACTIVITY 9

Testing Hearing Conduction Pathways

Materials for this Activity

Tuning fork

1. Hold the tuning fork by its handle and strike one tine from the side on the thick portion of the palm of your hand (near the pinky side of your palm).
2. Place the base of the tuning fork handle on your temporal bone about 1 inch behind your earlobe. If you strike the tuning fork hard enough, you should be able to hear the sound because you are directly vibrating the fluid in your cochlea, which is found within the temporal bone. If you do not hear a sound, repeat the procedure of striking the tuning fork and placing it on your temporal bone again until you do. When the sound disappears, move the tuning fork so that the ends of the tines are in front of your pinna. Do you hear a sound after moving the tines to the front of your pinna?

   ____

3. Repeat Step 2 in reverse order. That is, strike the tuning fork, place the tines in front of your pinna, and when the sound disappears, place the base of the tuning fork on your temporal bone. Do you hear a sound after moving the base of the tuning fork onto the temporal bone?

   ____

   What does this response suggest about the value of your conduction pathway (the eardrum, malleus, incus, and stapes)?

   ____

   ______________________________ ■

ACTIVITY 10

Testing for Hearing Localization

Materials for this Activity

Tuning fork

1. Work in pairs, with one partner closing his or her eyes. Strike the tuning fork. In random sequence, place the tuning fork in the following positions, asking your partner each time to identify the position of the tuning fork: near the right ear, near the left ear, near the front of the nose, near the top of the head, and a few inches behind the head. Prompt your partner by saying "now" when you have the fork in position. If he or she hears nothing, strike the tuning fork harder and try again. Do not tell your partner if he or she is right.
2. Repeat this exercise until a pattern becomes clear.

   At which location(s) was your partner able to most accurately and consistently judge the location of the tuning fork?

   ____

   At which location(s) was the subject most consistently *incorrect* in identifying the location of the tuning fork?

   ____

   Explain your results after considering the background information provided earlier regarding localization of sound.

   ____

   ______________________________ ■

The Ear and Balance

With slight modification, the very same kind of hair cell receptors that we identified in the cochlea are also used by the inner ear to heavily contribute to our sense of balance, or equilibrium. There are two main categories for our sense of equilibrium: **static equilibrium,** or our sense of equilibrium when stationary, and **dynamic equilibrium,** or our sense of equilibrium when in motion.

The **vestibule** is most responsible for our sense of static equilibrium (Figure 12.8a–c). Gravity pulls downward on hair cells embedded in a weighted, gelatinous material. Whether we are moving or standing still, we always know which way is down based upon the pull of gravity on these hair cells.

The **semicircular canals** provide our sense of dynamic equilibrium (Figure 12.9). When we move in any direction, the fluid in at least one of our semicircular canals will slosh backward in the opposite direction, in the same way the water in a container would seem to move when you start moving. Actually, the fluid is trying to remain stationary (the principle of inertia), but the effect is the same. The apparent sloshing back of fluid through the semicircular canals bends hair cells, which alerts our brains to movement along that plane. It is no coincidence that there are three dimensions in normal space and that we have three semicircular canals at right angles to each other. It is important to note that our

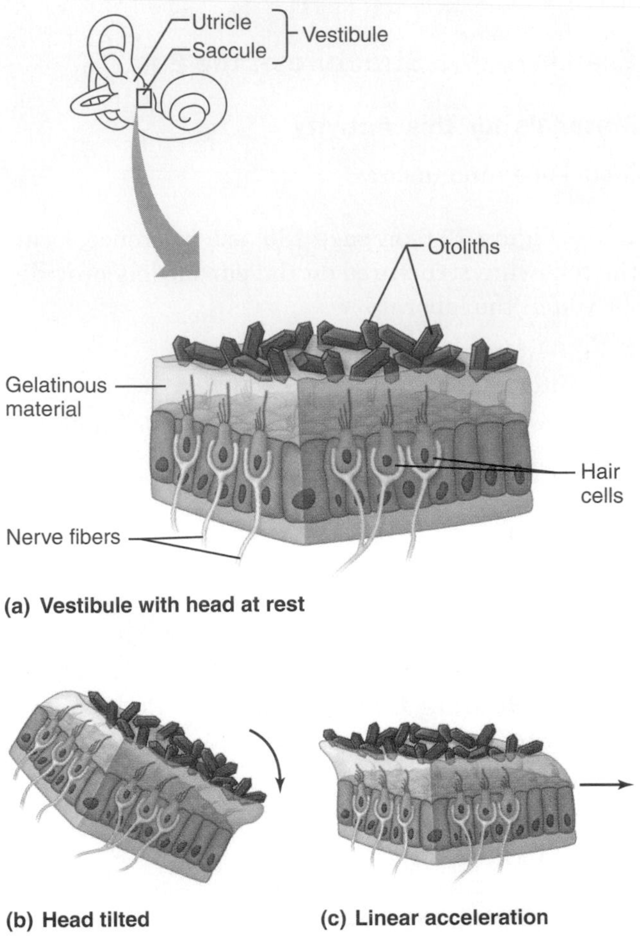

FIGURE 12.8 The role of the vestibule in balance.

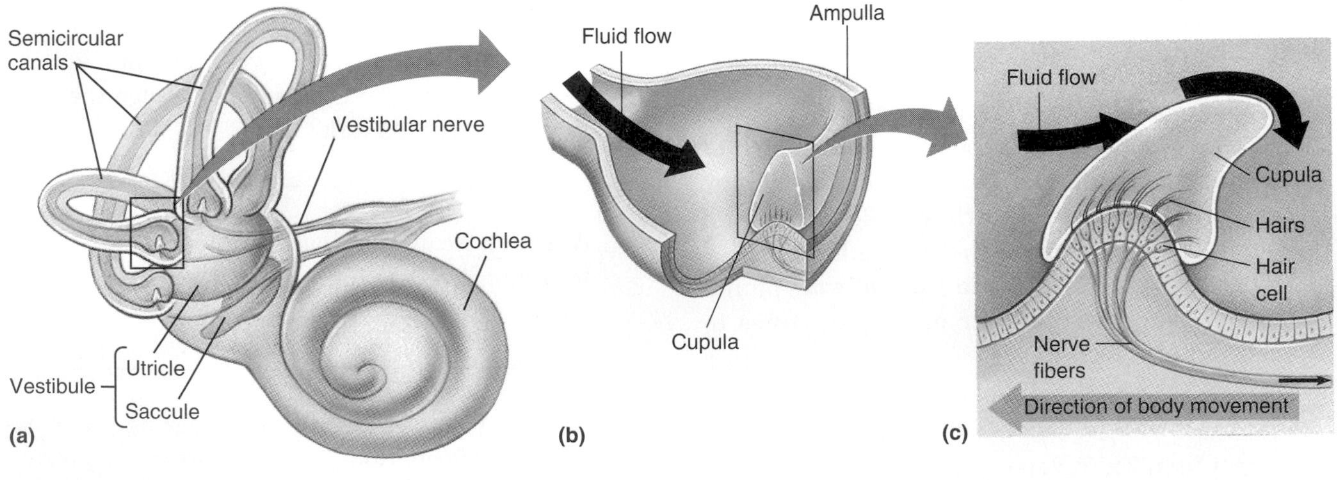

FIGURE 12.9 The role of the semicircular canals and vestibular apparatus in balance.

sense of balance is not restricted to input from the inner ear. You may recall that in our discussion of the cerebellum in Exercise 11, the inner ear's feedback is one of *several* pieces of sensory information used by the brain to help us maintain our balance.

ACTIVITY 11

Testing Equilibrium

Materials for this Activity

Chair with wheels
Stopwatch or clock

Part 1: Direction of Movement

1. Work in pairs. Have your partner sit in a chair with wheels in an area with some open space and *close his or her eyes.*
2. Slowly and carefully, move the chair a few feet randomly in the following directions, asking your partner each time to identify the direction of movement: right, left, forward, or backward. Was your partner able to consistently identify the direction of movement?

   ____

   Explain your results after considering the background information provided earlier regarding dynamic equilibrium.

   ____

   ____

Part 2: Maintenance of Equilibrium

Work in pairs. Have one person stand on one foot with his or her eyes open, and time how long he or she can maintain balance. Set a limit of 2 minutes. Repeat this activity with the subject's eyes closed, standing by to "spot" your partner if he or she loses balance. Record your results.

Eyes open ______________

Eyes closed ______________

Explain your results after considering the background information provided earlier regarding the role of the cerebellum and other factors involved in your sense of balance.

____

____

______________________________ ■

Section ______ Name ______

Date ______ Instructor ______

EXERCISE 12

THE SENSES

Review Questions

1. Match the receptor to its function or description:

____	Tactile receptors	a. temperature receptors
____	Stretch receptors	b. rods and cones
____	Thermoreceptors	c. touch receptors found mostly in the skin
____	Photoreceptors	d. taste receptors
____	Chemoreceptors	e. receptors found in muscles, tendons, joints, and lungs

2. Describe at least two examples of receptor adaptation.

______

______

______

______

3. Match the part of the eye to its function or description:

____	Cornea	a. adjustable light-bending structure
____	Lens	b. changes the tension on the lens
____	Iris	c. blind spot
____	Ciliary muscle	d. transparent portion of the sclera
____	Optic disk	e. modifies the size of the pupil

4. Explain what the term "20/30 vision" means.

______

______

______

5. Explain what the binocular vision experiment in Activity 7 taught you about depth perception.

______

______

6. Summarize the roles of the outer ear, middle ear, and inner ear in contributing to our sense of hearing.

______

______

7. Explain why it is difficult to localize a sound that originates from directly behind you.

8. Considering all that you have learned from this exercise, as well as the previous exercises on the nervous system, list and describe all of the structures that contribute to your sense of balance.

EXERCISE

13

THE ENDOCRINE SYSTEM

Objectives

After completing this exercise, you should be able to

1. List and describe the major endocrine glands.
2. List and describe the function of the major hormones of the endocrine system.
3. Identify the major endocrine glands in the fetal pig and/or human models.

Materials for Lab Preparation

Equipment and Supplies

- ❑ Sheep brain
- ❑ Compound light microscope
- ❑ Fetal pig
- ❑ Dissection equipment
- ❑ Dissection tray
- ❑ Disposable gloves (optional)
- ❑ String
- ❑ Cleaning disinfectant or detergent solution in a spray bottle
- ❑ Storage container or bag (to keep pig for later use)

Prepared Slides

- ❑ Pancreas
- ❑ Pituitary gland

Models

- ❑ Human torso
- ❑ Human brain
- ❑ Sheep brain

Introduction

You have already learned that the nervous system is the body's primary communication system. However, another system using chemical messengers is also important in maintaining homeostasis. The endocrine system uses the blood to carry its chemical messengers, called **hormones,** to target cells. The target cells have receptors capable of receiving the hormones and translating their messages. Because hormones are powerful chemical messengers, endocrine glands need not be particularly large or produce large amounts of hormone in order to have powerful effects on the human body.

Although there are numerous hormones, including some secreted by individual cells and tissues rather than a distinct gland, the scope of this exercise mainly focuses on the major endocrine glands and their hormones.

The Hypothalamus and the Pituitary Gland

The close relationship between the nervous system and the endocrine system is demonstrated by the hypothalamus and the pituitary gland (Figure 13.1). The hypothalamus uses the *posterior pituitary* as its release site for two hormones, **antidiuretic hormone (ADH)** and **oxytocin.** These hormones are actually made in cell bodies of neurons in the hypothalamus and transported to the end of the axons in the posterior pituitary, where they can be released. ADH stimulates water reabsorption by certain kidney tubules, and it is released as part of the hypothalamus's thirst response. Oxytocin stimulates smooth muscle contraction in the uterus and stimulates contraction of cells in the mammary glands.

The *anterior pituitary* makes and releases several hormones, some of which control other endocrine glands (Figure 13.1b). These **tropic hormones** (hormones that control other endocrine glands) give the pituitary its somewhat undeserved reputation as the "master gland." The hypothalamus tightly controls the release of hormones from the anterior pituitary through hormones it releases into a short network of blood vessels connecting the hypothalamus to the pituitary.

The tropic hormones released from the anterior pituitary are **thyroid stimulating hormone (TSH),** which stimulates the thyroid to produce thyroid hormone; **adrenocorticotropic hormone (ACTH),** which stimulates the adrenal cortex; and **follicle stimulating hormone (FSH)** and **luteinizing hormone (LH),** which stimulate the ovaries (and testes). The nontropic hormones are **growth hormone (GH),** which stimulates growth, particularly of bone and muscle tissue, and **prolactin (PRL),** which stimulates the mammary glands.

ACTIVITY 1

Observing the Hypothalamus and the Pituitary Gland

Materials for this Activity

Compound light microscope

Prepared slide of the pituitary gland

Sheep brain and/or human and sheep brain models

You should already be familiar with the hypothalamus from Exercise 11, which covers the brain.

1. Observe the pituitary gland, located directly under the hypothalamus, in the sheep brain and/or brain models provided by your instructor.
2. Next, observe a prepared slide of the pituitary gland using your microscope's 10× objective lens (100× total magnification). Your instructor may elect to set this up as a demonstration slide in order to ensure that both portions of the pituitary are in your field of view.

   You should notice that the two portions of the pituitary have very different textures. The *anterior pituitary* mostly consists of glandular tissue, and you should see it spotted with the nuclei of the cuboidal epithelial cells that produce its hormones. The *posterior pituitary* is mostly composed of axons of neurons that originate in the hypothalamus. So, instead of a spotted, glandular appearance, you should see many "streaks" running through the posterior pituitary.

   Which portion of the pituitary gland is to the left of your slide, the anterior or posterior?

   ____

   Which is to the right?

   ____

   Your instructor will verify your answer after you have had an opportunity to view this slide. ■

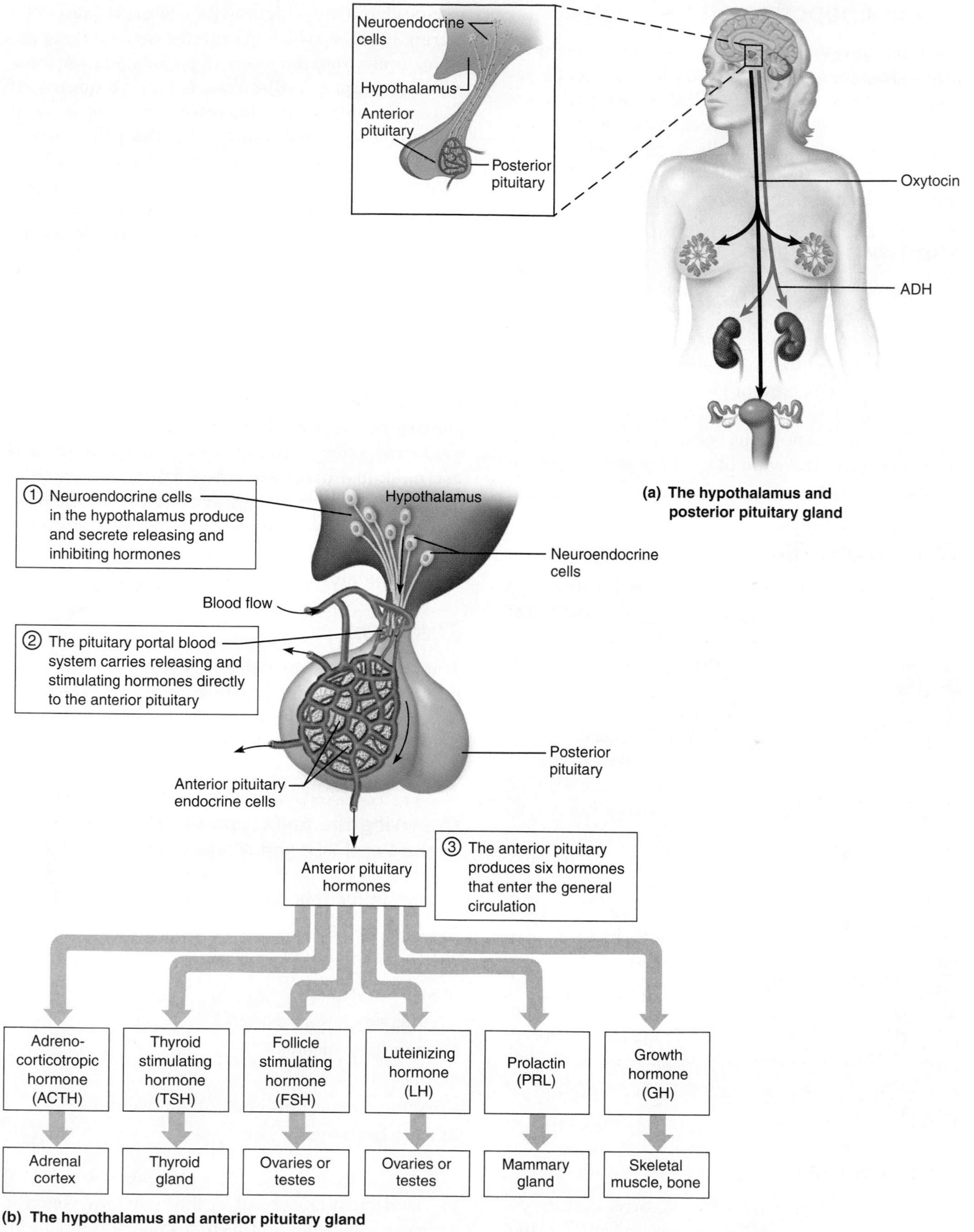
Neuroendocrine cells
Hypothalamus
Anterior pituitary
Posterior pituitary
Oxytocin
ADH
(a) The hypothalamus and posterior pituitary gland
① Neuroendocrine cells in the hypothalamus produce and secrete releasing and inhibiting hormones
Hypothalamus
Neuroendocrine cells
Blood flow
② The pituitary portal blood system carries releasing and stimulating hormones directly to the anterior pituitary
Posterior pituitary
Anterior pituitary endocrine cells
Anterior pituitary hormones
③ The anterior pituitary produces six hormones that enter the general circulation
Adreno-corticotropic hormone (ACTH)
Thyroid stimulating hormone (TSH)
Follicle stimulating hormone (FSH)
Luteinizing hormone (LH)
Prolactin (PRL)
Growth hormone (GH)
Adrenal cortex
Thyroid gland
Ovaries or testes
Ovaries or testes
Mammary gland
Skeletal muscle, bone
(b) The hypothalamus and anterior pituitary gland

FIGURE 13.1 The hypothalamus in relation to the posterior and anterior pituitary glands.

Other Endocrine Glands

The remaining endocrine glands are controlled by either hormones from the pituitary or simply by responding directly to the chemical or factor they are charged with regulating (Figure 13.2). The exception to this is the adrenal medulla, which is stimulated by the sympathetic nervous system as part of the "fight or flight response."

The Thyroid Gland

The thyroid produces two hormones (Figure 13.2). **Thyroid hormone (TH)** stimulates body cells to increase metabolism. Iodine is an essential component of the TH molecule—the main reason why salt is iodized in order to prevent iodine deficiency and dysfunction of the thyroid. **Calcitonin** helps the body store calcium in bones. This hormone lowers blood calcium levels and therefore plays an important role in calcium homeostasis.

The Adrenal Glands

Like the pituitary, each adrenal gland has two very different portions (Figure 13.2). The **adrenal medulla** is the inner part of the gland derived from nervous tissue. It releases **epinephrine** and **norepinephrine.** The medulla is directly connected to the hypothalamus by nerves for prompt release of these hormones during the "fight or flight" response.

The **adrenal cortex** is the outer part of the adrenal gland, and it is made from glandular epithelial tissue (Figure 13.2). It releases three main hormones. **Aldosterone** stimulates kidney tubules to reabsorb sodium and excrete potassium. The body releases aldosterone when potassium levels are too high, blood pressure is too low, and sodium levels are too low. The **glucocorticoids,** which include cortisone and cortisol, increase glucose production from other molecules and raise blood sugar levels. They also have anti-inflammatory properties and suppress the immune system. Small amounts of **androgens** (male sex hormones) and **estrogens** (female sex hormones) are also produced, regardless of gender.

The Pancreas

Although mainly a gland of the digestive system, the pancreas contains small clusters of cells called pancreatic islet cells (or islets of Langerhans) that produce hormones (Figure 13.2). Some of these cells secrete **insulin,** which stimulates many cells to absorb glucose from the blood. In particular, liver, muscle, and adipose cells absorb large amounts of glucose for processing. This results in a drop in blood sugar levels. Other pancreatic islet cells secrete **glucagon,** the antagonist of insulin, which stimulates the liver to break up glycogen and release glucose into the blood. Production of new glucose molecules is also encouraged. These activities result in a rise in blood sugar levels.

The Gonads

Both testes and ovaries are stimulated by the pituitary hormones LH and FSH (Figure 13.1). LH mainly stimulates hormone production in the gonads, whereas FSH is primarily involved in gamete production. **Testosterone** from the testes and **estrogen** and **progesterone** from the ovaries (Figure 13.2) have a wide variety of effects on the body associated with masculine and feminine characteristics, including proper development of the male and female reproductive systems.

The Thymus

The thymus secretes **thymus hormone,** or thymosin (Figure 13.2), which activates certain white blood cells of the immune system (T cells).

ACTIVITY 2

Observing the Endocrine Glands in the Fetal Pig and Models

Materials for this Activity

Fetal pig
Dissection equipment
Dissection tray
Disposable gloves (optional)
Human torso models
String
Cleaning disinfectant or detergent solution
Storage container or bag

This is the first exercise in which you use the fetal pig, so it must be opened so that you may view the structures specified in this and some of the remaining exercises.

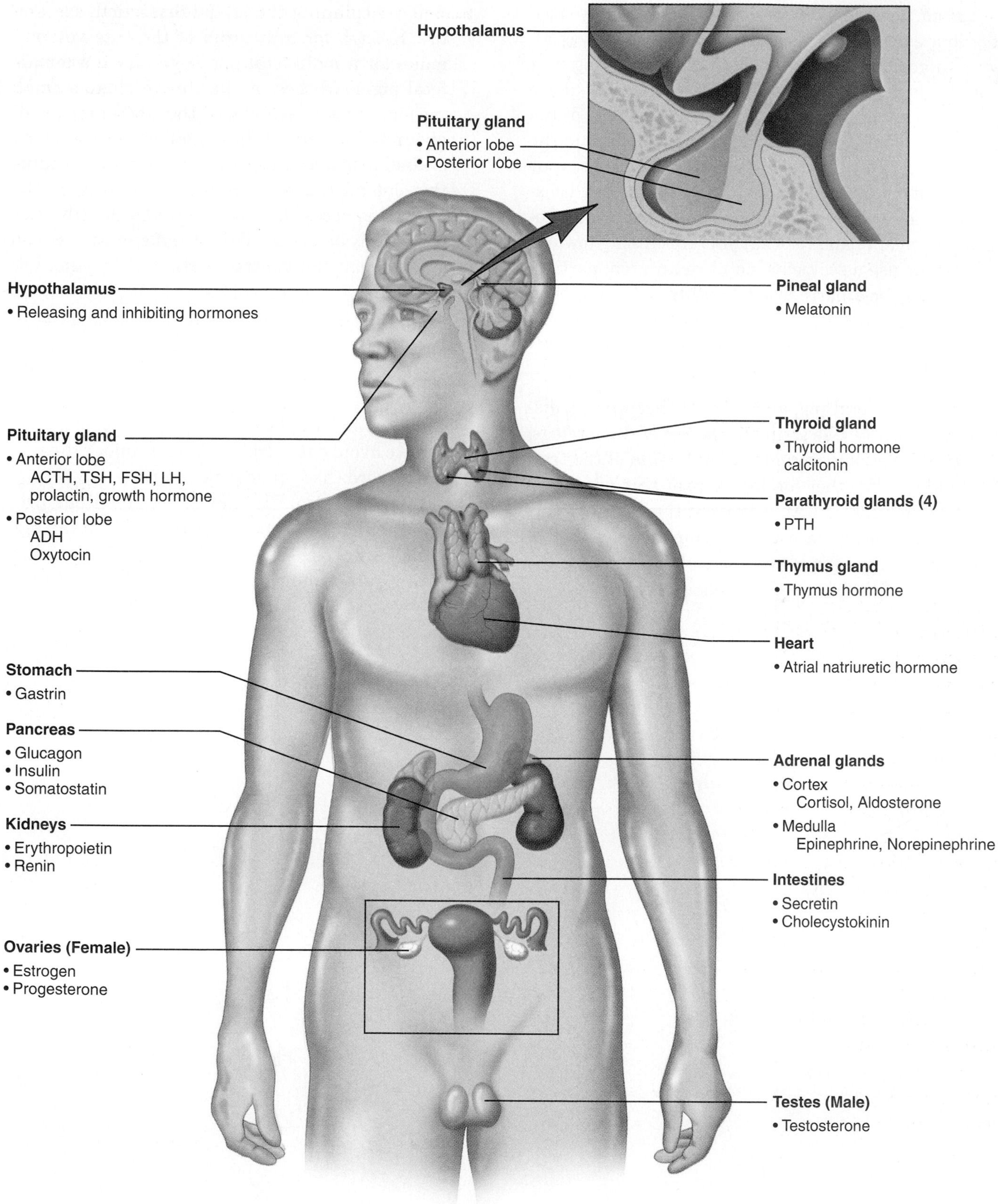
Hypothalamus
Pituitary gland
• Anterior lobe
• Posterior lobe
Hypothalamus
• Releasing and inhibiting hormones
Pineal gland
• Melatonin
Pituitary gland
• Anterior lobe
ACTH, TSH, FSH, LH,
prolactin, growth hormone
• Posterior lobe
ADH
Oxytocin
Thyroid gland
• Thyroid hormone
calcitonin
Parathyroid glands (4)
• PTH
Thymus gland
• Thymus hormone
Heart
• Atrial natriuretic hormone
Stomach
• Gastrin
Pancreas
• Glucagon
• Insulin
• Somatostatin
Adrenal glands
• Cortex
Cortisol, Aldosterone
• Medulla
Epinephrine, Norepinephrine
Kidneys
• Erythropoietin
• Renin
Intestines
• Secretin
• Cholecystokinin
Ovaries (Female)
• Estrogen
• Progesterone
Testes (Male)
• Testosterone

FIGURE 13.2 Endocrine glands.

First, consider a few words about safety. Eye protection, especially if you wear contact lenses, is always a good idea when working with anything that could splash in your eyes. Your instructor will advise you concerning proper handling and disposal of the fluid that comes with your specimen, including the policy regarding gloves and other precautions. Your dissection equipment, particularly the scalpel, is extremely sharp. Always make certain that your fingers are clear of an area that you are cutting toward, and never use a scalpel to point to an area, particularly if you or a partner are holding that area open with your fingers. Read and follow the directions below slowly and carefully—impatience and sharp cutting tools do not mix.

1. Obtain a fetal pig, and place it faceup on a dissection tray. Use string to tie one of the forelegs, run the string around the underside of the tray, and tie the other foreleg. Repeat this procedure for the hind legs. After you cut the pig open, this tying procedure and additional wrapping of loose string around the legs will help to keep the inside of the pig open and easily accessible.
2. Before beginning the actual dissection, see Figure 13.3a–b for a diagram of the cuts you will make for a male fetal pig. If you have a female fetal pig, read the note in Step 3. Make a small incision with a scalpel in the abdominal wall about ½ inch above the umbilical cord. Put the scalpel aside, and use your scissors to continue this cut all the way up into the middle of the neck. It is necessary to cut through the ribs (use bone snips here if available) and sternum as you cut through the ventral portion of the thoracic cavity.
3. Turn the tray around to make the next series of cuts easier. Continuing to use your scissors, make a "keyhole-shaped" cut around the umbilical cord of the male pig, leaving about ½ inch of material around the cord. This incision is important to avoid cutting blood vessels under the umbilical cord that you may be instructed to view in a later exercise. The keyhole-shaped cut prevents cutting the penis of the male fetal pig, which is not yet developed and barely visible as a thin tube running from the pelvic area to the

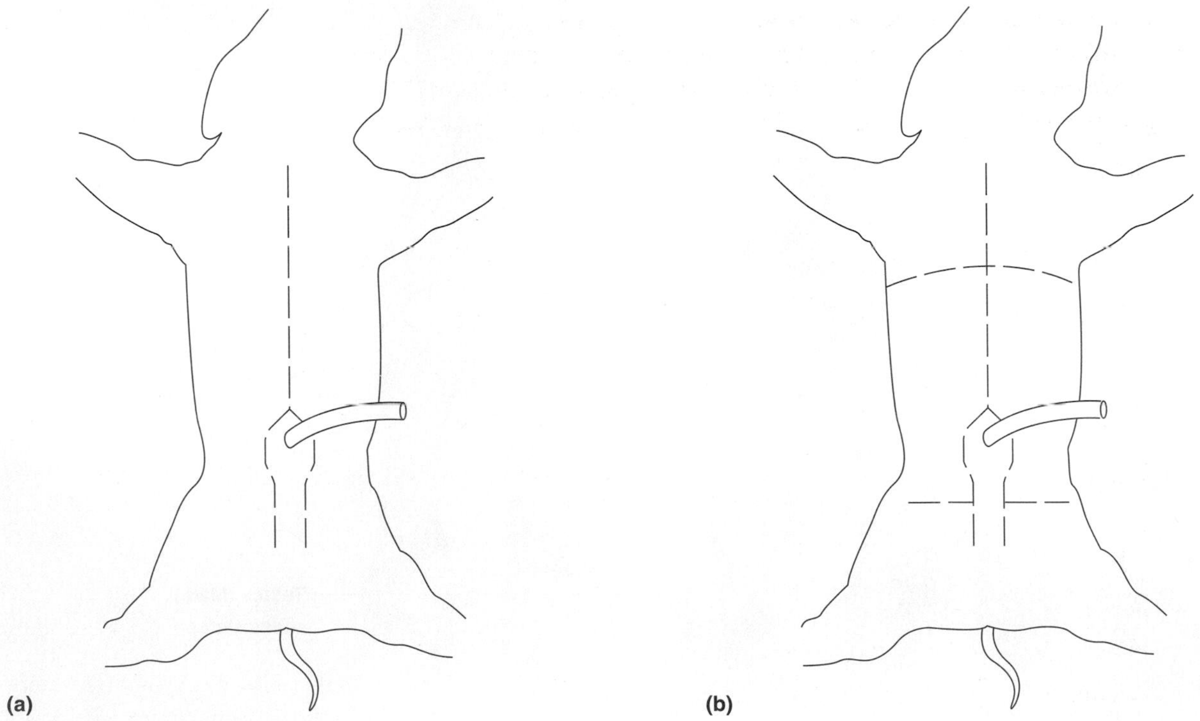

FIGURE 13.3 Fetal pig dissection. (a) Diagram outlining a midsagittal cut of the male fetal pig. (b) Diagram outlining transverse cuts of the male fetal pig.

umbilical cord. *Note:* If you have a female fetal pig, the keyhole-shaped cut is unnecessary, and you may simply circle around the umbilical cord, leaving a ½-inch border around it.

4. Next, make lateral cuts with scissors as shown in Figure 13.3b. These cuts should be just below the forelegs and just above the hind legs. So that you may more easily open up the body cavity, snip some of the diaphragm, which will be found inside the pig above the liver.
5. Observe Figure 13.4, and other fetal pig figures found in this manual (Figures 17.5, 18.3, and 19.6). Familiarize yourself with the major organs of the fetal pig. Although the focus of this exercise is the endocrine organs, you may find this step is helpful so that you will be able to follow the directions for locating the endocrine glands in the following steps.
6. Directly over and above the heart is a loose, tan-colored covering of glandular tissue. This is the **thymus,** which is proportionally much larger in the fetal pig than it would be in an adult pig. What is the function of the thymus?

   ____

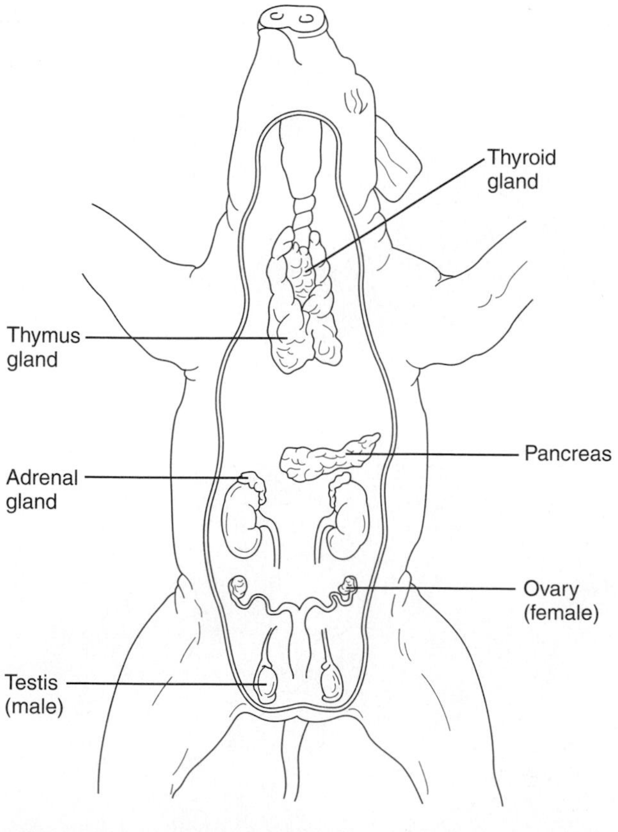

FIGURE 13.4 Major endocrine glands of the fetal pig.

7. Above the thymus in the lower neck area is the **thyroid,** which is often covered by the upper part of the thymus, muscle tissue in the neck, and possibly other tissues as well. Use your forceps to gently pull away the muscle tissue and other tissue covering the lower neck region until you see the small, brown, bean-shaped structure. What two hormones are produced by the thyroid?

   ____

   Do these hormones have similar or different functions?

   ____

8. Using your fingers and/or a blunt probe as needed, lift up (do *not* cut and remove) the liver on the pig's left side (your right). Then, lift up the saclike stomach, and use forceps to break and remove the thin, plastic-like material under the stomach. You should now be able to see the **pancreas,** which looks somewhat like a miniature ear of corn or many tiny beads stuck together. Why do you suppose the pancreas is so close to the stomach and intestines?

   ____

9. Lift, but again do *not* cut, the stomach and intestines on the *pig's* left side. Push them as far to the *pig's* right as you can without breaking them. You should be able to see the "kidney bean-shaped" kidney stuck into the back body wall. Use forceps to break and remove the thin, plastic-like material covering the kidney. Near the top and middle section of the kidney, you should see a lighter strip of material that might have become partially disconnected from the kidney. This is the **adrenal gland,** which is named for its close relationship to the kidney (*renal* = kidney). Which hormone of the adrenal gland is most closely associated with the function of the kidney?

   ____

10. Next, find the gonads. The **ovaries** are very small bean-shaped structures attached to the wavy uterus in the lower abdominopelvic cavity of the female fetal pig. To find the **testes,** use a scalpel to carefully cut the surface of the scrotum, which appears as two pockets of loose skin

in the groin area of the fetal pig. It is only necessary to cut the skin and connective tissue directly underneath the skin. Palpate the area, and find the small, brown, bean-shaped testes covered by a membrane. Find a classmate with a fetal pig of the opposite sex, and compare the two gonad types.

11. When finished, slide the string out from the bottom of the tray (do not cut or untie the string). Carefully clean and dry your dissecting tools. Spray your work area with the disinfectant and wipe it down. Your instructor will tell you how to prepare your pig for storage for use in future lab exercises.
12. Using Figure 13.2 as a reference, locate the following endocrine glands in the human torso models:
    - Adrenal glands
    - Ovaries
    - Pancreas
    - Testes
    - Thymus
    - Thyroid ■

ACTIVITY 3

Observing a Selected Endocrine Gland—The Pancreas

Materials for this Activity

Compound light microscope

Prepared slide of the pancreas

Observe a prepared slide of the pancreas, using your microscope's 10× objective lens (100× total magnification). Your instructor may elect to set this up as a demonstration slide in order to ensure that a pancreatic islet (islet of Langerhans) is in your field of view. Note that the islets are relatively small clusters of cells, and they make up a relatively small proportion of pancreatic cells.

What two hormones are produced by the pancreatic islets?

____

____

Considering the importance of these hormones, how is it that so few cells can do such a big job?

____

____

____ ■

Section ______ Name ______

Date ______ Instructor ______

EXERCISE

13

THE ENDOCRINE SYSTEM

Review Questions

1. Why can an endocrine gland be a great distance from its target tissue or organ?

______

______

2. Why do some people refer to the pituitary as the "master gland," and what part of the body controls the pituitary?

______

______

______

3. Match the hormone to its endocrine gland.

____	Insulin	a. adrenal medulla
____	Calcitonin	b. pancreas
____	Epinephrine	c. pituitary
____	Glucocorticoids	d. thyroid
____	Growth hormone	e. adrenal cortex

4. Match the hormone to its function.

____	Thyroid hormone	a. stimulates smooth muscle contraction in the uterus
____	Aldosterone	b. stimulates kidney tubules to reabsorb sodium and excrete potassium
____	Antidiuretic hormone (ADH)	c. stimulates the mammary glands
____	Oxytocin	d. stimulates body cells to increase metabolism
____	Prolactin (PRL)	e. stimulates water reabsorption by certain kidney tubules

5. Why is it best to minimize the use of your scalpel during dissection activities?

______

______

______

______

6. Which dissecting tools should be used for pointing to structures, removing connective tissue, and separating structures, and why?

____

____

____

____

7. Which two endocrine glands are made of a mixture of nervous and epithelial tissues?

____

8. Fill in the following chart, describing the location and hormones produced by each gland.

Gland	Location	Hormone(s)
Adrenal		
Ovary		
Pancreas		
Testis		
Thymus		
Thyroid		

EXERCISE

14

THE CARDIOVASCULAR SYSTEM I: BLOOD

Objectives

After completing this exercise, you should be able to

1. List the functions of blood.
2. Demonstrate the procedure for the hematocrit test.
3. Measure and calculate hematocrit test values.
4. Describe the function of the hematocrit test.
5. Identify red blood cells.
6. Identify five types of white blood cells.
7. Describe the functions of each type of blood cell.
8. Describe the antigen-antibody complex.
9. Describe the characteristics of the ABO blood type.
10. Describe the characteristics of the Rh blood type.
11. Demonstrate the procedure for determining blood type.
12. Properly dispose of blood-contaminated materials.

Materials for Lab Preparation

Equipment

- ❑ Compound light microscope
- ❑ Calculator

Prepared Slides

- ❑ Normal blood
- ❑ Pathological blood

Supplies

- ❑ Heparinized capillary tubes
- ❑ Microhematocrit centrifuge
- ❑ Microhematocrit reader
- ❑ Seal-ease or modeling clay
- ❑ Sterile lancets
- ❑ Alcohol prep pads
- ❑ Band-Aids
- ❑ Plastic centimeter ruler
- ❑ Disposable gloves
- ❑ "Sharps" container for sharp objects
- ❑ Autoclave bag
- ❑ Cleaning disinfectant or detergent solution in a spray bottle
- ❑ Paper towels
- ❑ Container of 10% bleach solution
- ❑ Freshly drawn animal blood (alternative choice for hematocrit test)
- ❑ Simulated human blood (alternative choice for blood typing)
- ❑ Anti-A, anti-B, and anti-D blood typing sera
- ❑ Disposable slides for blood typing
- ❑ Toothpicks
- ❑ Wax marker

Introduction

The cardiovascular system is composed of the heart, blood vessels, and blood. In this exercise, we will explore the different aspects of blood.

Blood serves several critical functions for the body. Our blood transports oxygen and nutrients in the form of blood glucose to all the cells of the body, and on the return trip, takes away the waste products of cells' metabolic processes. Hormones from the endocrine system, which serve to regulate growth and development, are also delivered by the blood. Other important functions of blood include the regulation of body temperature, fluid volume, and pH balance. Finally, blood defends us in two important ways: the white blood cells protect us against infections and diseases, and the platelets initiate a clotting mechanism to prevent excessive blood loss.

Components of Blood

There are four components of blood: red blood cells (RBCs), white blood cells (WBCs), platelets, and plasma (which consists of water, ions, proteins, hormones, gases, nutrients, and wastes). Table 14.1 lists the components of blood and their specific functions.

When anemia is suspected, a **hematocrit test** is performed to determine the amount of red blood cells in the blood. A blood sample is centrifuged, or spun, until a density gradient is established, with the lightest component at the top and the heaviest component at the bottom. The hematocrit test, also called **packed cell volume (PCV),** measures the volume of each blood component. Because the red blood cell volume is the largest component, it is often referred to as the **crit,** the **hematocrit,** or the **hematocrit value.** As indicated in Table 14.1, the main function of red blood cells is to carry oxygen, so the hematocrit is often used as a measure of the oxygen-carrying capacity of an individual.

Table 14.1 *Components of Blood*

Blood Component	*Examples and Functions*
Formed Elements (45%)	
Red blood cells	Transport oxygen and carbon dioxide to and from body tissues.
White blood cells	Defend body against invading microorganisms and abnormal cells.
Platelets	Initiate clotting to prevent blood loss.
Plasma (55%)	
Water	The primary component of blood plasma. Water serves as the major transport medium in plasma.
Electrolytes (ions)	Sodium, potassium, chloride, bicarbonate, calcium, hydrogen, magnesium, and others. Ions contribute to the control of cell function and volume, to the electrical charge across cell membranes, and to the function of excitable cells (nerve and muscle).
Proteins	Albumins maintain blood volume and transport electrolytes, hormones, and wastes. Globulins serve as antibodies and transport substances. Fibrinogens contribute to blood clotting.
Hormones	Insulin, growth hormone, testosterone, estrogen, and others. Hormones are messenger molecules that initiate metabolic processes.
Gases	Oxygen is needed for metabolism; carbon dioxide is a waste product of metabolism. Both are dissolved in plasma as well as carried by RBCs.
Nutrients and wastes	Glucose, urea, heat, and other raw materials. Nutrients and wastes (including heat) are transported by plasma throughout the body.

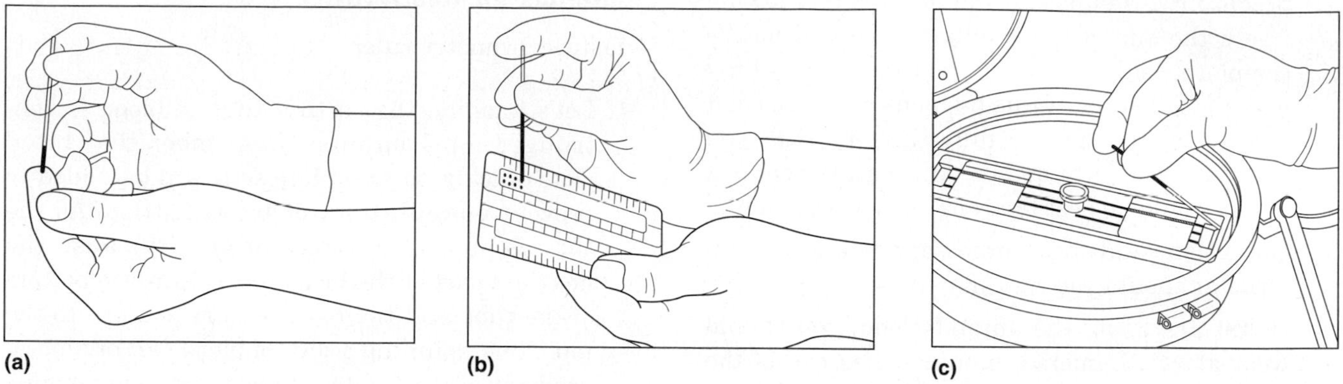

FIGURE 14.1 Steps in performing a hematocrit test. (a) Fill a heparinized capillary tube with blood. (b) Plug the capillary tube with clay. (c) Load the tube in a microhematocrit centrifuge.

ACTIVITY 1

Measuring Blood Components: The Hematocrit Test

Part 1: Performing a Hematocrit Test

Note to Instructor: Following are two activities that demonstrate the hematocrit test. Part 1 is recommended for students using labs with the proper equipment and supplies. Part 2 is designed for students whose lab instructors have decided to simulate this procedure rather than perform it.

Materials for this Activity

Heparinized capillary tubes
Microhematocrit centrifuge
Microhematocrit reader
Seal-ease or modeling clay
Sterile lancets
Alcohol prep pads
Band-Aids
Plastic centimeter ruler
Disposable gloves
"Sharps" container for sharp objects
Autoclave bag
Cleaning disinfectant or detergent solution
Paper towels
Container with 10% bleach solution
Freshly drawn animal blood (alternative choice)

Safety Precautions

- Please read the Review of Lab Safety Protocols section in Exercise 1.
- Remember that some infectious diseases are associated with the contact of blood. Protect yourself by following all instructions carefully. Perform only your own punctures and handle only your own blood. Wear gloves when you are in contact with slides containing other people's blood or animal blood and throughout this activity, including cleanup.
- Note that there are several different types of lancet and different methods of use. Please make sure you receive clear instructions on how to lance your fingertip to draw blood.
- Keep in mind that you are using a sharp instrument to draw your own blood. Concentrate and be alert.

Note: Your instructor will determine whether your own blood or animal blood will be used for this activity.

1. If you are using your own blood, obtain all of the supplies listed previously except for the animal blood, and return to your station.

   Clean one of your fingertips with the alcohol prep pad, and prick the fingertip with a lancet. When the blood is flowing freely, hold the unmarked end of the capillary tube to the blood drop, and fill the tube to the red line by capillary action (Figure 14.1a). Because blood clots quickly, the capillary tubes contain **heparin** to prevent **coagulation,** or clotting. If possible, prepare two samples. If the blood is not flowing freely, the end of the capillary tube will not be completely submerged in the blood during filling and air will enter. If this happens or you are not able to fill the tube to fully three-fourths, you will not be able to obtain an accurate reading, and it would be best to prepare another sample.

Use a Band-Aid and, if necessary, talk to your instructor should you continue to bleed.

If you are using the animal blood, you should wear gloves. Immerse the unmarked end of the capillary tube into the blood sample, and let it fill to the red line by capillary action.

2. Plug the blood-containing end of both capillary tubes by pressing it into the Seal-ease or modeling clay (Figure 14.1b).
3. Give your two tubes to the instructor, who will load and operate the microhematocrit centrifuge (Figure 14.1c). Be sure to record the numbers of the grooves in which your tubes are placed. The centrifuge will be on for approximately 5 minutes.

   During the operation of a centrifuge, what causes the heavier portions to settle to the bottom of the tube?

   ____

4. Upon getting your tubes back, measure the percentages of RBCs, WBCs, and plasma. There are two ways of making these measurements. You may use the microhematocrit reader, or use a centimeter ruler to measure the height of each component or layer. The RBCs are the red bottom layer, the WBCs and platelets are the white middle layer, and the plasma is the yellowish fluid on the top layer. Record your height measurements in Box 14.1 on page 161.
5. Cleanup: Place the used lancet in the "sharps" container. Place the blood-stained paper towels in the autoclave bag. Place the capillary tubes in the bleach jar. Spray some disinfectant on your counter area, and wipe it clean.
6. Go to Part 3: Calculating the Hematocrit Values.

Part 2: Simulating the Hematocrit Test

Note: If you have completed Part 1, please go to Part 3: Calculating the Hematocrit Values.

Materials for this Activity

Plastic centimeter ruler

1. Let's imagine that a thin tube of blood is centrifuged for 4 minutes. Remember that blood clots quickly, so some **heparin** will be added to prevent **coagulation,** or blood clotting. As the tube is spun, the forces of gravity cause the heaviest part of the blood to settle at the bottom of the tube and the lightest part to move to the top. Thus, spinning a tube of blood will produce a gradient of the lightest to heaviest components, from the top to the bottom of the tube. This procedure is called a **hematocrit test,** and it represents a common lab test for blood.
2. Figure 14.2a shows the original tube of blood before spinning. Figure 14.2b shows the result after it has been centrifuged. Notice there are three major components of blood: a large portion at the top, a very small portion in the middle, and another large portion at the bottom.
3. If you held the centrifuged tube in your hand, you would notice that the top portion was an amber fluid, the middle portion was a white-colored solid substance, and the bottom portion was a deep-red solid substance. These three sections represent the packed cell volume of blood: plasma at the top, platelets and white blood cells in the middle, and red blood cells at the bottom.
4. Obtain a plastic centimeter (cm) ruler. First observe the ruler. Notice each centimeter is divided into five sections. This means that each section represents ⅕ of a centimeter, in other words, each section represents 0.2 cm or 2 mm. If necessary, refresh your memory of the metric measurements for length in Exercise 1.
5. Measure the total height (in cm or mm) of the column of blood in the capillary tube shown in Figure 14.2a. Measure from the bottom of the test tube to the top of the blood sample. Measure the height as accurately as possible. This means that your reading may not be a whole number, but it may include fractions or decimals.
6. Using the same units as in Step 5 (cm or mm), measure the height of each element in the column of blood (Figure 14.2b). Due to the small size, the middle portion will be hard to read; in fact, it will be in the millimeter range. These respective measurements represent the plasma, platelet and WBC, and RBC heights.

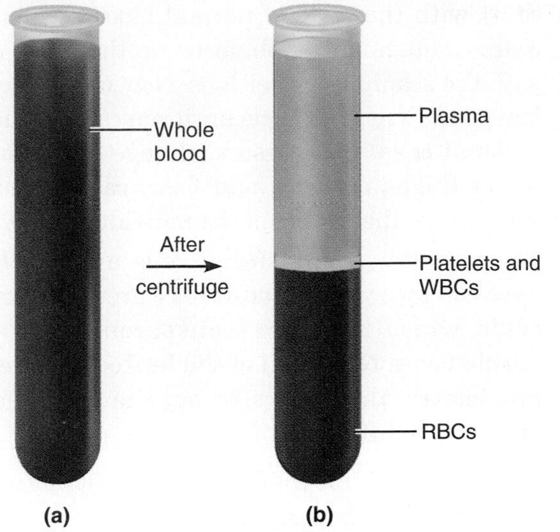

FIGURE 14.2 Hematocrit test results. (a) Whole blood before centrifuging. (b) Whole blood sample after centrifuging.

7. Record the four measurements under the Height column in Box 14.1.
8. Continue with Part 3: Calculating the Hematocrit Values.

Part 3: Calculating the Hematocrit Values

Materials for this Activity

Calculator

1. Using the height measurements you recorded in Box 14.1, calculate the percentage of each of the blood elements as follows:

   plasma% = plasma layer height ÷ total height × 100

   platelet and WBC% = platelet and WBC layer height ÷ total height × 100

   RBC% = RBC layer height ÷ total height × 100

2. Record the three calculations under the PCV% column in Box 14.1.

3. You have just calculated the **PCV%** for each of the three components of blood. Your values should be close to that of normal blood. The PCV% values for normal blood are as follows:

   plasma%: 55%

   platelet and WBC%: 1%

   RBC%: 44% (also called the hematocrit)

4. The **hematocrit** is the PCV% for RBCs. This is a simple way of expressing your oxygen-carrying capacity. In the United States, average hematocrit values for males range from 42% to 52%; for females, the range is from 37% to 47%. If your hematocrit is lower than these values, it may mean that you are anemic, a disorder of insufficient RBCs. Speculate on the reason for higher hematocrit values in males.

   Females contain A monthly cycle that decreases the hematocrit.

5. Not all shifts from these reference values are due to disorders. If you moved to a city situated at a high elevation, for example, Denver, where the air is "thinner" with less oxygen per volume of air, what immediate physical changes would you expect to experience?

   How you breathe.

   After several weeks in this environment, what kind of shift in hematocrit would you expect to see?

   Increase.

Box 14.1 Hematocrit Calculations

Blood Elements	Height (cm or mm)	PCV%
RBC	2.5 CM	80% to 45%
WBC and Platelets	.5 CM	10%
Plasma	2.5	45%
Total	5.5 CM	**100%**

Blood Cells

The proper term for red blood cells is **erythrocytes.** There is only one type of erythrocyte. The function of erythrocytes is to transport gases. Earlier, you learned that the hematocrit, which is the erythrocyte percentage of blood, is a measure of the oxygen-carrying capacity of the body.

The proper term for white blood cells is **leukocytes.** There are five types of leukocytes. The general function of leukocytes is defense, and each type of leukocyte has a different way of defending the body.

Let's look at the actual structure of the blood cells by examining slides of normal and pathological blood.

ACTIVITY 2

Observing Erythrocytes and Leukocytes

Materials for this Activity

Compound light microscope

Prepared slides

Normal blood

Pathological blood

1. Obtain a microscope and a sample of each type of blood slide.
2. Start with the slide of normal blood. Using the coarse-adjustment knob, focus on the blood cells with the scanning-power lens. Now change to the low-power lens, and focus until you can see the individual cells. If necessary, change to the high-power (high-dry) lens, and focus carefully until you can see the details of the individual cells.
3. The numerous pinkish-red cells are **erythrocytes.** Scattered here and there are a few leukocytes, which look like transparent cells with purple lobes. Notice all of the leukocytes are approximately the same size, but the purple lobes are shaped differently.

   Why do you think there are so many more erythrocytes than leukocytes?

   ____

4. Observe the round, disklike appearance of the erythrocytes. In Box 14.2, draw a few erythrocytes as they appear on the slide. Remember to record the magnification of your samples.
5. Observe the differences between the five **leukocytes** illustrated in Figure 14.3. The agranular leukocytes are composed of monocytes and lymphocytes. What is the main structural difference between monocytes and lymphocytes?

   ____

Box 14.2 Blood Cell Sketches

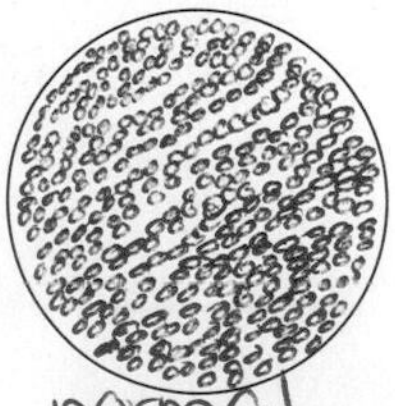

normal
~~Erythrocytes~~
40 ×

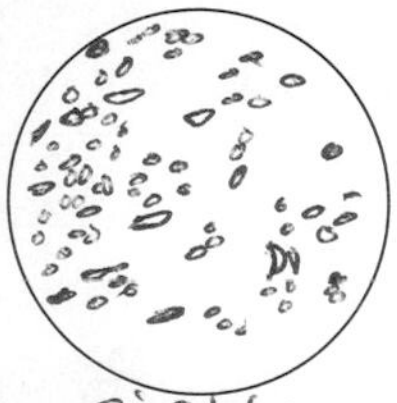

sickle
~~Monocytes~~
40 ×

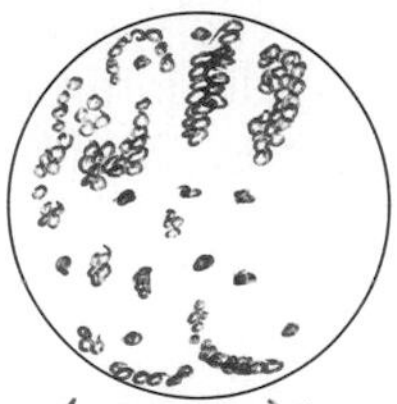

leukemic
~~Lymphocytes~~
40 ×

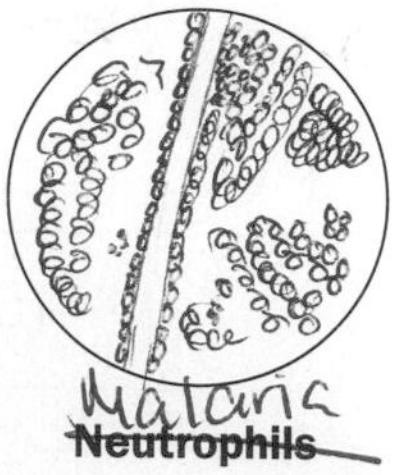

Malaria
~~Neutrophils~~
____ ×

Eosinophils
____ ×

Basophils
____ ×

The granular leukocytes are composed of the neutrophils, eosinophils, and basophils. What are the main structural differences between neutrophils, eosinophils, and basophils?

____

____

Examining your blood slide, find and draw a few of each type of leukocyte, and write the total magnification used in the space provided in Box 14.2. You may have to move the slide around a bit to find each type of leukocyte. What appears to be the *main* structural difference between the agranular and granular leukocytes?

____

____

6. Focus on a slide of pathological blood. Because each campus has a different collection of pathological slides, we will only make some general observations. Infections involve the proliferation of specific leukocytes, so depending on the type of disease, certain types of leukocytes will be more prevalent. For example, in sickle cell anemia, the red blood cells have a crescent or sickle shape, unlike the usual round, concave disk shape.

   Compare the cells of the normal blood slide with the cells of the pathological blood slide. Describe some differences.

   ____

   ____

   ______________________________________ ■

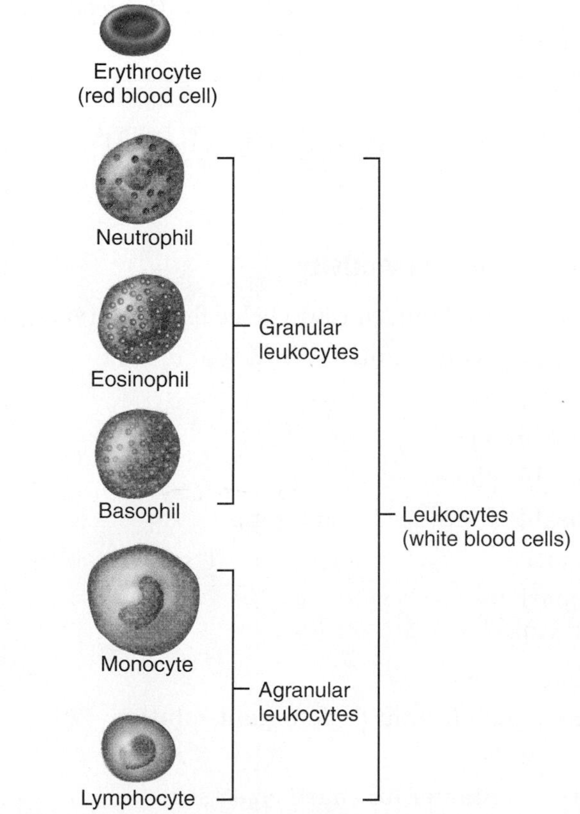

FIGURE 14.3 Erythrocytes and leukocytes.

Blood Types

Successful transfusions of blood from one person to another depend upon the compatibility of the people's blood types. The right transfusion can save lives; the wrong one can result in severe illness or even death for the recipient.

Your blood type is determined by your genes, and it does not change during your lifetime. Before you can understand about blood type, you must learn some new terms: "antigen," "antibody," and "antigen-antibody complex."

Antigen

The surfaces of your erythrocytes have specific proteins called "self" surface proteins, which allow your immune system to identify your erythrocytes as your own cells (Figure 14.4a). The surfaces of erythrocytes from other blood types contain foreign surface proteins, called antigens, which allow your immune system to identify these erythrocytes as foreign cells (Figure 14.4a). There are different antigens, depending on your blood type.

Antibody

Your plasma contains specific plasma proteins called antibodies, which mount an attack on cells identified as foreign cells (Figure 14.4a). There are different antibodies, each one specialized to attack a particular antigen.

Antigen-Antibody Complex

Once the antigens have been identified, the antibodies mount an attack on foreign erythrocytes by binding to the antigen and forming an **antigen-antibody complex** (Figure 14.4b). Incompatible

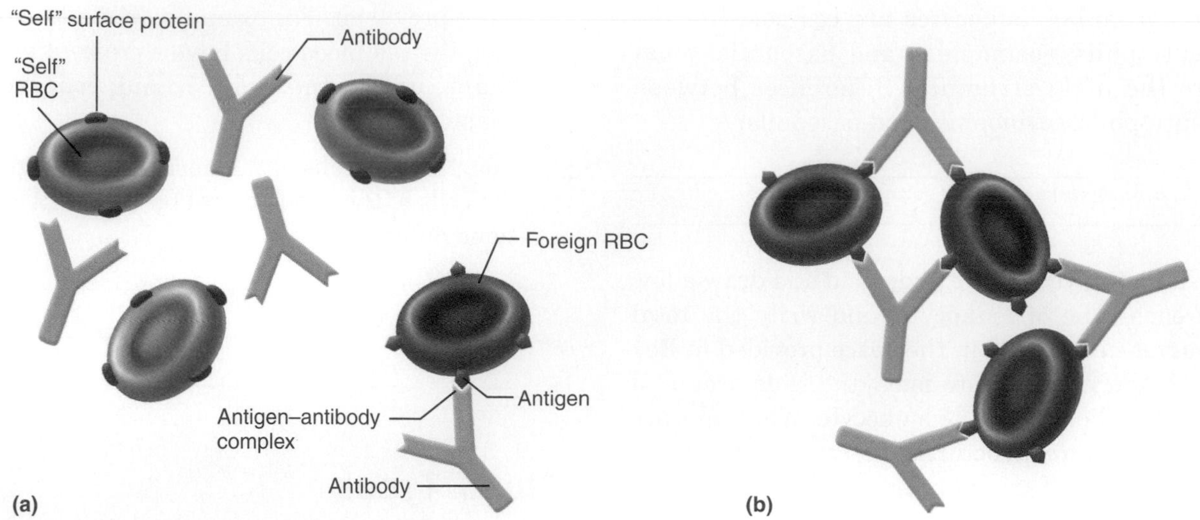

FIGURE 14.4 The role of antibodies in recognizing and inactivating foreign RBCs. (a) The antibody recognizes the foreign RBC and binds to its antigen. (b) The formation of the antigen-antibody complexes causes foreign RBCs to clump, thus rendering them inactive.

transfusions lead to the formation of an antigen-antibody complex, which then leads to agglutination (clumping together of RBCs) and hemolysis (rupture of RBCs). In essence, incompatible transfusions are fatal because the RBCs are critically damaged.

Figure 14.5 shows the different antigens and antibodies found in each blood type. Notice that in the United States, the distribution of blood types in Caucasians, African-Americans, and Native Americans differs significantly.

There are two classifications for blood type: ABO and Rh. Figure 14.5 presents the characteristics of the ABO blood type. The Rh blood type is also important because in the United States, 85% of the population is Rh+. These people have erythrocytes with antigens called Rh factor, and their plasma does not contain Rh antibodies. The other 15% of the population is Rh−; these people have erythrocytes without the Rh antigen.

Normally, Rh− individuals do not have Rh antibodies in their plasma, but once there is an initial exposure to Rh+ blood, they will begin to produce Rh antibodies. This means that the first time an Rh− individual receives an Rh+ transfusion, there may be no danger. But the second Rh+ transfusion may cause a severe reaction. A special situation arises for an Rh− mother carrying an Rh+ fetus. The leakage of the fetal Rh+ blood into the mother's Rh− bloodstream may cause the mother to begin producing Rh antibodies. The result may be a form of fetal anemia, which is potentially life threatening.

In the following activity, we will determine blood type using anti-sera or chemical solutions with antibodies. The way an anti-sera reacts with blood indicates blood type.

ACTIVITY 3

Blood Typing

Materials for this Activity

Simulated blood (alternative choice for blood typing)
Anti-A, anti-B, and anti-D blood typing sera
Sterile lancets
Alcohol prep pads
Disposable gloves
Disposable slides for blood typing
Toothpicks
Wax marker
Band-Aids
Autoclave bag
Cleaning disinfectant or detergent solution
Paper towels
"Sharps" container for sharp objects
Container with 10% bleach solution

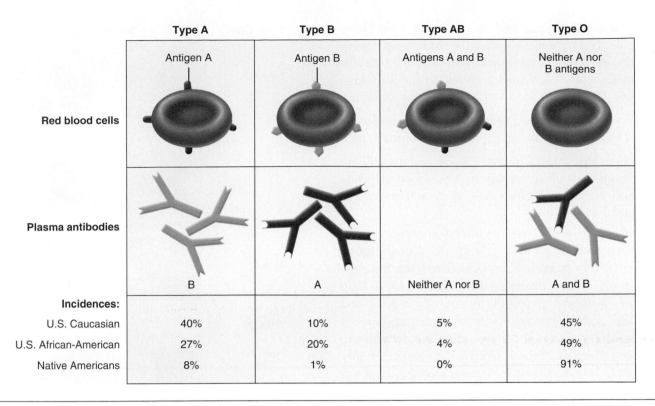

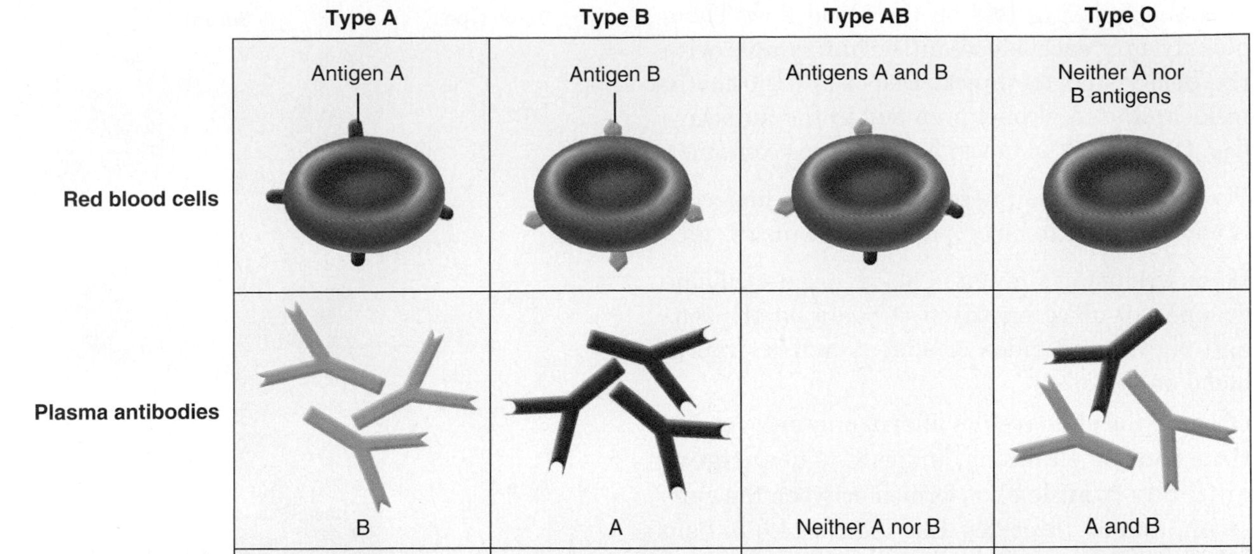

Incidences:	Type A	Type B	Type AB	Type O
U.S. Caucasian	40%	10%	5%	45%
U.S. African-American	27%	20%	4%	49%
Native Americans	8%	1%	0%	91%

FIGURE 14.5 Characteristics of ABO blood types.

Safety Precautions

- Please read the Review of Lab Safety Protocols section in Exercise 1.
- Remember that some infectious diseases are associated with the contact of blood. Protect yourself by following all instructions carefully. Perform only your own punctures and handle only your own blood. Wear gloves when you are in contact with slides containing other people's blood, and throughout this activity, including cleanup.
- Note that there are several different types of lancet and different methods of use. Please make sure you receive clear instructions on how to lance your fingertip to draw blood.
- Keep in mind that you are using a sharp instrument to draw your own blood. Concentrate and be alert.

Note: Your instructor will determine whether your own blood or simulated blood will be used for this activity.

1. Obtain the indicated supplies from the supply area.
2. Cover your work area with paper towels, on which you can place blood-contaminated instruments and slides.
3. Mark slide 1 with a wax pencil as follows: Draw a line in the middle, mark the left upper corner "anti-A" and the right upper corner "anti-B." Place a drop of anti-A sera on the middle of the left side and a drop of anti-B sera on the middle of the right side. Place a toothpick on the left side and another toothpick on the right side.
4. Mark slide 2 with a wax pencil as follows: Write "anti-Rh" on the top of the slide. Place a drop of anti-Rh sera in the middle of the slide. Then place a toothpick on the slide.
5. If you are using your own blood, thoroughly wash your hands with soap, and dry them with paper towels. Clean your fingertips with an alcohol prep pad.
6. Pierce your fingertip with the lancet, and wipe away the first drop of blood. Then place one drop of freely flowing blood on each of the three marked sections of the slides. Press your thumb

over the fingertip to stop the blood flow. Then, quickly mix each blood-antiserum sample with the designated toothpick. Dispose of the toothpicks and used alcohol prep pad in the autoclave bag. Dispose of the lancet in the sharps container.

Place a Band-Aid, if necessary, on your fingertip. If you continue to bleed, talk to your instructor.

If you are using simulated blood, use a medicine dropper to place one drop of blood on the left- and right-hand sides of slide 1 and a drop of blood on slide 2.

7. Observe the slide results after 2 minutes. **Agglutination,** or clumping, indicates an **antigen-antibody complex** has formed between the antiserum and erythrocytes. Use Figure 14.6 to help you interpret your ABO results. Type AB blood will agglutinate with both the anti-A and anti-B sera. Type A blood will agglutinate only with the anti-A serum. Type B blood will agglutinate only with the anti-B serum. Type O blood will not agglutinate with either anti-A or anti-B sera. Record your results in Box 14.3.

   Sometimes, it is clear that agglutination has or has not occurred. It may be hard, however, to tell the difference between dried blood and agglutination. Ask your instructor to confirm your interpretations of agglutination.

8. Figure 14.6 does not illustrate agglutination for the Rh factor. Rh+ blood will agglutinate with the anti-Rh serum. Rh− blood will not. Agglutination in Rh+ blood is sometimes rather fine and hard to observe. If necessary, observe the slide under the microscope, or ask your instructor for assistance in determining agglutination.

   The convention for labeling the Rh blood type is to use the plus symbol (+) to indicate Rh+ blood and the minus symbol (−) to indicate Rh− blood. Record your results as (+) or (−) in Box 14.3 below.

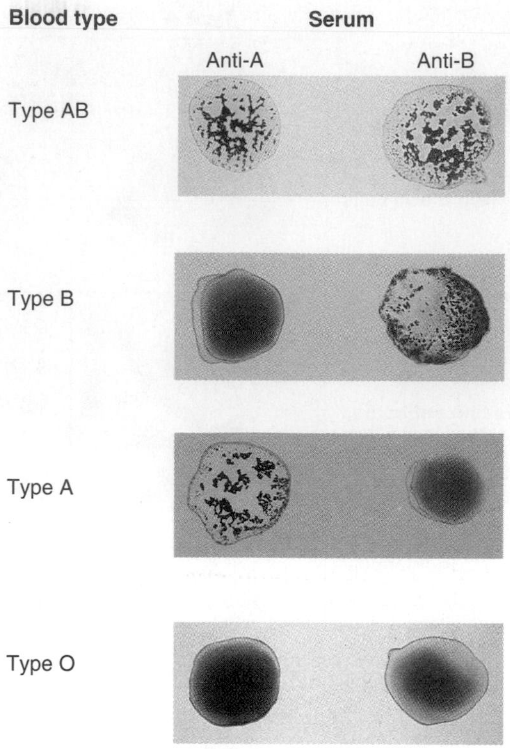

FIGURE 14.6 Agglutination in ABO blood typing.

Box 14.3 Blood Type Results

Results	Observed	Not observed
Clumping with anti-A	Blood type A clumps while in B it doesn't. The RH is clump which makes the Blood A+. AB is clumped in all three making it AB+	
Clumping with anti-B	O is O− B=B−	
Clumping with anti-Rh		

9. To determine your blood type, first determine if it is A, B, AB, or O, and then add (+) or (−) as appropriate. For example, if your blood type is AB and Rh+, you would record it as AB+. If your blood type is O and Rh−, you would record it as O−. Record your blood type.

____

10. The instructor may select certain student slides against which all students may view and compare their results. The instructor may also focus certain student slides in a microscope to demonstrate the difference between agglutination and dried blood. It is useful to observe a range of slides, and to see how blood agglutinations may look very different from slide to slide and between the naked eye and microscope.

    Caution: Look without touching. Remember to use gloves if you handle any microscopes or slides containing other people's blood.

    If you were able to view other slides, what were your observations about the different types of agglutination?

    ____

    ____

11. Cleanup: Wear gloves. Place the used slides in the jar of bleach. Place gloves, toothpicks, and the blood-stained paper towels in the autoclave bag. Spray some disinfectant on your counter area and wipe it down. If the microscope was used, wipe it down. Remember that the next student to use the counter is relying on you for a clean, safe work area and instruments. ■

EXERCISE

15

THE CARDIOVASCULAR SYSTEM II: HEART AND BLOOD VESSELS

Objectives

After completing this exercise, you should be able to

1. Describe the location of the heart in the chest.
2. Identify the anatomy of the human heart.
3. Trace the flow of blood through the heart.
4. Become familiar with the internal structure of the heart by dissecting a sheep heart.
5. Demonstrate the procedure for measuring heart rate.
6. Define an artery and a vein.
7. Describe the cardiac cycle.
8. Demonstrate the procedure for measuring blood pressure.

Materials for Lab Preparation

Models and Charts

- ❑ Torso model with removable heart
- ❑ Human heart models (dissectible)
- ❑ Sheep heart (preserved or fresh)
- ❑ Anatomical charts of the systemic circuit
- ❑ Models of the systemic circuit

Equipment and Supplies

- ❑ Dissection equipment
- ❑ Dissection tray
- ❑ Disposable gloves
- ❑ Autoclave bag
- ❑ Paper towels
- ❑ Blunt sticks
- ❑ Disinfectant in spray bottle
- ❑ Stopwatch or timer with second hand
- ❑ Stethoscope
- ❑ Sphygmomanometer
- ❑ Alcohol prep pads

Introduction

This exercise continues our examination of the cardiovascular system, which consists of three components: blood, which we studied in Exercise 14, and the heart and blood vessels, which we will focus on in this exercise. The heart provides the pressure to pump blood through the network of blood vessels throughout our body. Because there are literally miles and miles of blood vessels, the extreme ends of the body, for example, the fingers and toes, would be deprived of blood without sufficient blood pressure. The blood vessels transport blood to and from each cell of our body through arteries and veins.

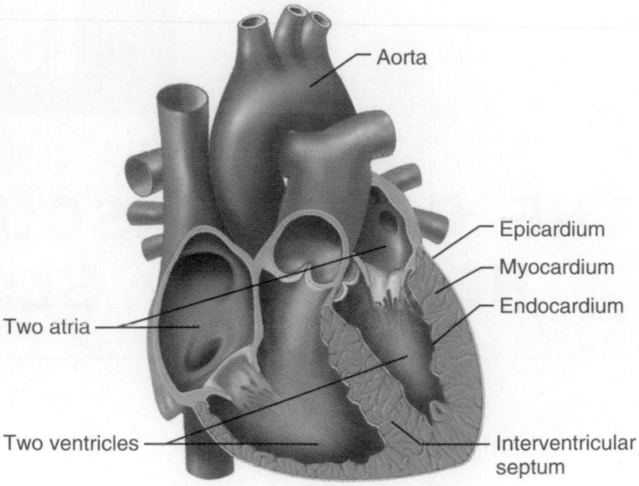

FIGURE 15.1 Layers of the heart wall. In this cross section, we can see the septum, which separates the right and left ventricles of the heart.

The Heart

The human heart, like that of any mammal, is primarily muscle. In Exercise 5, Tissues, we examined three kinds of muscle, one of which was cardiac muscle. Cardiac muscle is special because (1) it does not connect to bone like skeletal muscle, and (2) it contracts continuously to propel blood through the blood vessels to every region of the body.

The heart's natural position is in the chest, or thoracic cavity. A tough, fibrous sac, called the **pericardium,** protects the heart on the outside, while anchoring it to surrounding structures. There are three layers to the wall of the heart: the **epicardium,** consisting of epithelial and connective tissue; a middle layer, the **myocardium,** which is thick and consists mainly of the cardiac muscle that forms the bulk of the heart; and the innermost layer, the **endocardium,** which consists of a thin endothelium resting on a layer of connective tissue (Figure 15.1).

The heart's structure is relatively simple. It consists of four chambers, two at the top of the heart called **atria** (singular *atrium*), and two at the bottom called **ventricles.** There is a muscular partition that separates the right and left sides of the heart, which is called the **septum** (Figure 15.1). When blood returns to the heart from the body's tissues, it enters the heart at the *right atrium* and then passes through a valve into the *right ventricle.* The right ventricle pumps blood through a second valve and into the artery leading to the lungs. Blood returning from the lungs to the heart enters the *left atrium* and then passes through a third valve into the *left ventricle,* which pumps blood through a fourth valve into the **aorta,** the body's largest artery (Figure 15.1).

In close proximity to the heart are the lungs (Figure 15.2). The lungs are intimately related to the heart, which sends blood to the lungs for oxygenation.

We will look at the position of the heart inside the thoracic cavity to see the physical relationships between the heart, lungs, trachea, and esophagus, and then we will examine the internal structure of the heart.

ACTIVITY 1

Locating the Heart and Examining its Internal Structure

Materials for this Activity

Torso model with removable heart
Human heart models (dissectible)

Part 1: Locating the Heart

1. Using the torso model, locate the **heart,** which is near the center of the **thoracic cavity.** Notice that the right side of the heart rests on the diaphragm. The **lungs** surround most of the heart, which means the heart is well situated to send oxygen-poor blood to the lungs and to receive oxygen-rich blood in return (Figure 15.2).

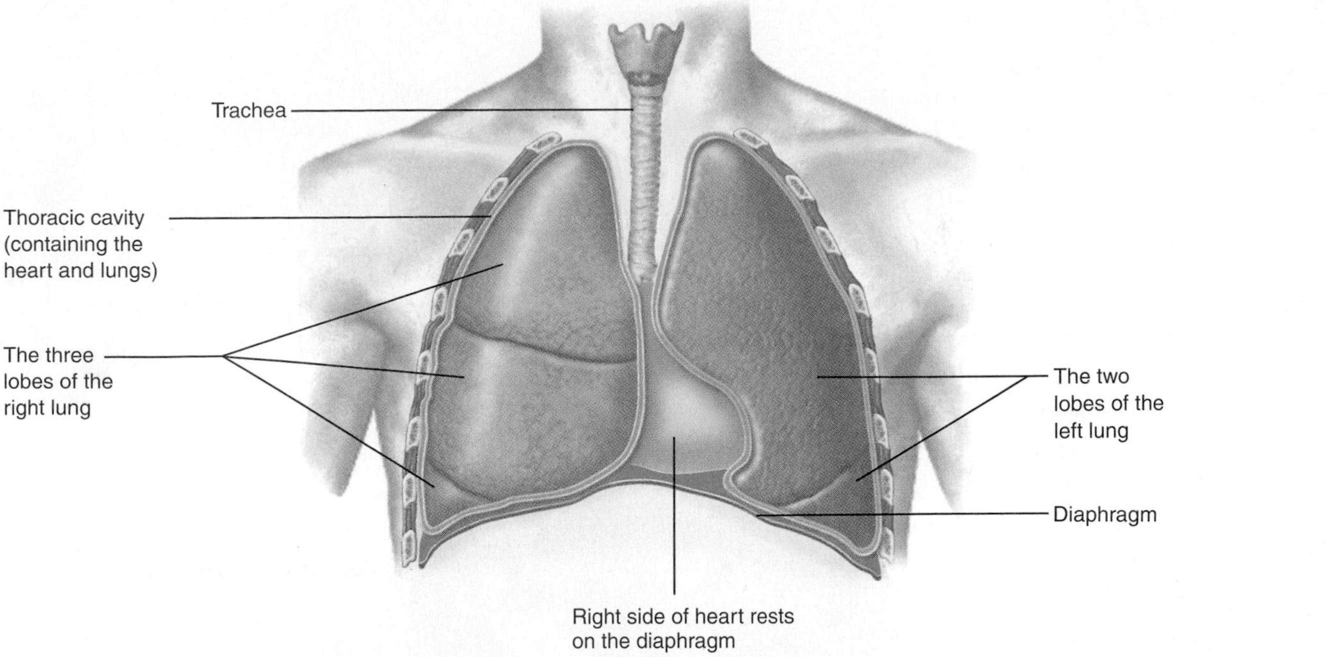

FIGURE 15.2 The heart as it is situated in the thoracic cavity.

2. Remove the heart from the torso model. Notice that both the **trachea,** or wind pipe, and the **esophagus,** or food tube, are located near the major blood vessels on the upper portion of the heart. As you look at the tightly packed and entwined arrangement of the trachea, esophagus, and blood vessels of the heart, you can appreciate the efficient use of a small space.
3. With the heart removed, look at the point of the entry into the diaphragm for the esophagus. Notice that the stomach is underneath the heart, separated by the diaphragm. Can you see the basis for calling acid reflux "heartburn"?
4. Look for the cut rib bones in the **rib cage** on the edges of the thoracic cavity. Recall what you learned about the axial skeleton in Exercise 8.

   What function does the rib cage play as it relates to the heart and lungs?

   ____

Part 2: Examining the Internal Structure of the Heart

1. Use the dissectible model of the heart to examine each of the internal structures illustrated in Figure 15.3. Notice that there are three large blood vessels. The one in the center, the **aorta,** is situated between the **superior vena cava** on the right side and the **pulmonary trunk** on the left side of the heart.
2. As mentioned earlier, there are four chambers to the heart. The upper chambers are the **right** and **left atria.** The lower chambers are the **right** and **left ventricles,** which are separated by the thick **interventricular septum.** Notice that the two lower chambers are more muscular. This is a consequence of their individual functions: The right ventricle pumps blood through a second valve into the artery that leads to the lungs; the left ventricle pumps blood through a fourth valve into the aorta.
3. The aorta arches over the pulmonary trunk. At the top of the **aortic arch,** there are three **aortic arteries,** which provide oxygenated blood for the head and arms. The aorta then descends to provide oxygenated blood for the lower part of the body. The aorta supplies the **systemic circuit,** which is the system of blood vessels for the entire body. We will discuss the systemic circuit in detail later.
4. The **inferior vena cava,** which brings blood to the heart from the inferior or lower part of the body, is almost directly underneath the

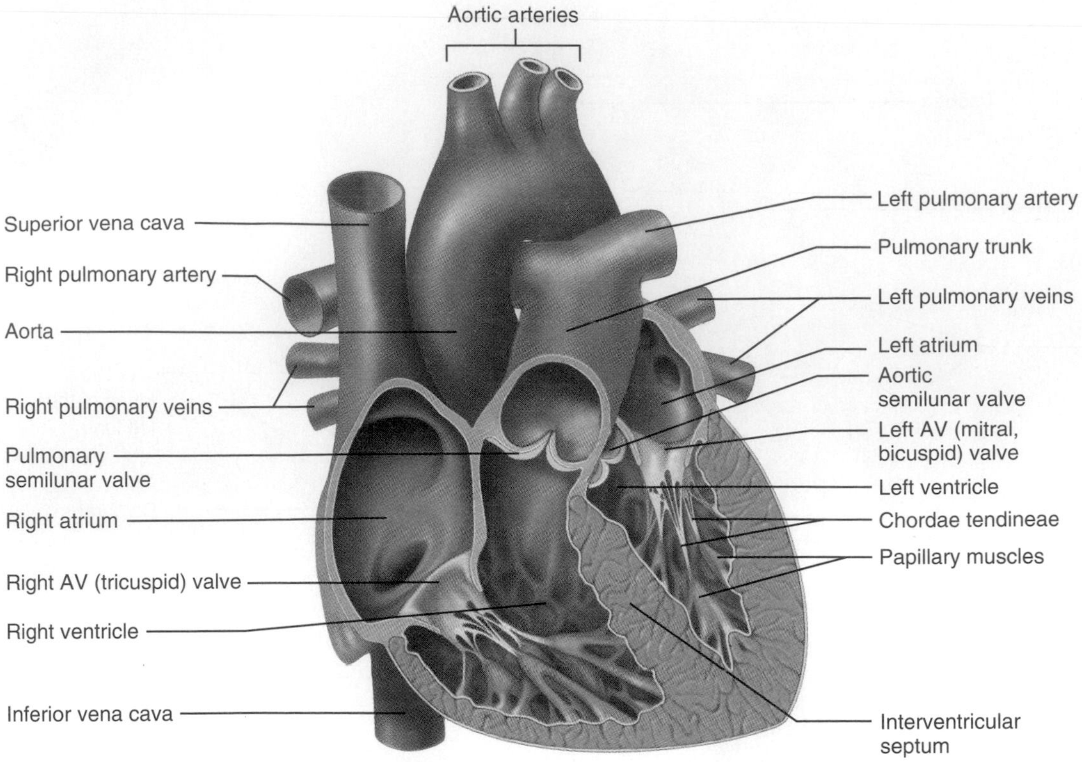

FIGURE 15.3 An internal view of the heart.

superior vena cava, which brings blood to the heart from the superior or upper part of the body. These two major veins are generally referred to as the venae cavae.

5. The pulmonary trunk separates into two branches, the **right and left pulmonary arteries,** which bring oxygen-poor blood to the lungs. These arteries supply the **pulmonary circuit,** which is the system of blood vessels in the lungs and which we will discuss later. The **right and left pulmonary veins** bring oxygen-rich blood from the lungs to the heart. Notice there are four pulmonary veins.
6. Identify the four heart valves. There are two semilunar valves to regulate the flow of blood from the heart. On the right, the **pulmonary semilunar valve** regulates the flow of blood into the pulmonary trunk. On the left, the **aortic semilunar valve** regulates the flow of blood into the aorta.

   Two valves regulate the flow of blood from the atrial chambers to the ventricular chambers. On the right side, the blood flow is regulated by the **right atrioventricular (AV) valve,** also known as the **tricuspid valve.** On the left side, the blood flow is regulated by the **left atrioventricular (AV) valve,** also known as the **mitral** or **bicuspid valve.**
7. The **chordae tendineae** anchor the AV valves to the **papillary muscles** of the ventricles. The contractions of these powerful muscles provide the blood pressure to transport blood from the heart to the extremities: the fingers and toes. ■

ACTIVITY 2

Tracing Blood Flow Through the Heart

1. Study Figure 15.4 a moment. Notice the arrows, which represent the flow of blood through the heart. Now, refer to Box 15.1, where you will see two sections: A and B. Fill in the blank beside each structure listed in Section A with the correct number from Figure 15.4. Then, list the structures through which blood passes as it travels through the pulmonary and systemic

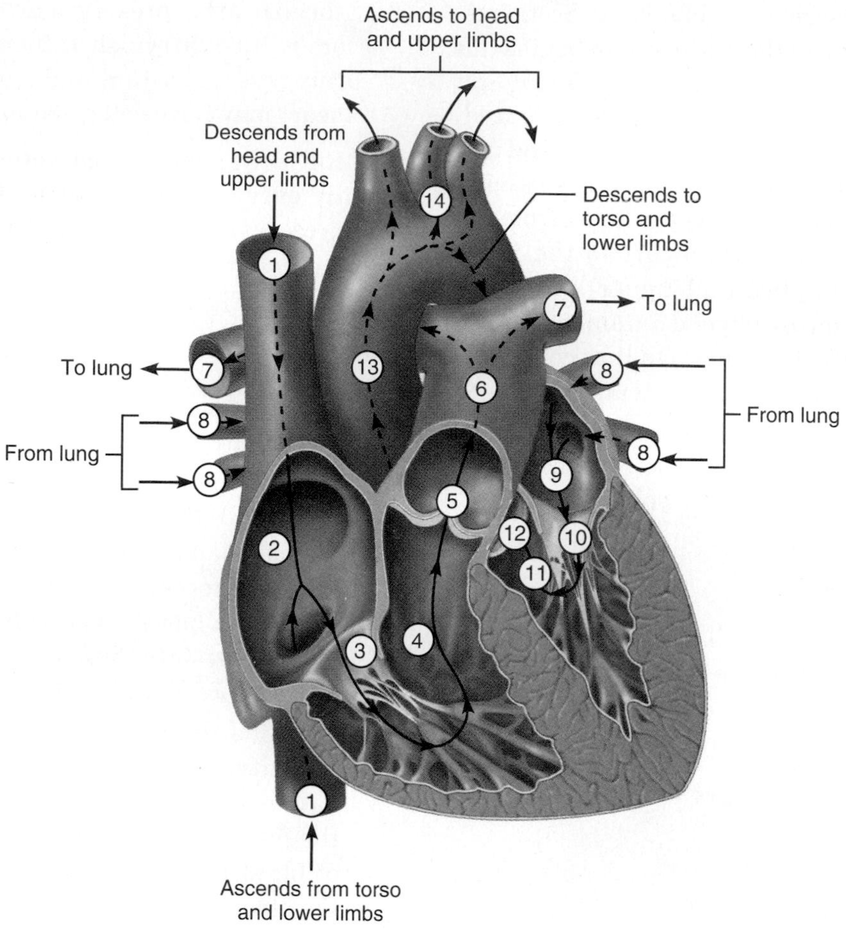

FIGURE 15.4 Blood flow in the heart.

Box 15.1 Tracing Blood Flow Through the Heart

Section A

____ Pulmonary arteries	____ Aorta
____ Pulmonary semilunar valve	____ Pulmonary trunk
____ Aortic arteries	____ Aortic semilunar valve
____ Right atrium	____ Left atrium
____ Right AV valve	____ Left AV valve
____ Right ventricle	____ Left ventricle
____ Pulmonary veins	____ Venae cavae

Enter the numbers of the structures above that constitute the pulmonary circuit: ____________

Enter the numbers of the structures above that constitute the systemic circuit: ____________

Section B

Oxygen-Poor Blood	Oxygen-Rich Blood
1. ____________	8. ____________
2. ____________	9. ____________
3. ____________	10. ____________
4. ____________	11. ____________
5. ____________	12. ____________
6. ____________	13. ____________
7. ____________	14. ____________

circuits in the appropriate blanks in Section A. There are 14 steps in the route, which begins (for our purposes) at the venae cavae ①. You'll first trace the flow of oxygen-poor (deoxygenated) blood through the right side of the heart and into the lungs. You'll continue as oxygen-rich (oxygenated) blood returns to the heart from the lungs and flows out of the heart to the head, torso, internal organs, and upper and lower limbs. Once you have assigned a number to each structure, identify the structures that constitute both the pulmonary and the systemic circuits.

2. In Section B on Box 15.1, enter the names of structures from Section A in their correct order, ①–⑭, to trace the pathway of blood.

   Is this a correct definition? "Arteries carry oxygen-rich blood."

   ______________________________ ■

ACTIVITY 3

Dissecting the Sheep Heart

Materials for this Activity

Sheep heart (preserved or fresh)
Dissection equipment
Dissection tray
Disposable gloves
Autoclave bag
Paper towels
Blunt sticks
Disinfectant in spray bottle

The sheep heart is used for dissection because, like the human heart, it has four chambers and is about the same size. Because this activity involves a dissection, please refresh your memory on the dissection protocol in Exercise 1.

1. Pair up with a lab partner. Supply yourself and your partner with dissection equipment, dissection tray, disposable gloves, and paper towels. Don your gloves, and obtain a sheep heart from the supply area.
2. Compare your sheep heart to the photos in Figure 15.5 which show fresh sheep hearts. If your sheep heart is preserved instead of fresh, the colors will be brownish rather than pink, the texture will be stiffer, and the blood vessels and heart may have collapsed in certain areas.

   Rinse the heart in cold water, and squeeze it gently over a sink a few times to remove trapped preservatives and blood clots.
3. Examine the external features of the sheep heart (Figure 15.5a, b). Observe the **pericardium,** if it is still present. Slit it open. Notice there is quite a bit of adipose tissue, which needs to be removed. The two earlike **auricles** are projections from the **atria.**
4. Identify the major blood vessels on the top (Figure 15.5a). The **pulmonary trunk** is anterior to the aorta and the **pulmonary arteries** (if they are still present; they may already be removed). The **aorta** has thicker walls. Insert one or more blunt sticks into these blood vessels in the sheep heart to prevent them from collapsing.
5. Examine the back of the heart (Figure 15.5b). Identify the **superior and inferior venae cavae,** and insert blunt sticks into both veins in the fresh sheep heart. Notice that the insertion of blunt sticks provides additional firmness to the structure of the dissected heart organ.

   Compare the thickness and diameter of the aorta and venae cavae. Which one is larger? Which one has thicker walls?

   ____
6. Insert a blunt stick into the superior vena cava, and use scissors to cut through to the **right atrium.** Observe the **right AV valve,** which is anchored with thin strands of **chordae tendineae** to the muscles of the ventricle (Figure 15.5c).

   How does the size of the right atrium compare to the left atrium?

   ____
7. Continue the cut into the **right ventricle.** Identify the **papillary muscles** of the ventricle. The **ventricles** do much more work and are correspondingly much larger than the atria.

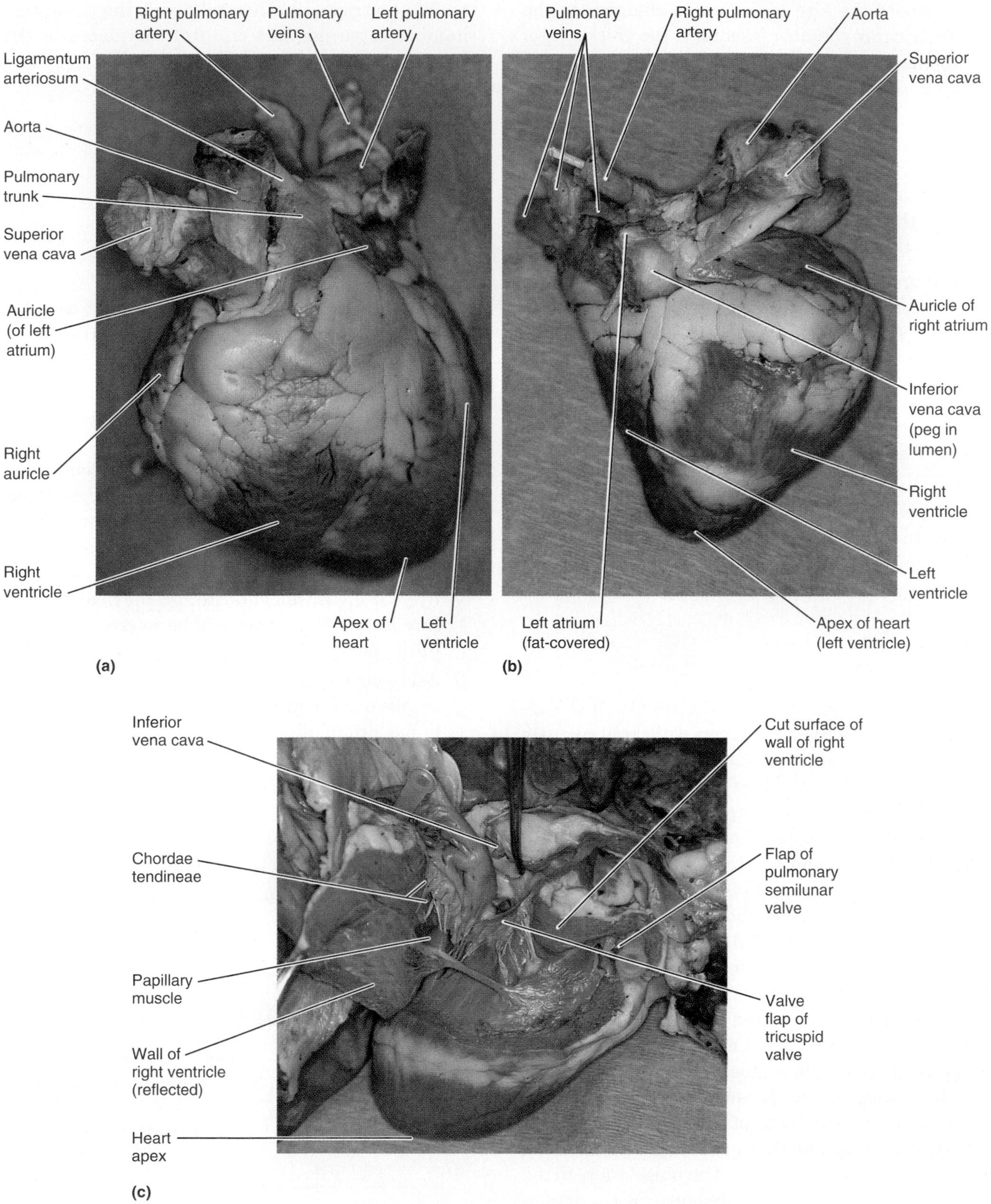

FIGURE 15.5 Sheep heart. (a) Anterior view. (b) Posterior view. (c) Internal view.

Compare the size of the atrial chamber to the ventricular chamber. Compare the thickness of the heart wall. Which chamber is larger? Which wall is thicker?

____

8. Insert a blunt stick into the aorta, and use scissors to cut through to the **left ventricle.** Notice that the heart wall there is substantially thicker than that of the right ventricle.
9. Continue the cut from the left ventricle to the left atrium. Compare the **left AV valve** to the right AV valve. Notice that the right AV valve has three sections. How many sections does the left AV valve have?

____

____

10. Properly dispose of the dissected sheep heart and blunt sticks in the autoclave bag. Clean and dry the dissection tools and tray. Spray and wipe down your counter with a disinfectant. ■

Heart Rates

Your heart is about the size of your fist. It is constructed of living cells, yet it can work continuously for 60 to 100 years without stopping for repairs. There are only about 5 liters of blood in the adult body, yet the heart can pump 5 to 25 liters of blood per minute! This means that it is possible to move your entire blood supply through the heart every minute!

Another amazing fact about your heart is that it never stops pumping blood. At rest, the typical heart rate is 75 beats per minute. During exercise, the heart rate can exceed 125 beats per minute. During a heart attack, it can exceed 200 beats per minute. Over the course of a typical lifetime, your heart will beat more than 3 billion times!

The easiest way to measure the heart rate is by measuring the pulse rate of the radial artery. The pulse point for the radial artery is located on the side of the wrist, above the thumb. The pulse refers to the alternating surges of the pressure in an artery, which correlates to the contraction and relaxation of the left ventricle. In a normal heart, the pulse rate (number of surges per minute) correlates to the heart rate (beats per minute).

ACTIVITY 4

Measuring Heart Rate

Materials for this Activity

Stopwatch or timer with second hand

1. Have your lab partner sit comfortably, with his or her arm and hand on the counter. Have your timer or stopwatch easily accessible. Place the fingertips of the first two fingers of one hand over the pulse point, and press gently. Remember not to press too hard.
2. Count the number of pulses in 15 seconds, and then multiply that number by four. For example, if you measure 20 pulses over 15 seconds, then $20 \times 4 = 80$ pulses per minute, or a heart rate of 80 beats per minute. Repeat the count, and verify your measurements. Record the heart rate in Box 15.2. This number will be referred to as the initial heart rate.
3. Ask your partner to do 10 jumping jacks or run in place for 1 minute. Measure the heart rate immediately after the exercise, and record it in Box 15.2. This will be the exercise heart rate.
4. Measure the heart rate 3 minutes after the exercise, and record it in Box 15.2. This will be referred to as the final heart rate.
5. Now, reverse your roles. Have your lab partner measure your heart rate under the three conditions.

Box 15.2 Heart Rate Measurements

	Your Heart Rate	Lab Partner's Heart Rate
Initial heart rate		
Exercise heart rate		
Final heart rate		

Checking the results in Box 15.2, answer the following:

Use this formula to determine the percentage change over initial heart rate: Initial HR/Exercise HR × 100. Write the percentage in the space below.

___

What was the percentage change between the initial heart rate and the exercise heart rate ?

___

If there was a decrease, instead of the expected increase, try to determine the reason. You may repeat the measurement, and you may also explore the medical reasons for the decrease.

___

___

What was the percentage change from the initial to the final heart rate (Initial HR/Final HR × 100)? What was the change from the exercise to the final heart rate (Exercise HR/Final HR × 100)?

___

___

If the initial heart rate was approximately the same as the final heart rate, what does this information say about the health of the individual?

___

At 75 beats per minute, how much time does a heartbeat take? (*Hint:* There are 60 seconds per minute. Calculate the time in terms of seconds.)

___

At 75 beats per minute, how many times will your heart beat in 1 day? (*Hint:* There are 60 minutes per hour, and 24 hours per day.)

___ ■

The Blood Vessels

There is an extensive network of blood vessels to deliver blood to all the tissues of your body. The arteries and veins that are located near the heart have the largest diameters. Remember that arteries are blood vessels that transport blood away from the heart, whereas veins transport blood back to the heart. As they extend to the extremities, arteries and veins begin to narrow and branch off into the arterioles and venules. The capillaries are the smallest and thinnest blood vessels, and they deliver the blood at the cell level. There is an inverse relationship between the diameter and length of blood vessels. The larger the diameter, the shorter the length. Thus, the arteries and veins have the largest diameter and the shortest length, while capillaries have the smallest diameter and the longest length.

As pointed out earlier in this exercise, there are two major circulatory circuits in the human body. The **pulmonary circuit** is located in the lungs. It transports oxygen-poor blood through the lungs, where the blood gives up carbon dioxide and picks up a fresh supply of oxygen from the air we inhale. The **systemic circuit** is located throughout the body. The systemic **arteries** transport oxygen-rich blood from the heart to every part of the body (Figure 15.6). The arteries divide into arterioles, which divide again into the capillaries that serve the individual cells of the tissues. At the cells, the blood gives up oxygen and picks up a fresh supply of carbon dioxide, which is a product of the metabolic work of the tissues. The systemic veins transport this oxygen-poor blood back to the heart (Figure 15.6).

Blood Pressure and its Relationship to the Cardiac Cycle

Keep in mind that the right side of the heart (the right atrium and the right ventricle) pumps blood to the lungs. The left side of the heart (left atrium and left ventricle) pumps blood to the head, torso, and upper and lower limbs. This pumping action consists of two distinct pulses: **contraction** and **relaxation.** Contraction in the two atria forces blood into the ventricles. The subsequent contraction of the two ventricles pumps blood into the pulmonary artery and the aorta. The repeated sequence of contraction and relaxation is known as the **cardiac cycle.**

As blood surges into the arteries, the artery walls are stretched by the higher pressure to accommodate the extra volume of blood. These same artery walls recoil as blood flows out of them. **Blood pressure** is the force that blood exerts on the wall of a blood vessel, and its maintenance is crucial

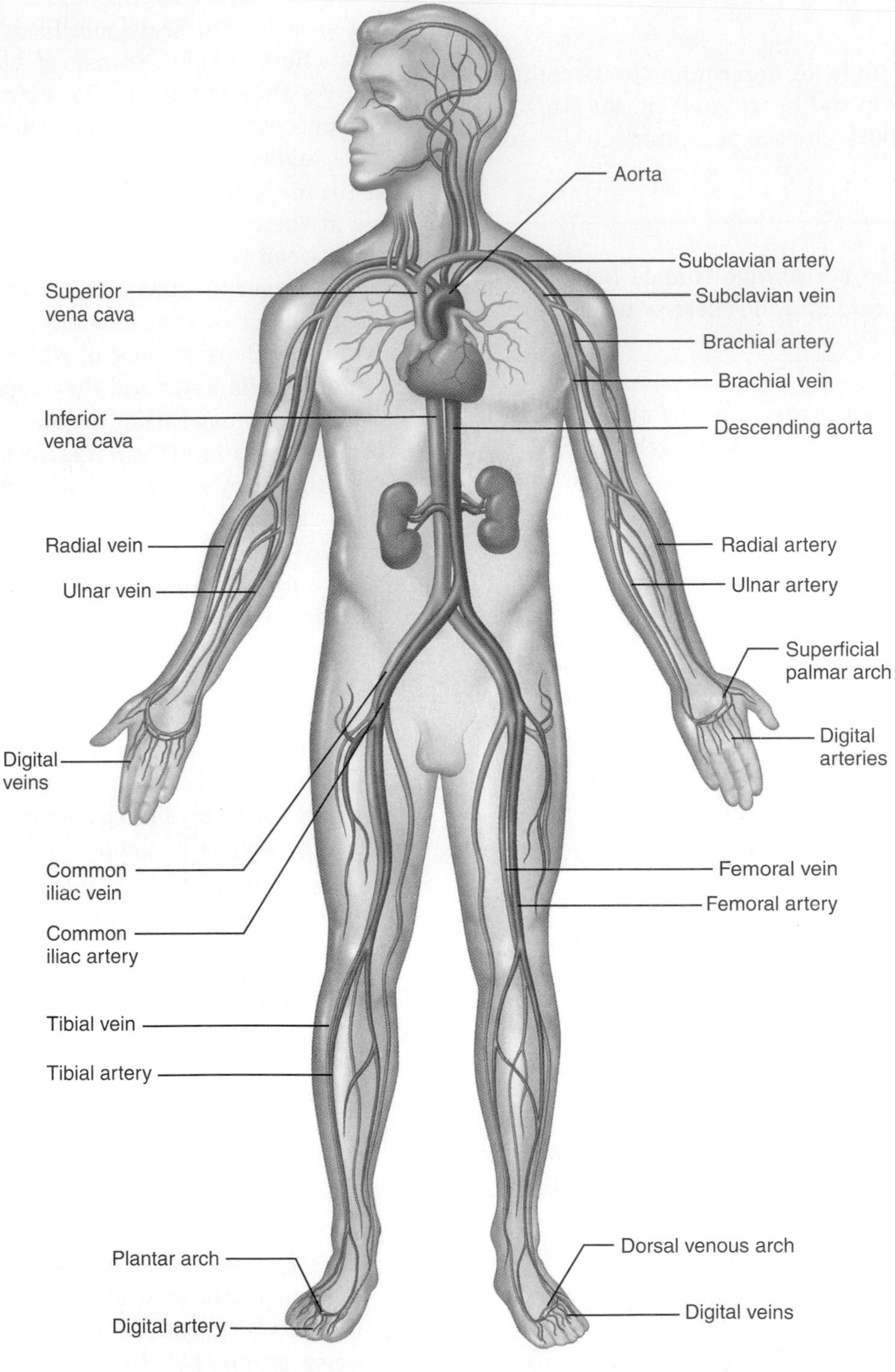

FIGURE 15.6 Simplified systemic circuit, showing arteries and veins.

to drive the flow of blood throughout the body and force the return of blood to the heart. **Systolic pressure** is the highest pressure of the cardiac cycle, which is reached when the ventricles contract to eject blood *from* the heart. The **diastolic pressure** occurs when both the atria and the ventricles relax and the blood pressure in the chambers of the heart is lower than the blood pressure in the veins.

Measuring blood pressure helps health care providers track trends in the blood pressure of an individual over time, and it is a good means for tracking changes in the cardiovascular system, as well as its overall efficiency.

ACTIVITY 5

Measuring Heart and Blood Pressure

Materials for this Activity

Stethoscope
Sphygmomanometer
Alcohol prep pads
Stopwatch or timer with second hand

The pressure of body fluids is measured in millimeters of mercury (mm Hg). A **sphygmomanometer** measures the changing blood pressure associated with the pulse.

We learned earlier that when measuring the pulse rate, we use the pulse point for the radial artery at the wrist. In this exercise, you will be introduced to a new pulse point: the elbow for the brachial artery. A stethoscope will be used to enable you to listen to the sounds in the artery as it experiences blood pressure changes.

1. Obtain a stopwatch, sphygmomanometer, and stethoscope. With an alcohol prep pad, clean the earpieces of the stethoscope, let them dry, then place them in your ears. Examine the stethoscope sensors to see if they need cleaning as well. Examine the tubing of the sphygmomanometer to make sure there are neither tears nor holes in it. Also, be sure all the air is pressed out to avoid inaccurate readings. Return damaged equipment to your instructor.
2. Place your fingertips over the pulse point for the radial artery on the wrist. Using a watch, count the number of pulses in 15 seconds. Multiply your result by four. This will be the pulse rate. For young adults, the pulse rate is usually about 75 pulses per minute.
3. Keep your fingertips lightly pressed on the pulse point to assess the pulse strength. For young adults, the pulse strength is usually rated as strong, regular, and easily detected. Are the pulse waves strong or weak, regular or spotty, easily or hardly detectable?

   ____

4. Have your lab partner sit comfortably and place his or her arm on the counter. Wrap the sphygmomanometer air cuff around his or her left upper arm, just above the elbow. Place the stethoscope sensor over the pulse point for the brachial artery. Compare your setup to the depiction in Figure 15.7. Listen to the pulse in the brachial artery.
5. Close the valve of the sphygmomanometer's air bulb, and pump the cuff to a pressure of about 180 mm Hg.
6. While listening to the artery and watching the mercury column, open the valve slightly and slowly release the air. This action will slowly release the pressure and cause the blood pressure in the artery to decrease. As this happens, tapping sounds, the **heart sounds,** can be heard. The first heart sound is caused by the closure of the AV valves. Note the pressure of the first heart sound. This is the **systolic pressure.** Continue to decrease the pressure until the second heart sound. The second heart sound is caused by the closure of the semilunar valves. Note the pressure of the second heart sound. This is the **diastolic pressure.**

   Blood pressure is expressed as systolic pressure *over* diastolic pressure. Average values are "120 over 80," which means 120 mm Hg / 80 mm Hg. These two values are labeled as points 1 and 2 on the graph in Figure 15.7.
7. Let your lab partner rest a minute, and then measure the blood pressure again to verify your measurements. Record these numbers in Box 15.3. This recording will be referred to as the initial blood pressure.

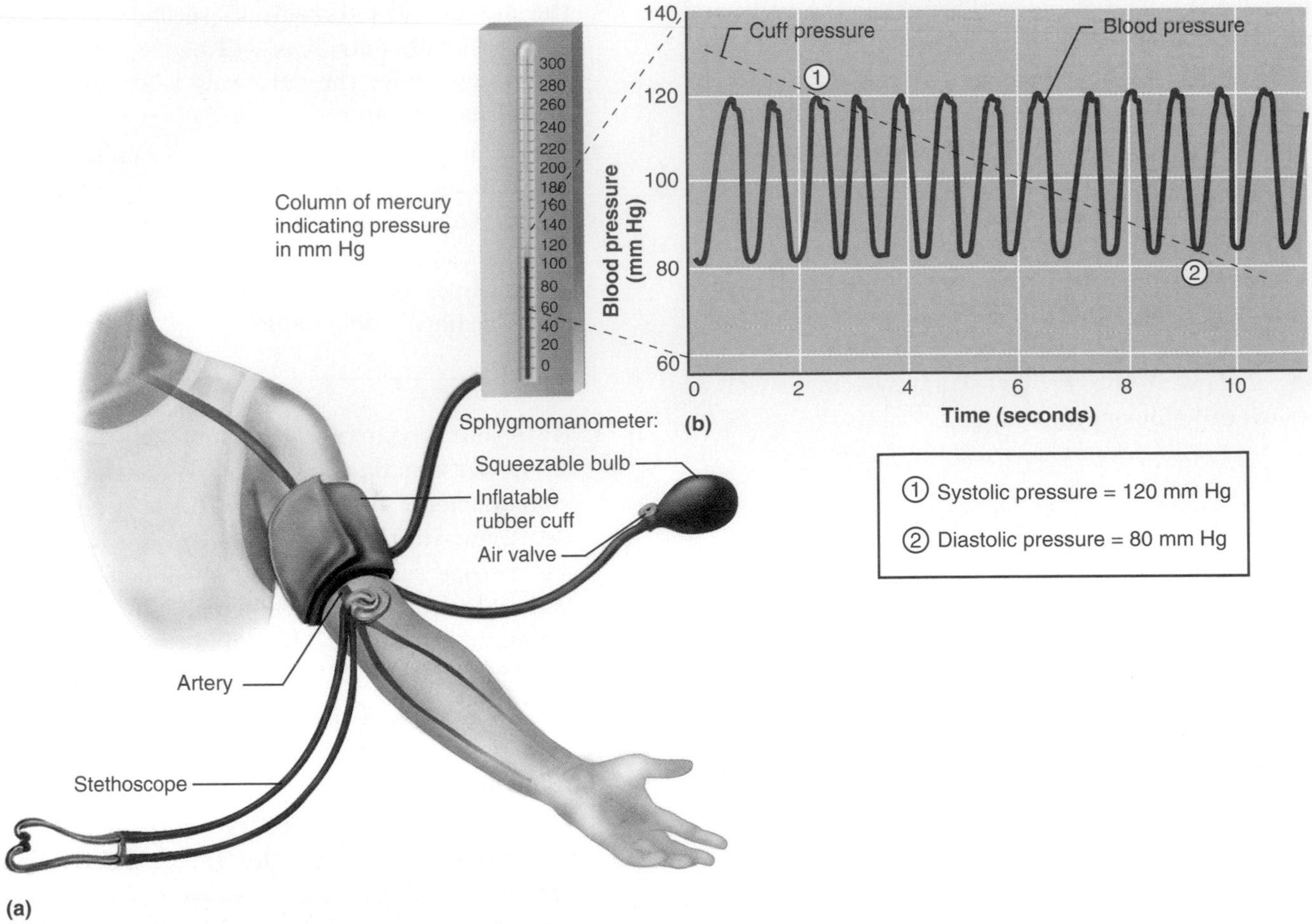

FIGURE 15.7 Using a sphygmomanometer. (a) The sphygmomanometer air cuff is wrapped around the upper arm over the brachial artery. (b) In this schematic representation, blood pressure pulses are superimposed over declining cuff pressure. Point 1 indicates when sounds are first heard (systolic pressure); point 2 indicates when sounds cease (diastolic pressure).

Box 15.3 Blood Pressure Measurements

	Your Blood Pressure	Lab Partner's Blood Pressure
Initial blood pressure		
Exercise blood pressure		
Percent change		
Final blood pressure		

8. Ask your partner to do 10 jumping jacks or run in place for 1 minute. Measure his or her blood pressure immediately after the exercise. These numbers will be referred to as the exercise blood pressure.
9. Measure his or her blood pressure 3 minutes after the exercise. These numbers will be referred to as the final blood pressure.
10. Now reverse your roles. Have your lab partner measure your blood pressure at rest, immediately after exercise, and 3 minutes after exercise.
11. Compare your respective values for initial blood pressure to those in Table 15.1. If you find that your blood pressure values are in stage 3 or 4, repeat your measurements again.

Table 15.1 ***Systolic and Diastolic Blood Pressures***

Systolic Pressure	
Normal	129 and lower
High-normal (borderline hypertension)	130–139
Stage 1 (mild hypertension)	140–159
Stage 2 (moderate hypertension)	160–179
Stage 3 (severe hypertension)	180–209
Stage 4 (very severe hypertension)	210 and higher
Diastolic Pressure	
Normal	84 and lower
High-normal (borderline hypertension)	85–89
Stage 1 (mild hypertension)	90–99
Stage 2 (moderate hypertension)	100–109
Stage 3 (severe hypertension)	110–119
Stage 4 (very severe hypertension)	120 and higher

12. Refer to your measurements in Box 15.3 to answer the following questions.

What was your initial blood pressure? ________________

What was your exercise blood pressure? ________________

What was the percentage change from your initial blood pressure (Initial BP/Exercise BP × 100)? ________________

If there was a decrease instead of the expected increase, try to determine the reason.

____

What was the final blood pressure? ________________

What was the percentage change from the initial blood pressure (Initial BP/Final BP × 100)? ________________

What was the percentage change from the exercise blood pressure (Exercise BP/Final BP × 100)? ________________

If the initial blood pressure was approximately the same as the final blood pressure, what does this statistic say about the health of the individual?

____

What was your pulse rate and strength? ________________

Let's say that your blood pressure was 180/120. Refer to Table 15.1. Is this a normal or hypertensive value?

____

EXERCISE

16

THE RESPIRATORY SYSTEM

Objectives

After completing this exercise, you should be able to

1. Describe the anatomy of the respiratory system.
2. Trace the route of inhaled air.
3. Understand chest volume changes during breathing.
4. Understand the components of total lung capacity.
5. Demonstrate the use of a nonrecording spirometer.
6. Compare lung volume differences between smokers and nonsmokers.

Materials For Lab Preparation

Equipment and Supplies

- ❑ Bell jar model of the lungs
- ❑ Metric tape measure
- ❑ Nonrecording spirometer (wet or dry)
- ❑ Disposable mouth pieces for spirometers
- ❑ Alcohol prep pads
- ❑ Nose clips

Models

- ❑ Human torso
- ❑ Human half head
- ❑ Human lung
- ❑ Human alveolar sac
- ❑ Human thoracic organs

Introduction

When someone chokes, whether from a piece of food or strangulation, he or she has less than 1 minute to remove the obstruction to the air supply before he or she suffers a loss of consciousness. This scenario points out our absolute dependency on a constant air supply.

All living cells require a constant supply of oxygen to drive their normal cell processes. Oxygen deprivation is very dangerous, and the brain cells are the most vulnerable. When they "shut down," many body activities become uncoordinated, so our body also "shuts down" by becoming unconscious. Carbon dioxide removal is also critical because as the carbon dioxide level in the blood rises, carbonic acid is formed, which acidifies the blood. As blood pH drops, many body functions are affected.

The main function of the respiratory system is to provide oxygen to cells and remove carbon dioxide, which is the waste product that cells release upon oxygen use. The exchange of these gases with the air is called **aerobic metabolism.**

Four processes, collectively referred to as respiration, must occur in order for the respiratory system to fulfill this main function. We will be primarily concerned in this exercise with two of these processes: **pulmonary ventilation,** known as breathing, and **external respiration,** which involves gas exchange in the lungs.

Respiratory Anatomy

The respiratory system is divided into two tracts. The upper respiratory tract consists of the following structures: nose, pharynx, and larynx (Figure 16.1), through which air flows to the lungs. The lower respiratory tract consists of the lungs themselves. We will study the anatomy of both respiratory system tracts by tracing the route of inhaled air as it travels from the nose to the lungs. The route for exhaled air is the same but occurs in the reverse order.

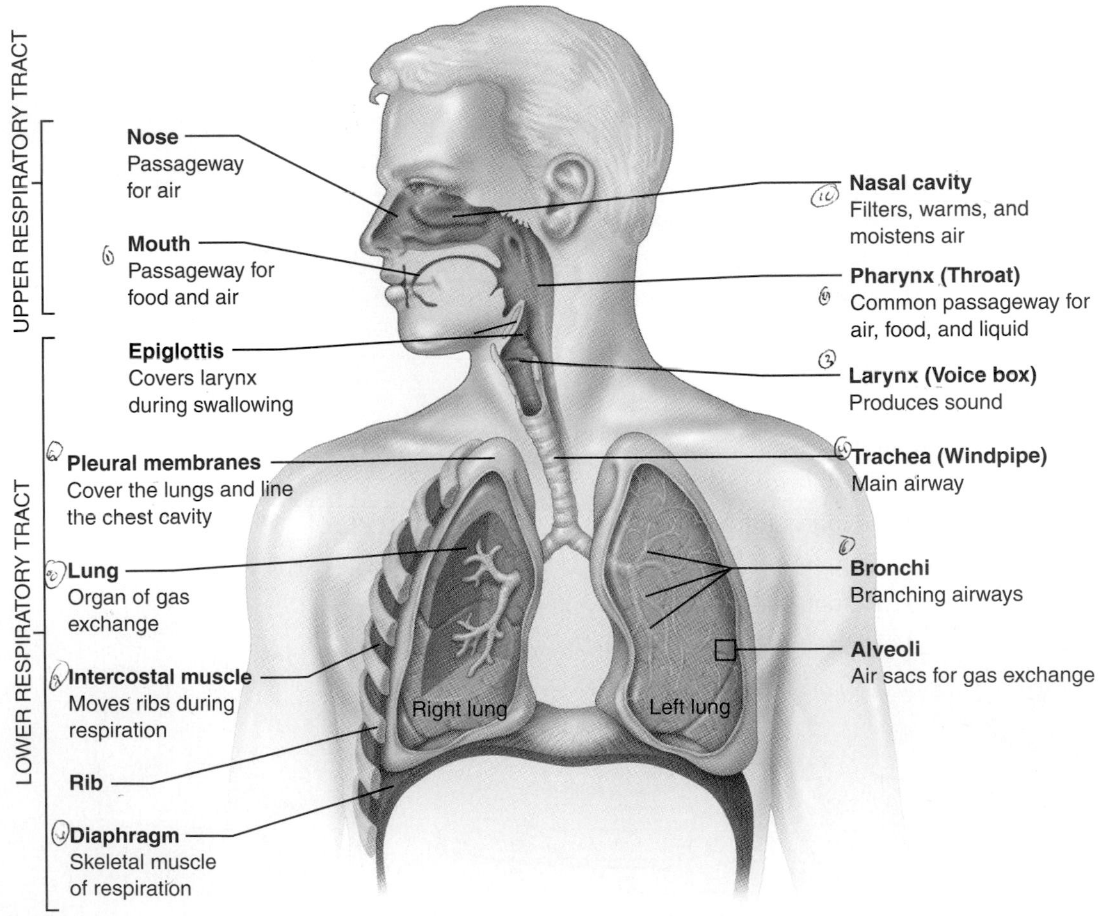

FIGURE 16.1 Human respiratory system.

ACTIVITY 1

Tracing the Route of Inhaled Air

Materials for this Activity

Models

Human torso
Human half-head
Human lung
Human alveolar sac
Human thoracic organ

1. Using various models and referring to Figures 16.1–16.3, we will trace the route of inhaled air from the nose to the lungs. If the torso model is too heavy, bring your lab manual to that lab station. *Note:* Remember these models are used for many other lab exercises, so do not touch them with pencils or pens; use your fingers or the eraser ends of pencils.
2. During inhalation, nostril hairs filter particles from the air entering the **nose.** The air is then warmed and moistened in the **nasal cavity** before traveling down to the throat, or pharynx (Figure 16.1). The **pharynx** is the common area for air traveling from either the nose or the mouth. Notice that air may also be inhaled through the **mouth,** but this is a less desirable method of inhalation.

   What are some disadvantages of using the mouth to breathe?

   Bacteria can enter easily
3. The cleaned, warmed, and moistened air now travels past the **epiglottis** (Figure 16.1). This little flap of cartilage is located at the opening to the larynx. It remains open when air is flowing into the larynx, but it temporarily covers up the larynx when it is pulled up as we swallow. Place your hand midway on the anterior surface of your neck and swallow. Can you feel the larynx rise as you do so? This "switching mechanism" routes food and drink into the esophagus and air into the trachea. If anything other than air enters the larynx, a cough reflex is triggered to expel the substance and prevent it from moving into the lungs.

   What else besides food might cause a blockage in your air supply?

   trapped air

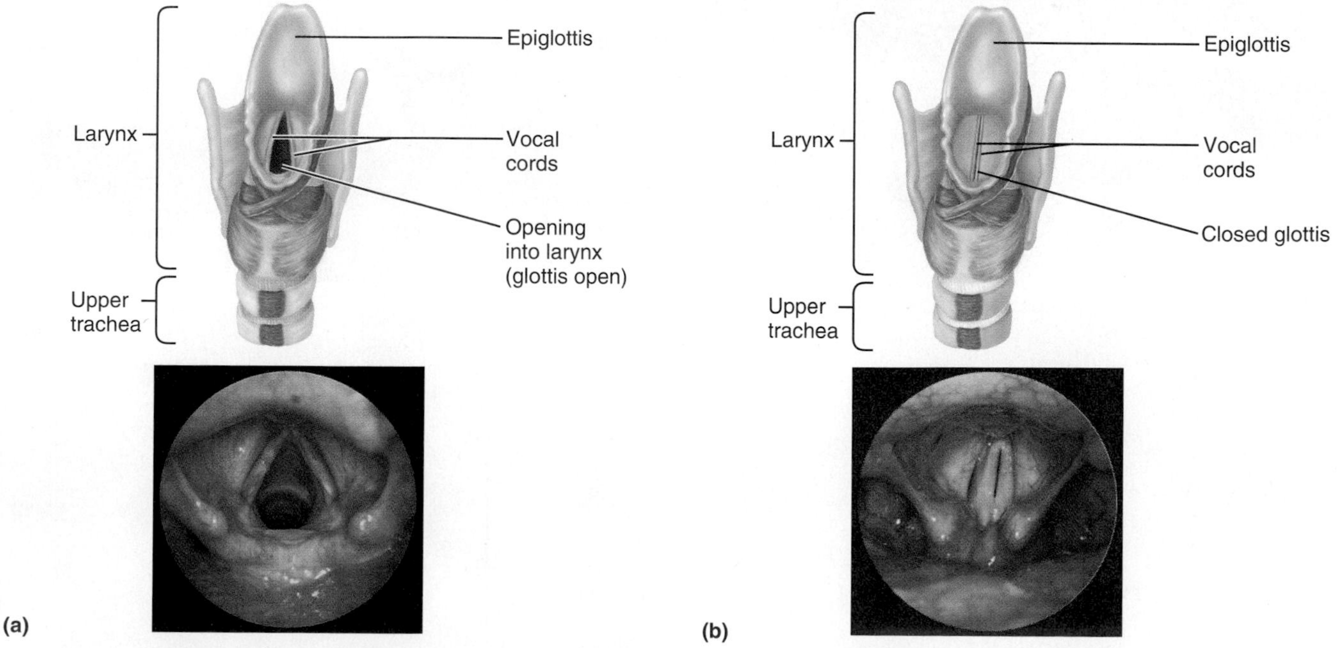

FIGURE 16.2 Vocal cords: production of sound. (a) The position of the vocal cords during quiet breathing. (b) The position of the vocal cords during sound production.

4. Within the **larynx,** or voice box, are a pair of folds called **vocal cords,** which vibrate when air is expelled, thus giving us the ability to produce sound (Figure 16.2). When we are not talking, the vocal cords are relaxed and open. When we talk, the vocal cords are stretched tightly by skeletal muscles. The tones of our speech depend on how tightly the vocal cords are stretched.

   In the event of a prolonged blockage of the upper respiratory tract, a tracheotomy is performed. This involves making an opening in the trachea and inserting a tube to assist in breathing.

   How would a tracheotomy affect the tonal quality of our speech?

   The ~~trachea went~~ tube can't be "controled".

5. The inhaled air travels past the larynx and enters the **trachea,** or windpipe (Figure 16.1). The trachea forms two major branches called the **bronchi,** which enter the right and left lung cavities. The bronchi divide into a system of smaller branched airways inside the lungs. The smallest bronchi, which lack cartilage, are called **bronchioles,** a term that means "little bronchi."
6. The bronchioles are too small to be seen by the naked eye, but refer to Figure 16.3 as we continue our trace of the route of inhaled air. The smallest bronchioles terminate in clusters of **alveoli,** or air sacs. With proper breathing, the alveolar air maintains a high level of oxygen and a low level of carbon dioxide. The alveoli are the sites for gas exchange. They are surrounded by

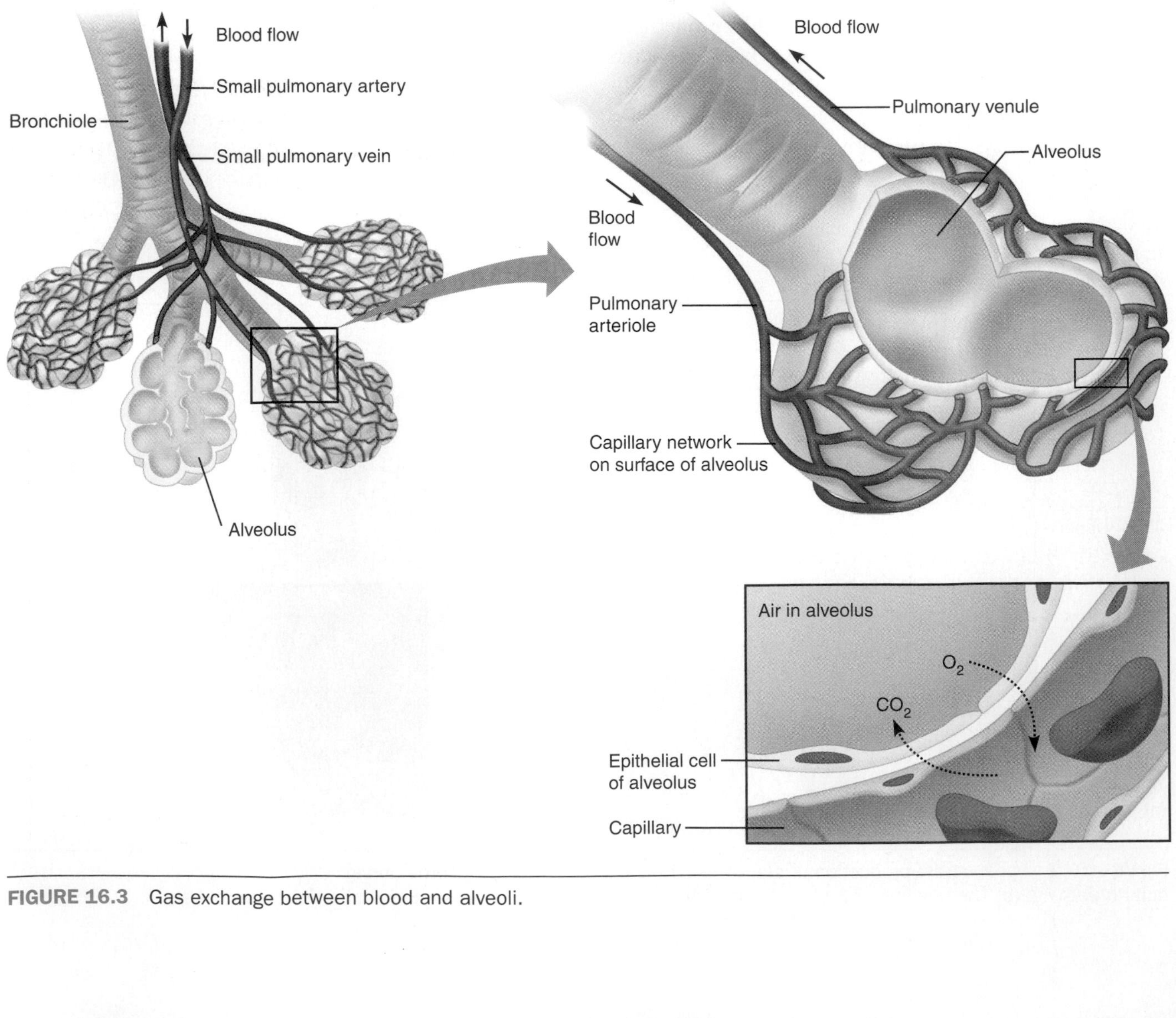

FIGURE 16.3 Gas exchange between blood and alveoli.

capillaries, which carry blood low in dissolved oxygen and relatively high in carbon dioxide. Diffusion of oxygen and carbon dioxide between the alveoli and capillaries reduces the oxygen level and increases the carbon dioxide level in the alveoli. To maintain the high oxygen and low carbon dioxide content of the alveolar air, the deoxygenated air must be exhaled, and new oxygen must be inhaled.

7. Now that we have traced the route for inhaled air, trace the route of exhaled air, and list the structures along this route in the appropriate order. Remember there should be about 10 structures.

____

____

____

The inhaled air mixes with the exhaled air in the respiratory tract. Compare the oxygen content of the atmosphere to the air in the trachea. ■

Breathing

Breathing, also called pulmonary ventilation, is the process of moving air through our lungs. We usually are only aware of our breathing when there is a respiratory problem, but as we will come to see, breathing is not a simple process.

There are two phases to breathing: **inspiration** (commonly known as inhalation), during which our chest expands and air flows into our lungs, and **expiration** (also known as exhalation), during which our chest contracts and air flows out of our lungs. This process is illustrated in Figure 16.5 on page 192.

During inspiration, the diaphragm, which is dome shaped in a relaxed state, will contract and flatten. At the same time, the intercostal muscles between the ribs pull them up and out, thus expanding the thoracic cavity. As a result of these two actions, the volume of the pleural cavity increases and the pressure within the pleural space lowers. When the pressure surrounding the lungs falls relative to the atmosphere, the lungs expand; this reduces air pressure within them relative to the atmosphere, which allows airflow into the lungs.

During expiration, the same muscles—the diaphragm and intercostals—relax, and the lungs contract. As the lungs contract, the volume within them decreases. As a result, the pressure within the lungs increases relative to atmospheric pressure, which causes air to flow out of the lungs.

We can observe these two processes and demonstrate the principles involved by using the Bell jar model in our next activity.

ACTIVITY 2

Demonstrating Airflow into and out of the Lungs

Materials for this Activity

Bell jar model of the lungs

We have learned that in a closed container, increasing the volume will lead to a decrease in the air pressure. This condition establishes a pressure gradient, which results in air from the outside flowing into the lungs.

1. Obtain a Bell jar model of the lungs and compare it to Figure 16.4. The Bell jar represents the chest or thoracic cavity. The rubber membrane represents the diaphragm, which when pushed up or down, changes the volume within the thoracic cavity. The Y-shaped tube represents the trachea. The two balloons represent the lungs.
2. Firmly hold the Bell jar across your front but do not cover up the balloons. Work the model of the lungs by slowly pushing the button on the

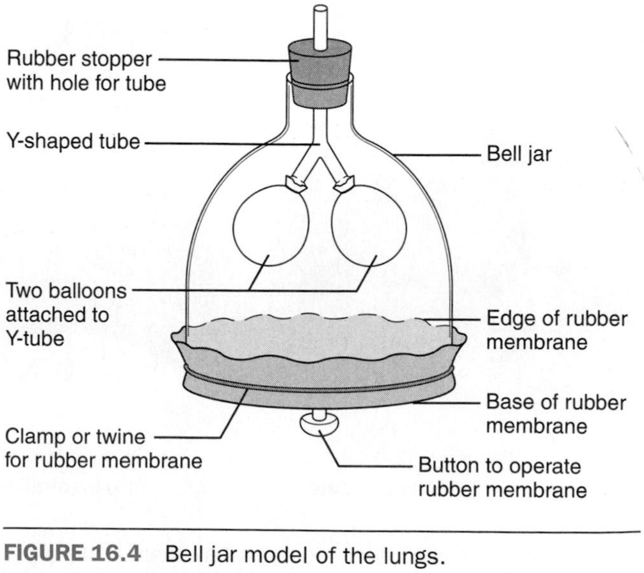

FIGURE 16.4 Bell jar model of the lungs.

rubber membrane up and down a few times, while observing the changes in the balloons (lungs).

Caution: Remember that the rubber membrane is easily damaged through punctures from long fingernails or through tears caused by pushing the button too hard.

3. Slowly push the button up, which moves the membrane inward. Answer the first three questions with the words "increase" or "decrease."

   What happens to the volume of the thoracic cavity?

   ~~inc~~ decrease

   What happens to the pressure of the thoracic cavity?

   increase

   What happens to the size of the lungs?

   increase

   Does the air flow into or out of the lungs?

   both

4. Return the Bell jar model of the lungs to the proper storage area. ■

ACTIVITY 3

Measuring Chest Volume Changes

Materials for this Activity

Metric tape measure

From the Bell jar experiment we have learned that expanding the volume of the closed container will cause air to flow into the balloons (lungs). In this activity, we will examine the volume changes of our own lungs by measuring the volume changes of the chest during three states: relaxed state, quiet breathing, and forced breathing. Notice the movement of the ribs and the diaphragm with inspiration and inhalation, as illustrated in Figure 16.5.

1. In this exercise, you will measure the chest volume changes of your lab partner. Obtain a metric tape measure. Measure the chest circumference by placing the tape around your partner's chest, as high up under the armpit as possible. Make the following measurements in centimeters.

   Quiet breathing

   Hold breath after normal inhalation ~~[illegible]~~ 97cm

   Hold breath after normal exhalation ~~[illegible]~~ 94cm

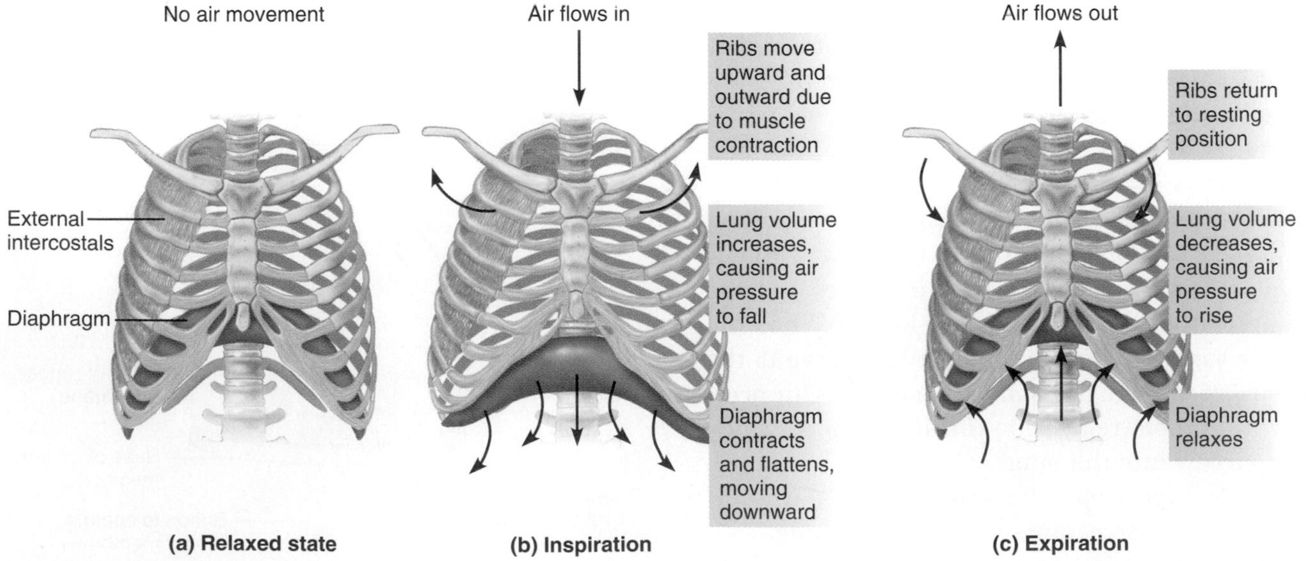

FIGURE 16.5 Chest volume changes. (a) Relaxed state. (b) Inspiration. (c) Expiration.

Forced breathing

Hold breath after forced inhalation ~~3742~~ 94.5

Hold breath after forced exhalation ~~3737~~ ~~3714~~ 95.75

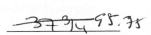

2. Repeat your measurements to verify your results.
3. Subtract your quiet breathing normal *exhalation* measurement from your quiet breathing normal *inhalation* measurement and record 3.
4. Subtract your forced breathing *exhalation* measurement from your forced breathing *inhalation* measurement and record 1.25.
5. Considering that a larger difference indicates a larger volume change, what can you conclude about the relationship between chest volume changes and air movement?

____

____

Lung Capacity

Depending on age, gender, and physical condition, there can be great variation in the chest volume changes that were measured for normal and forced breathing. These variations make sense, considering the lungs are geared to accommodate personal oxygen needs, which are different for each individual.

Measurement of these **lung volumes** is useful because it allows medical personnel to track the changes in your lung capacity during the course of a disease or in the diagnosis of a respiratory disease.

There are four individual components to your **total lung capacity (TLC)** (Figure 16.6):

1. Inspiratory reserve volume (IRV)
   Volume of forcibly inhaled air after a normal inhale
2. Tidal volume (TV)
   Volume of air in a normal inhale or normal exhale
3. Expiratory reserve volume (ERV)
   Volume of forcibly exhaled air after a normal exhale
4. Residual volume (RV)
   Volume of air in the lungs at all times

In summary,

$$TLC = IRV + TV + ERV + RV$$

Vital capacity (VC) represents the sum total of IRV, TV, and ERV. Notice it includes every lung volume except for RV, which is always part of the lung. Vital capacity refers to the total exchangeable air of the lungs. There are two ways of calculating VC:

$$VC = IRV + TV + ERV$$
$$VC = TLC - RV$$

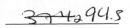

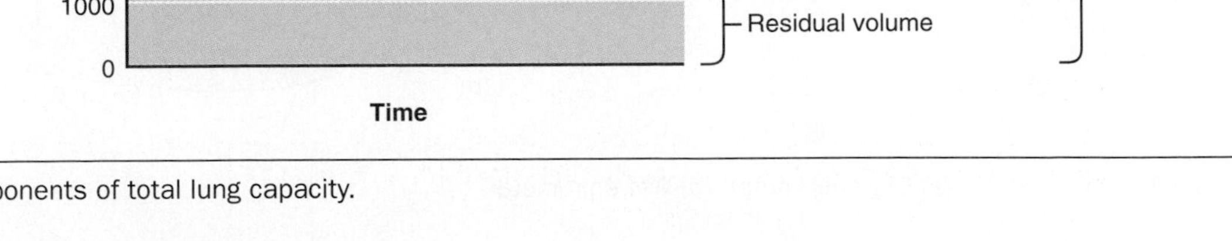

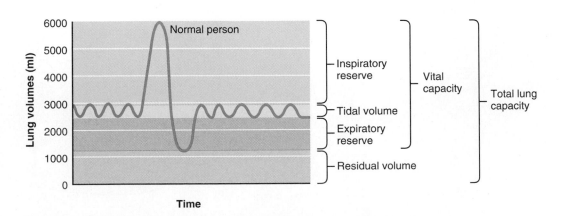

FIGURE 16.6 Components of total lung capacity.

ACTIVITY 4

Measuring Lung Volumes

Materials for this Activity

Nonrecording spirometer (wet or dry)
Disposable mouthpieces for spirometers
Alcohol prep pad
Nose clips

1. Obtain a wet or dry spirometer (Figure 16.7), mouthpiece, and alcohol prep pad. Upon return to your station, immediately write your name on your mouthpiece to avoid using the wrong one. If you obtained a dry spirometer, untwist the top and clean out the interior chambers with an alcohol prep pad. Replace the top of the dry spirometer. If you obtained a wet spirometer, prepare a clear work space in case of water spillage.
2. Take a look at your spirometer. They are called nonrecording spirometers because there is no record kept of the data. After each use, you need to reset the gauge back to zero. Note that the volumes are to be measured in milliliters or liters. The typical values quoted for each measurement are based on a young, healthy male.
3. Your job is to provide the following instructions to your partner, record the data, and reset the gauge of the spirometer. As the subject, your lab partner should handle the mouthpiece of the spirometer, inserting it into his or her mouth.
4. Count the number of normal breaths in 1 minute.

   Normal respiratory rate: 12

   A typical value for a normal respiratory rate is about 12 breaths per minute.
5. Practice exhaling through the mouthpiece, without exhaling through the nose. If this is not possible, clean and use the nose clips.
6. Measure **TV** as follows: Inhale a normal breath, place the spirometer in your mouth, and exhale normally into the mouthpiece. Take three measurements, and calculate the average of the three measurements. Remember to set the gauge to zero before making the next measurement.

   First TV 1200

   Second TV 13..

   Third TV 1300

   Average TV 126.67

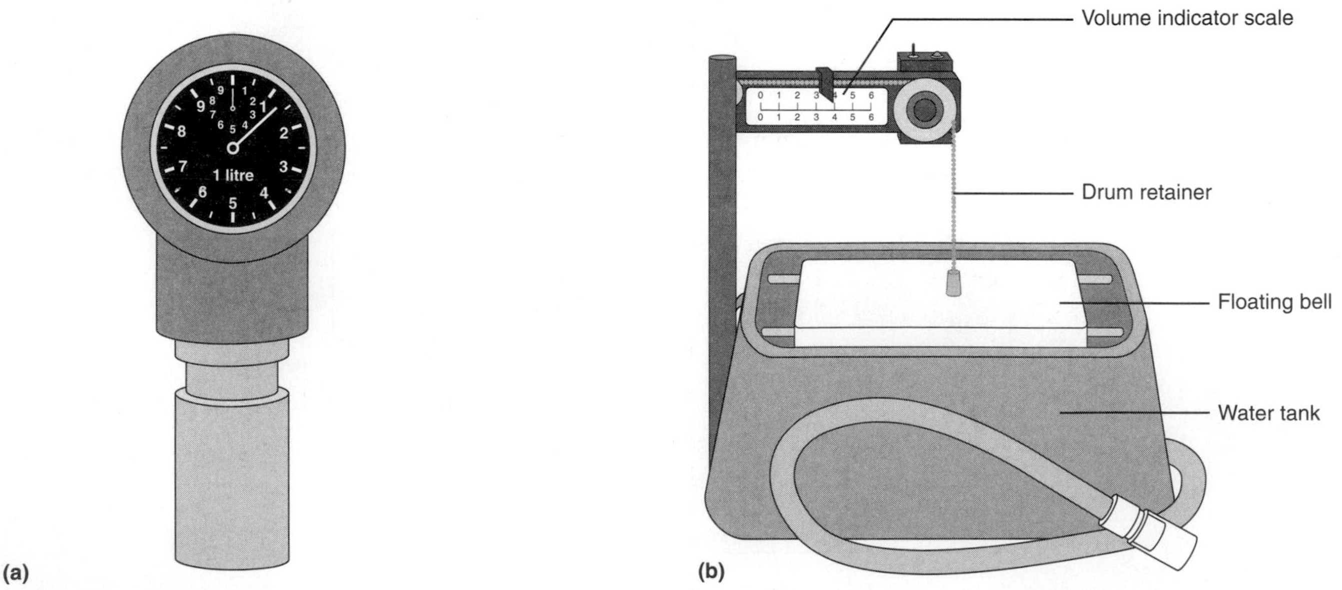

FIGURE 16.7 Spirometers. (a) Dry spirometer. (b) Wet spirometer.

Typical values for TV range from 300–500 ml. About one-third of the air stays in the respiratory tract, and two-thirds reaches the alveoli to become involved in gas exchange.

7. Calculate the **minute respiratory volume (MRV)** as follows:

MRV = TV × respiratory rate

MRV 1520.04 ~~2533.4~~

Typical values for the MRV range are determined as follows:

300 ml × 12 breaths per min = 3600 ml or 3.6 l

500 ml × 12 breaths per min = 6000 ml or 6 l

Range: 3600–6000 ml, or 3.6–6 l

8. Measure ERV as follows: Take a few normal inhalations and exhalations. After your last inhalation, place the spirometer in your mouth and forcibly exhale. Take three measurements, and calculate the average of the three measurements. Remember to set the gauge to zero before making the next measurement.

First ERV	200
Second ERV	230
Third ERV	260
Average ERV	230

Typical values for ERV are about 1300 ml or 1.3 l.

ERV is greatly reduced in respiratory disorders such as emphysema. Because energy must be used to deflate the lungs, exhalation is also exhausting.

9. Measure VC as follows: Take a few normal inhalations and exhalations. Bend over and exhale as completely as possible. Raise yourself and inhale as completely as possible. Quickly insert the mouthpiece and exhale forcibly. Take three measurements, and calculate the average of the three measurements. Remember to set the gauge to zero before making the next measurement.

First VC	200
Second VC	190
Third VC	180
Average VC	190

The typical range for VC is 3600–4800 ml, or 3.6–4.8 l.

10. Calculate IRV as follows:

IRV = VC − TV − ERV

Average IRV ~~166.6 700~~

The typical range for IRV is from 2000–3000 ml, or 2–3 l. ■

ACTIVITY 5

Comparing Nonsmoker to Smoker Data

Materials for this Activity

Lung volume data from Activity 4

1. The instructor will divide the class into two groups: smokers and nonsmokers.
2. On the board, the instructor will put up two tables for students to record the individual data for the smoker and nonsmoker categories.
3. Obtain the averages of your personal data from the previous activity. Reorganize the data as follows:

MRV	1520.04	15199.24
IRV	-166.67	-1666.6
TV	126.67	1266.66
ERV	230	2300
VC	190	1900

4. Record your individual data on the board under smoker or nonsmoker.
5. Calculate the average for each category on the board. The average is the sum of the values divided by the number of values.

Average smoker data:

MRV	8400
IRV	467
TV	600
ERV	1933
VC	2000

Average nonsmoker data:

MRV ____same____

IRV ____a____

TV ____min____

ERV ________

VC ________

6. How do your data compare to the appropriate group—smoker or nonsmoker?

   Compare your average to the group average. Are your statistics higher or lower?

   MRV ____High____

   IRV ____higher____

   TV ____higher____

   ERV ____lower____

   VC ____~~higher~~ lower____

7. The range is the simplest measure of the dispersion of the data. Many public reports, for example, are presented in the form of ranges. Weather forecasts typically state the high and low temperature projections for each day of the week. Stock reports track the highs and lows of stock prices over a certain time period.

   Calculate the range of statistics for each category on the board.

   Range of smoker data:

   MRV ________

   IRV ________

   TV ________

   ERV ________

   VC ________

   Range of nonsmoker data:

   MRV ________

   IRV ________

   TV ________

   ERV ________

   VC ________

8. How do your data compare to the appropriate group—smoker or nonsmoker? Compare your data to the group range. Are you in the low, middle, or high end?

   MRV ________

   IRV ________

   TV ________

   ERV ________

   VC ________

9. How do the nonsmoker data compare to the smoker data?

   Compare the averages. Are the nonsmoker numbers higher or lower than the smoker numbers?

   MRV ________

   IRV ________

   TV ________

   ERV ________

   VC ________

   If the data are not what you expected, what might be some reasons?

   ____

   ____

   ____

   Compare the ranges. Is the middle of the nonsmoker range higher or lower than the middle of the smoker range?

   MRV ________

   IRV ________

   TV ________

   ERV ________

   VC ________

   If the data are not what you expected, what might be some reasons?

   ____

   ____

   ____

EXERCISE

17

THE DIGESTIVE SYSTEM AND NUTRITION

Objectives

After completing this exercise, you should be able to

1. List, identify, and describe the major organs of the digestive system.
2. Describe the functions of the major organs of the digestive system.
3. List the major enzymes of the digestive system and describe their functions.
4. Explain the basic concept of metabolism and how it is connected to nutrition.

Materials for Lab Preparation

Equipment and Supplies

- ❑ Compound light microscope
- ❑ Fetal pig
- ❑ Dissection equipment
- ❑ Dissection tray
- ❑ Disposable gloves
- ❑ 1-liter beaker filled about halfway with tap water heated to 38°C
- ❑ 1-liter beaker filled about halfway with tap water heated to 70°C
- ❑ 1-liter beaker filled about halfway with ice water
- ❑ Tap water
- ❑ Distilled or deionized water
- ❑ 250-ml flask
- ❑ Five large test tubes
- ❑ pH paper
- ❑ China marker
- ❑ Standard personal weighing scale

Prepared Slides

- ❑ Small intestine

Models

- ❑ Small intestine
- ❑ Human torso

Solutions

- ❑ Phenol red solution
- ❑ Dropper bottle with vegetable oil
- ❑ Dropper bottle with 2% sodium carbonate
- ❑ Dropper bottle with 1% bile solution (which has been adjusted to a pH of 7.5–8.0)
- ❑ Dropper bottle with 1% pancreatin or pancreatic lipase solution

Introduction

There are three major parts to the study of the digestive system: digestion, absorption, and metabolism (how the absorbed nutrients are used by the body). Indirectly related to all three is the concept of nutrition. As always, it is important to first learn the "parts" of the system before we can understand what they do and why being aware of nutrition issues is so important. It is also important to understand the role played by enzymes in the digestive system.

The structures of the digestive system are usually divided into two sections: the organs of the **gastrointestinal (GI) tract** and the **accessory organs** located outside it (Figure 17.1). After the mouth, the GI tract consists of the pharynx (throat), esophagus, stomach, small intestine, large intestine, rectum, and anus. The main accessory organs are the pancreas, liver, gallbladder, and salivary glands (Figure 17.1). We will examine each of these structures and the tissues they are made from in the next section.

Digestive System Anatomy and Basic Function

The Layers of the Gastrointestinal (GI) Tract

All of the tubular structures of the organs composing the GI tract have the same basic tissue organization. There is an outer **serosa,** a thick layer of muscle called the **muscularis,** an inner layer of connective tissue called the **submucosa,** and a membrane with epithelial cells lining the inside called the **mucosa** (Figure 17.2, see page 202).

These tissues are modified in each part of the digestive tract to fit the special needs and functions of each organ. For example, the stomach has three layers of muscle so that it can churn the food being digested. The rest of the tract has two layers so that food can simply be pushed along the length of the tract. The mucosa of some areas produces less mucus in places where nutrient absorption is the highest priority, and it produces a great deal of mucus where protection from acid or friction is the main concern.

ACTIVITY 1

Observing the Tissue Layers of the GI Tract

Materials for this Activity

Compound light microscope
Slide of the small intestine
Model of the small intestine

1. Observe a slide and/or model of the small intestine, and identify the four tissue layers shown in Figure 17.2.

   Note the outer *serosa* (which is often mostly removed during slide preparation and may appear as one small area of adipose tissue). Next, observe the two layers of smooth muscle in the *muscularis.* The deep layer is circular, while the superficial layer runs longitudinally.
2. Observe the connective tissue of the *submucosa,* and note the blood vessels, nerves, and other structures embedded in this layer. The submucosa provides most of the blood supply for the other layers and protects the delicate nerves and vessels that pass through its connective tissue.
3. Finally, observe the *mucosa,* and note its surface epithelium, connective tissue, capillaries, and lacteals (lymphatic capillaries). Its major functions are secretion (e.g., of hormones, mucus, and enzymes), absorption of digested food, and protection against bacteria. Into which structure in the mucosa would water-soluble nutrients be absorbed?

   ______________________________ ■

The Mouth, Pharynx, and Esophagus

You probably already know that the mouth, pharynx, and esophagus are structures that allow us to chew and swallow food. But how are they specialized for this job? All three are lined with *stratified squamous epithelium* to protect the lining from friction. You may recall that this is the same epithelium as the skin; however there is one big difference. The skin cells are filled with keratin for extra protection and prevention of water loss. Because water loss is not really an issue in these structures, and a softer, more flexible lining is desirable, the lining of these structures is **nonkeratinized.**

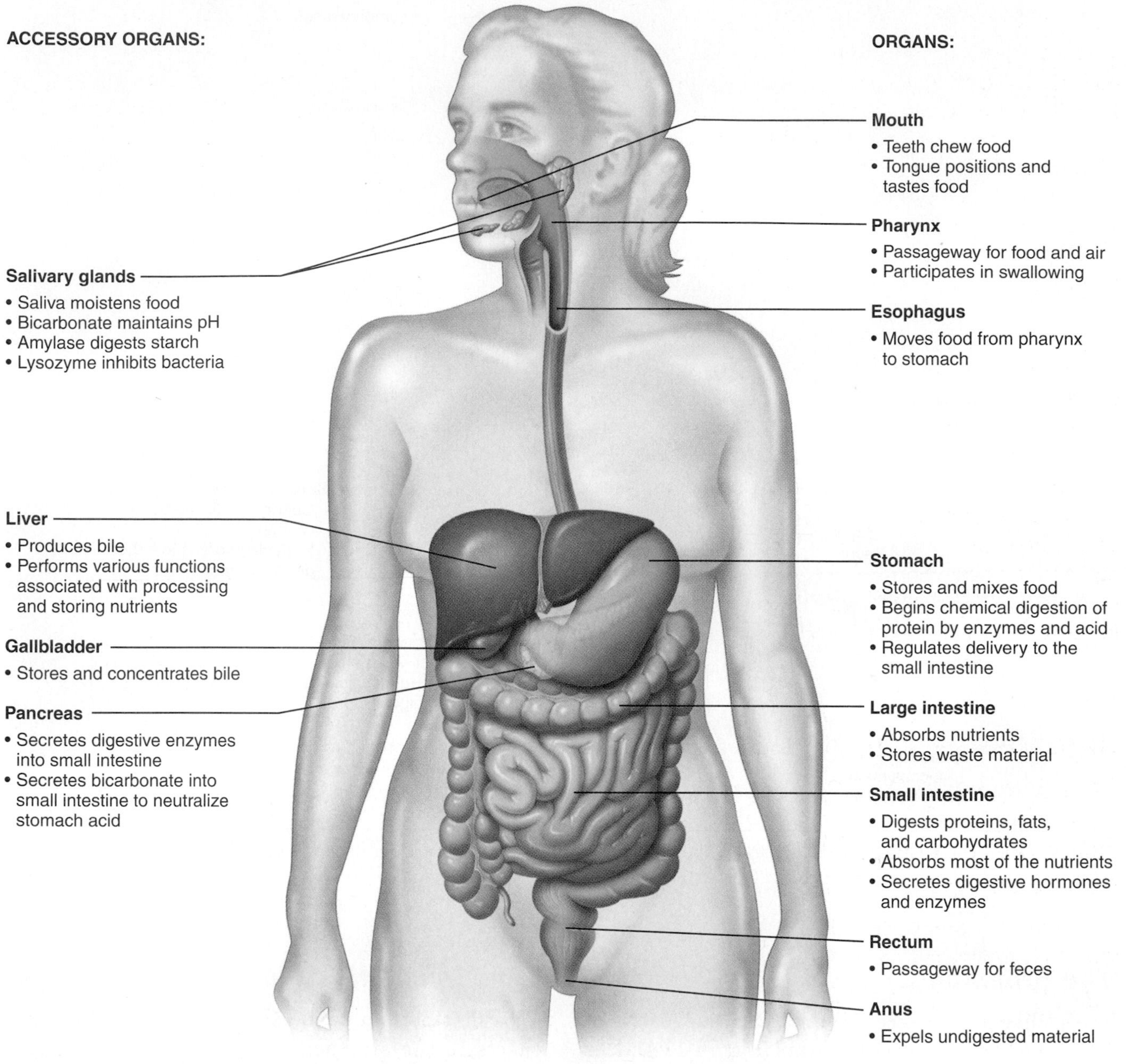
ACCESSORY ORGANS:
ORGANS:
Mouth
• Teeth chew food
• Tongue positions and tastes food
Pharynx
• Passageway for food and air
• Participates in swallowing
Esophagus
• Moves food from pharynx to stomach
Salivary glands
• Saliva moistens food
• Bicarbonate maintains pH
• Amylase digests starch
• Lysozyme inhibits bacteria
Liver
• Produces bile
• Performs various functions associated with processing and storing nutrients
Gallbladder
• Stores and concentrates bile
Pancreas
• Secretes digestive enzymes into small intestine
• Secretes bicarbonate into small intestine to neutralize stomach acid
Stomach
• Stores and mixes food
• Begins chemical digestion of protein by enzymes and acid
• Regulates delivery to the small intestine
Large intestine
• Absorbs nutrients
• Stores waste material
Small intestine
• Digests proteins, fats, and carbohydrates
• Absorbs most of the nutrients
• Secretes digestive hormones and enzymes
Rectum
• Passageway for feces
Anus
• Expels undigested material

FIGURE 17.1 Organs and accessory organs of the digestive system. The digestive organs, from the mouth to the small intestine, share a common function: getting nutrients into the body. The rectum and anus provide the means for expelling waste. Accessory organs aid digestion and absorption.

Another aspect common to all three structures of the upper GI tract is muscle tissue. Skeletal muscle is found in and around the mouth, including the tongue, in order to chew and manipulate food. Skeletal muscle is also found in the pharynx and upper portion of the esophagus so that we may voluntarily initiate swallowing. The muscle tissue of the esophagus gradually transitions to smooth muscle, which explains why initiating swallowing is voluntary, but the continuation of the swallowing process is not.

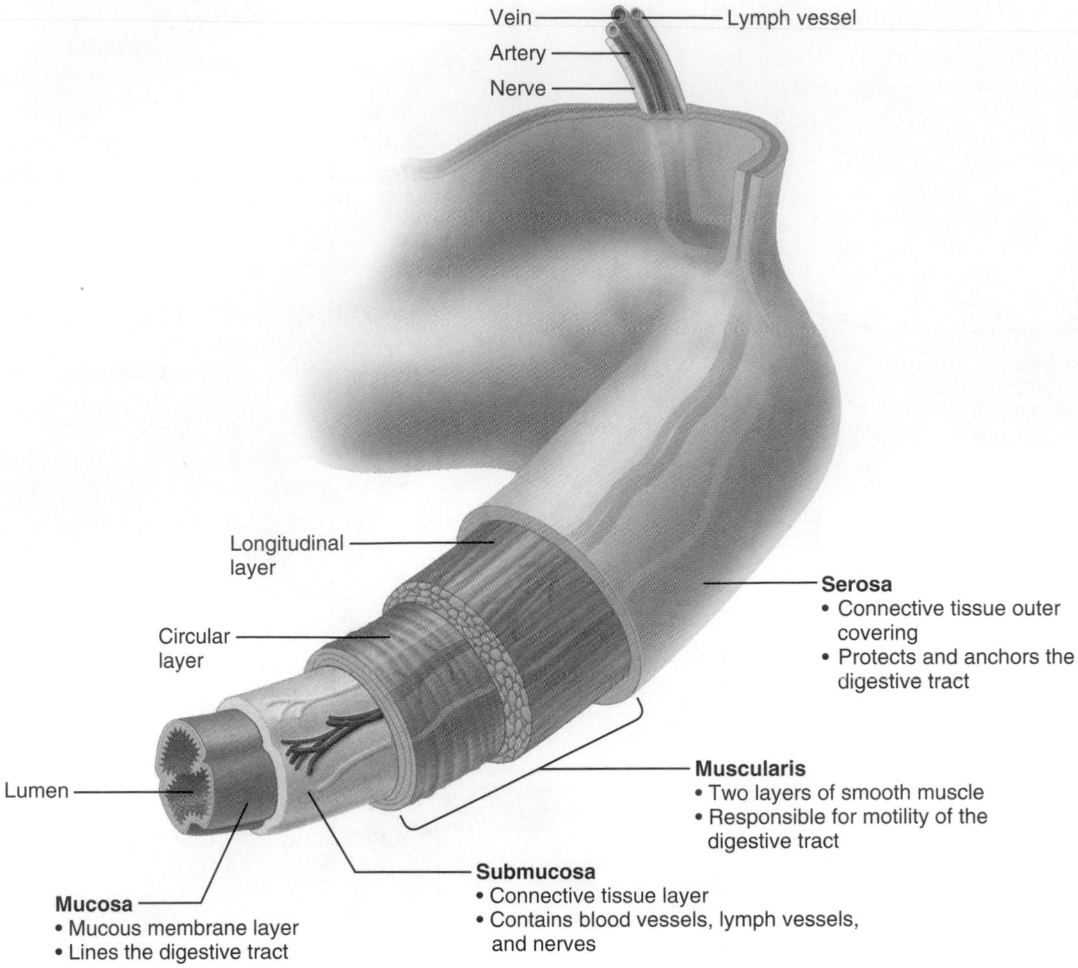

Vein
Lymph vessel
Artery
Nerve
Longitudinal layer
Circular layer
Serosa
• Connective tissue outer covering
• Protects and anchors the digestive tract
Lumen
Muscularis
• Two layers of smooth muscle
• Responsible for motility of the digestive tract
Submucosa
• Connective tissue layer
• Contains blood vessels, lymph vessels, and nerves
Mucosa
• Mucous membrane layer
• Lines the digestive tract

FIGURE 17.2 The gastrointestinal (GI) tract. The GI tract walls are composed of four tissue layers.

The Stomach

The saclike appearance of the stomach suggests that it plays a role as a food reservoir (Figure 17.1). It is also a major site of digestion, as solid food is liquified by the enzymes and acid produced by the stomach. A pair of **sphincter** muscles help to hold food in the stomach during this part of the digestive process. The cardiac (or gastroesophageal) sphincter, near the top of the stomach where the esophagus and stomach connect, prevents the acidic food matter from moving back up into the esophagus. The pyloric sphincter, where the small intestine and stomach connect, prevents food from moving into the small intestine until it is fully liquified. What common ailment is caused when the cardiac sphincter allows a small amount of material to leak into the esophagus?

____

The Small Intestine

The first portion of the small intestine, which connects to the stomach, is called the **duodenum.** Ducts from the pancreas, liver, and gallbladder also connect to this busy crossroads of the digestive tract. Bicarbonate ions from the pancreas neutralize the acid from the stomach. Bile from the liver and gallbladder and enzymes from the pancreas are also added to the duodenum so that complete digestion of our food can occur during the long trip through the small intestine.

The remainder of the small intestine is divided into two parts, the **jejunum** and **ileum.** Not only is food digested down to simple molecules in these structures, but absorption of nutrients is a major function as well. Numerous folds and microscopic folds called villi increase surface area for absorption

(Figure 17.3a, b). For the most part, water-soluble nutrients such as glucose are absorbed into the capillaries, and fat-soluble nutrients are absorbed into the lacteals, which are lymphatic capillaries in the mucosa of the villi (Figure 17.3c).

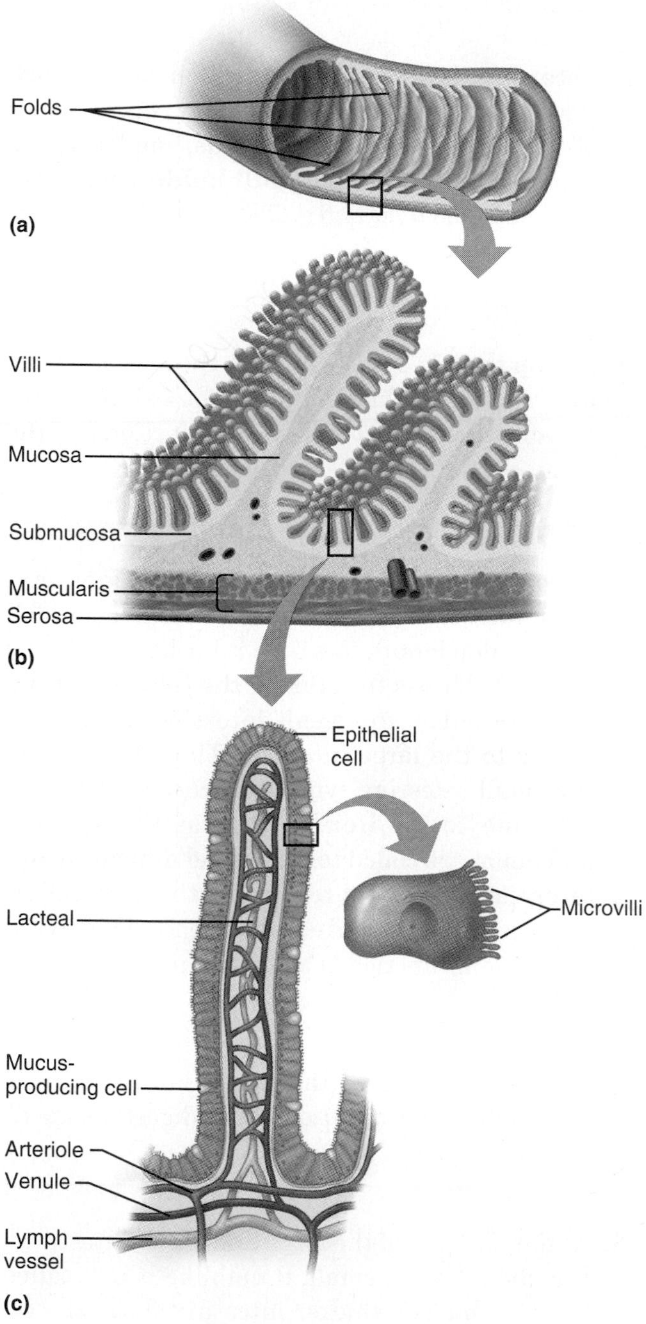

FIGURE 17.3 The walls of the small intestine. (a) The numerous folds of the small intestine increase its surface area. (b) Each fold is covered with microscopic folds called villi. (c) This view of a single villus shows a single layer of mucosal cells, a lacteal (lymph vessel), and blood vessels.

The Large Intestine

The ileum of the small intestine plugs into the large intestine slightly above its starting point. The short blind alley of large intestine below this entry point is called the **cecum.** In humans, the cecum has a wormlike "tail" called the **appendix** (Figure 17.4).

The **colon** begins just above the cecum, and it is named for the direction in which the digestive wastes move: the **ascending colon** on the right side of the abdominal cavity, the **transverse colon** stringing across from right to left in the upper portion of the abdominal cavity, the **descending colon** down the left side, and the squiggly **sigmoid colon** connecting to the rectum in the pelvic cavity. We refer to the material in the large intestine as waste because essentially all of the nutrients and most of the water should have been absorbed from the food before it left the small intestine.

It is important to note that a significant amount of water and some salts are absorbed by the large intestine, a process that helps to solidify the wastes. Bacteria in the large intestine also provide a service to us by producing vitamins such as vitamin K, which plays an important role in blood clotting.

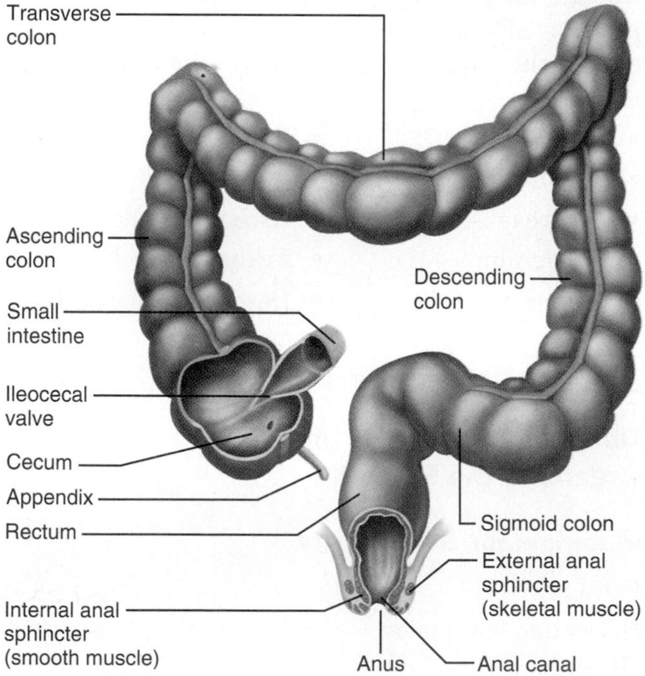

FIGURE 17.4 The large intestine. The large intestine absorbs the last of the remaining water, ions, and nutrients in the digestive tract and stores the remaining near-solid waste until it can be expelled.

Accessory Organs

Although not considered part of the digestive *tract*, the pancreas, salivary glands, liver, and gallbladder play critical roles in the digestive process (Figure 17.1). The **pancreas** and **salivary glands** produce enzymes and other materials needed by the digestive system. As mentioned previously, the pancreas helps to neutralize the acid from the stomach by secreting bicarbonate ions, and it produces enzymes that help digest many different kinds of food molecules. The salivary glands produce mucus and amylase, an enzyme that helps to break down starches into sugars. The **liver** produces bile, which is then stored and concentrated in the **gallbladder.** Although it is not an enzyme, bile plays an important role in fat digestion by breaking up or emulsifying large fat globules into much smaller ones.

ACTIVITY 2

Identifying Human Digestive System Organs

Materials for this Activity

Human torso models

Using Figures 17.1 and 17.4 as references, trace the pathway of food through the GI tract by identifying each of the following structures on the human torso models:

- Esophagus
- Stomach
- Pancreas
- Duodenum
- Jejunum
- Ileum
- Cecum
- Ascending colon
- Transverse colon
- Descending colon ■

ACTIVITY 3

Observing the Major Digestive Organs in the Fetal Pig

Materials for this Activity

Fetal pig

Dissection equipment

Dissection tray

Disposable gloves

Because your fetal pig should already be open from previous lab exercises, you will now simply need to examine the inside of the pig in order to find these organs. Some of the organs will probably be familiar, considering they were pointed out previously as landmarks to help you find other structures. Please review the safety procedures for dissection provided by your instructor and those that appeared in Exercise 13. Figure 17.5 provides a reference for the dissection.

1. Obtain a fetal pig, place it faceup on a dissection tray, and reattach the string as described in Exercise 13.
2. Locate the large dark **liver,** which covers much of the upper abdominal cavity. Lift up the lobes of the liver on the *pig's* right side, and locate the small, greenish, saclike **gallbladder.** What secretion do the liver and gallbladder add to the digestive tract?

   ____

3. Lift up the lobes of the liver on the *pig's* left side, and locate the large, pale, saclike **stomach.** What are the most important functions of the stomach?

   ____

4. Follow the stomach toward the pig's right as it tapers down to join the **duodenum.** Then, follow the duodenum to where it hooks back to the pig's left. This is the start of the **jejunum.** Continue to follow the small intestine until it attaches to the large intestine. The latter half of the small intestine, which attaches to the large intestine, is the **ileum.** Because the jejunum and ileum are coiled together and difficult to differentiate, most sources refer to them collectively as the **jejuno-ileum** in the fetal pig. What is the main function of the small intestine?

   ____

   Why do you suppose that the small intestine is the single longest portion of the digestive tract?

   ____

5. At the junction of the large and small intestines, you should see a small, thumblike pouch called the **cecum.** The thicker intestine above the cecum is the **colon,** which is typically coiled together in the pig. For this reason, it is usually referred to as the *spiral colon* in the fetal pig. The colon straightens out to run along the back of the pelvic cavity and form the **rectum.**

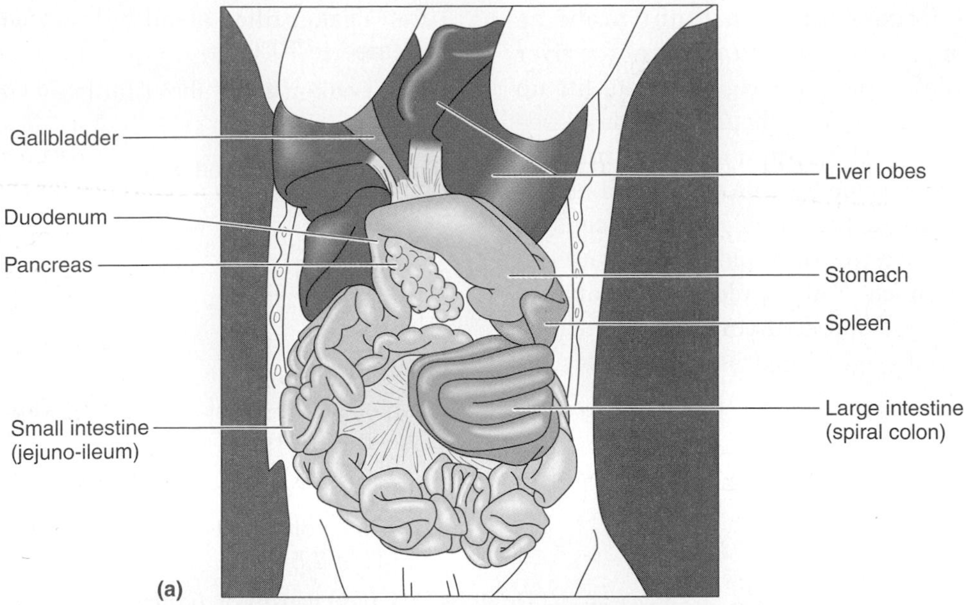

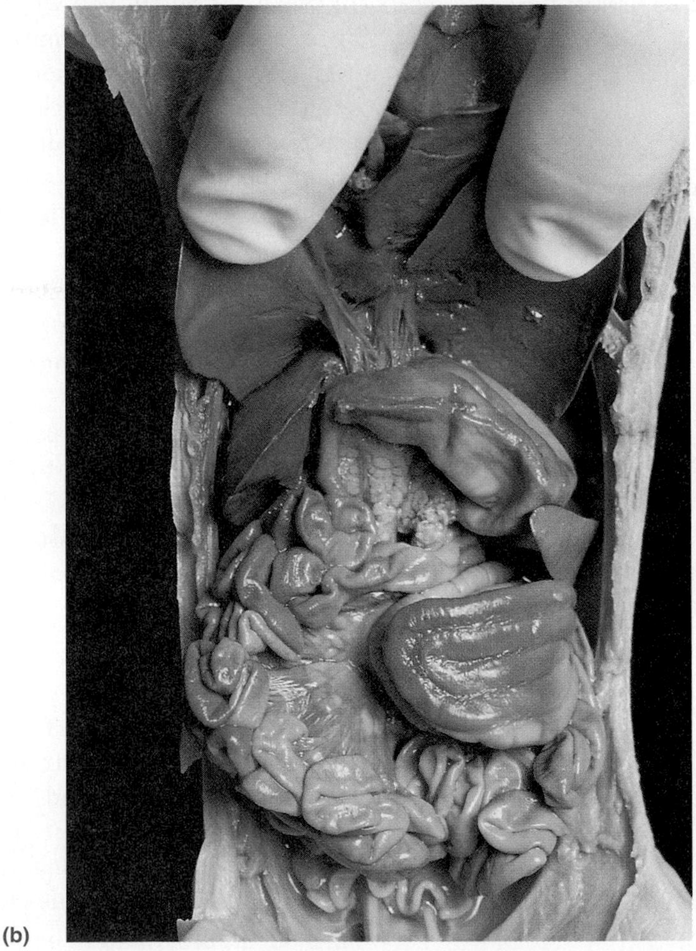

FIGURE 17.5 The digestive organs of the fetal pig. (a) This diagram illustrates the photograph in part (b).

6. Using your fingers and/or a blunt probe as needed, lift up (do *not* cut and remove) the liver on the pig's left side (your right). Then, lift up the saclike stomach. You should be able to see the **pancreas,** which you first observed in Exercise 13, The Endocrine System. If you did not locate the pancreas previously, use forceps to break and remove the thin, plasticlike material under the stomach. You should now be able to see the pancreas. What structure is the pancreas attached to on its right side (remember, the *pig's* right side)?

______________________________ ■

Enzymes

Enzymes are the helper molecules, made of protein, that make many chemical reactions occur much easier and faster. They are used by the digestive system to break the large food macromolecules down to their more easily absorbed building blocks.

The enzymes of the digestive system are generally named for the substance they break down. For example, proteases break down proteins, lipases break down lipids, and amylases break down starches such as amylose. Within these general categories, there are enzymes with specific jobs. Pepsin is a protease produced by the stomach, which is mainly responsible for breaking big proteins down into smaller ones. Other proteases, such as trypsin produced by the pancreas, then break down proteins further.

Because they are made of delicately folded proteins, enzymes require fairly specific environmental conditions to maintain their shape and function effectively. Although some enzymes are more tolerant than others, a slight change in temperature or pH can render an enzyme useless. Not surprisingly, each enzyme is designed for the environment in which it must work. For example, pepsin works best in the acidic conditions we find in the stomach, while trypsin only works at higher pHs found in the small intestine.

ACTIVITY 4

→changed to animal fat (milk)

Observing Enzyme Function

Materials for this Activity

1-liter beaker filled about halfway with tap water heated to 38°C

1-liter beaker filled about halfway with tap water heated to 70°C

1-liter beaker filled about halfway with ice water

Tap water

Distilled or deionized water

250-ml flask

Five large test tubes

Phenol red solution

pH paper

China marker

Dropper bottles containing the following:

- Vegetable oil
- 2% sodium carbonate
- 1% bile solution (which has been adjusted to a pH of 7.5–8.0)
- 1% pancreatin or pancreatic lipase solution

In this experiment, you will be testing the performance of an enzyme, pancreatic lipase, under a variety of environmental conditions. You will be preparing five test tubes: one serves as a control and each of the remaining tubes is subjected to a different variable. The lipid that pancreatic lipase will be breaking down is called a **triglyceride.** When broken down, the triglyceride is separated into a glycerol molecule and three fatty acid molecules.

1. Add about 120 ml of distilled water to the 250-ml flask. The exact amount is not important. Test your solution with pH paper. The pH of distilled water should be approximately 7.0. Add 1–2 drops of sodium carbonate to the flask, and swirl the flask to mix the solution. Test your solution with pH paper. If the solution has a pH between 7.5–8.0, proceed to Step 2.

   If the pH of the solution is lower than 7.5, repeat the step of adding sodium carbonate and testing the solution with pH paper until the pH is in the 7.5–8.0 range. *Note:* If the pH of your solution is already above 8, your source of distilled water may not be pure and has an abnormally high pH. You may substitute tap water or bottled water if necessary, as long as you end up with a pH in the slightly alkaline range (7.5–8.0).
2. Pour approximately equal amounts of the solution you prepared in Step 1 into each of the five large test tubes. Label the tubes 1, 2, 3, 4, and 5. Add five drops of phenol red solution to each test tube. Phenol red is somewhat like liquid litmus

paper; it turns a bluish color at a high pH and an orange color at lower pHs. Because you made your solution slightly alkaline in Step 1, your tubes should all be about the same color—a fuchsia or magenta color.

3. Add five drops of pancreatin or pancreatic lipase solution to each test tube.
4. Add nothing to test tube 1; add five drops of vegetable oil to test tubes 2, 3, 4, and 5. Add three drops of bile solution to test tube 3. Propose a reason why you added nothing further to test tube 1. (What function is test tube 1 serving in this experiment?)

   ____

5. Place test tubes 1, 2, and 3 in the beaker serving as the 38°C water bath; place test tube 4 in the beaker serving as the 70°C water bath; place test tube 5 in the beaker serving as the ice-water bath.
6. Every 10 minutes for 1 hour, check to make sure the temperature has not changed more than 2–3 degrees from the starting temperature. If it has, place the 38°C or 70°C beaker on a hot plate for a minute or two in order to return it to its starting temperature. The ice-water bath is unlikely to require adjustment in 1 hour's time, but you may add more ice if necessary. Use this 10-minute check as an opportunity to remix the contents of the test tube by swirling or shaking the test tube from side to side. (Your instructor may demonstrate the best procedure for your test tubes.)
7. After 1 hour, compare the color changes, if any, that occurred in your test tubes. Use test tube 1 as your point of comparison, considering it should not have changed color during this experiment.

   Recalling that the enzyme should break down the vegetable oil into glycerol and fatty *acids,* a change in pH should occur in any tube in which the enzyme was working.

   Recalling that phenol red is like liquid litmus paper, a color change to pinkish-red or red is a strong positive result. A slight color change to pink would still be considered a positive reaction, but a weaker one.

   Record your results in Box 17.1. *Note:* It is helpful to hold the tubes against a white background when judging your color changes. A dark background makes it difficult to notice slight color changes.

8. The bicarbonate ions, which are secreted by the pancreas along with enzymes such as pancreatic lipase, raise the pH of the small intestine to approximately the same slightly alkaline range used in this experiment. However, the temperature of the human body in the abdominal cavity is more likely to be in the 37–38°C range. Which were the only test tubes to give significant positive reactions?

   ____

   Which test tube demonstrated the strongest reaction?

   ____

   What ingredients were initially placed in the test tube of your answer above?

   ____

   ____

   Explain your results.

   ____

   ____

9. The temperatures experienced by test tubes 4 and 5 should have been too extreme for the enzyme to work. Propose a reason why it might be possible to get a slight color change in these test tubes anyway. *Hint:* Review your procedure for this activity for potential sources of experimental error.

   ____

   ____

Box 17.1 Results of Enzyme Activity as Measured by Color Change

Test Tube	Color Change (insignificant, slight, moderate, substantial)
1	
2	
3	
4	
5	

Speculate as to why either a high fever or hypothermia can be lethal.

____

______________________________ ■

Metabolism and Nutrition

Metabolism mainly describes how the nutrients we absorb are used by the body. These nutrients can be used to build molecules, cells, and tissues, or to repair existing ones. Most of the molecules we absorb can also be used for energy. Because humans are warm-blooded creatures, an enormous amount of the nutrient molecules we take in are simply used to maintain our body temperature and keep our organs functioning.

Basal metabolism, or **basal metabolic rate (BMR),** is the measure of the energy required for essential life-support functions such as generating body heat. A calorie is a very small heat measurement unit and is simply the amount of energy needed to raise 1 gram of water 1°C. Because this tiny unit would be impractical in discussing human nutritional needs, the kilocalorie (1000 calories) is used. We use the capitalized term, **Calorie,** when referring to the larger kilocalorie in discussions of nutrition.

For the average-sized person, the Calories of energy required for basal metabolism would be approximately 2000 calories per day. This may vary greatly depending upon numerous factors such as gender, size, and muscle density. The total number of Calories we expend depends on our BMR plus our daily activities. A sedentary person will clearly expend a number of Calories that is not substantially higher than his or her BMR, while a triathlete in training will expend many more Calories in a day. A simple fact of weight gain versus weight loss is that excess Calories are stored as fat, and reducing body weight requires burning more calories than we consume.

It is an unfortunate fact of nutrition that fatty foods tend to be significantly higher in Calories, ounce for ounce, than other food sources. They also tend to taste better to most people—probably a leftover adaptation from a time in human history when food was far less abundant, so the more Calories, the better.

ACTIVITY 5

Learning About Metabolism, Nutrition, and You

Materials for this Activity

Standard personal weighing scale

In this activity, we will examine a person's Caloric needs, compare that to a proposed menu, and examine the amount of additional activities this person would need to undertake to avoid weight gain.

1. To calculate your BMR, it is necessary to determine your weight in kilograms and adjust the number if you are female, because females burn slightly fewer Calories pound for pound than males. Weigh yourself, divide the number of pounds by 2.2, and record this number:

   71.36 ______________________________

   If you are female, multiply this number by 0.9, and record it:

   64.22 ______________________________

2. Multiply the number you obtained in Step 1 by 24 and record it:

   ~~1712.64~~ 1541.28 ______________________________

   This is your BMR, or the number of Calories you burn every day just to maintain body temperature and organ function.

3. Use Table 17.1, or other references provided in the lab, to choose items for a hypothetical menu for typical daily food intake. *Note:* This activity will not work if you make food choices that do not truly represent your food intake on most days. Enter these items and their number of Calories in Box 17.2. Add the Calories in Box 17.2 to calculate total Calories for the day. Record this total.

   1490 ______________________________

4. Subtract the BMR you recorded in Step 2 from the Calorie total you recorded in Step 3. If this is a negative number, you may have forgotten to include all the items you would truly eat in a typical day, or you may have planned a menu that you would follow if you were on a diet. If you are certain that you did plan a representative menu and still obtained a negative number, this result

Table 17.1 *Common Foods and Their Estimated Caloric Content*

Meal	Food item	Calories
Breakfast foods	Cereal with milk	200
	Pancakes (2) with syrup	450
	Doughnut	300
	Bagel, medium (with margarine or cream cheese, add 40)	250
	Nonfat yogurt	150
	Medium muffin (with margarine, add 50)	250
	Fruit, medium size (apple, orange)	80
	French toast (2 slices) with syrup	450
	Toast, one slice, with margarine and jam	170
	Scrambled eggs (2)	200
	Glass of milk	150
	Orange juice	100
	Black coffee (with milk, add 25)	5
Lunch and dinner items	Sandwich*	400
	Fish (fried, add 150)*	200
	Medium hamburger (fast food, add 150)*	350
	Steak*	350
	Chicken breast (fried, add 100)	150
	Pizza, 1 slice (each meat topping, add 50)*	250
	Pasta with tomato sauce	250
	Soup*	150
	Potato, baked or mashed (with margarine, add 50)	150
	Mixed vegetables (with margarine, add 50)	50
	Garden salad (with dressing, add 40)*	100
	French fries (fast food, add 150)	200
	Average soft drink (non-"diet")	130
Dessert and snack items	Cake*	300
	Potato or corn chips*	150
	Popcorn*	70
	Apple (or similar) pie slice*	300

Note: The listed Calorie numbers are simplified, rough estimates provided only for the purposes of this activity. They should not be used for dietary planning. The estimates will vary greatly depending on the exact preparation, portion size, and ingredients used.

*The Calories in these items are especially variable depending upon the exact preparation and ingredients. For example, some species of fish are much higher in fat content than others, and a lean steak can easily be half the calories of a fatty one.

suggests that your current eating habits *may* result in weight loss over time. If you obtained a positive number (the expected outcome), proceed to Steps 5 and 6.

5. There is no need for concern if the number of Calories you calculated from your menu is only 10–15% higher than your BMR. Unless a person's lifestyle is very sedentary, it is likely that he or she will burn at least 10% more Calories above his or her BMR. But if the difference between the BMR and Caloric intake from the proposed menu is great, he or she would need to reconsider dietary choices or increase his or her activity level in order to avoid weight gain over time.
6. Walking and most light activities burn Calories at a rate of about 250 Calories per hour for a person of average size. Jogging and more intense physical

Box 17.2 Proposed Menu and Calorie Totals

	Food items	Calories
Breakfast	~~cereal w/milk~~	~~300~~
	coffee	30
	Bagel w/cream	290
	Total	590
Lunch	sandwich	400
	soup	150
	Total	550
Supper	Mix vegetables	50
	chicken breast	150
	Total	200
Snack	Chips	150
	Total	180
	Total for all food items	1490

activities burn approximately 600 Calories per hour or considerably more depending on a person's size and the intensity of the activity performed.

How many hours of walking would you need to do to burn the Calories above your BMR? ________

How many hours (or minutes) of jogging would you need to do? ________

It is important to conclude by mentioning that numerous factors may affect your metabolic rate and Caloric needs; your doctor or other health professional is in the best position to assess your actual dietary needs. Also, Calories are not the only consideration in a proper diet. Other factors, such as vitamins, fiber, and fat content are important too. The purpose of this activity was simply to get you off to a good start in thinking about the Calories you burn versus the Calories you consume. ■

EXERCISE

18

THE URINARY SYSTEM

Objectives

After completing this exercise, you should be able to

1. List, identify, and describe the major organs of the urinary system.
2. List the main parts of the nephron and describe their functions.
3. Explain how urine is formed and the factors that affect its composition.

Materials for Lab Preparation

Equipment and Supplies

- ❑ Dissection equipment
- ❑ Fetal pig
- ❑ Disposable gloves
- ❑ Urine specimen containers
- ❑ Clinistix or other suitable testing strips that test urine pH and specific gravity
- ❑ Drinking cups (8-ounce and 12-ounce)
- ❑ Salted pretzels
- ❑ Distilled water
- ❑ Baking soda (sodium bicarbonate)
- ❑ Caffeinated coffee, tea, or diet soda
- ❑ Cleaning disinfectant or detergent solution

Models

- ❑ Human torso
- ❑ Nephron

Introduction

The digestive tract absorbs nutrients, as well as substances that we cannot metabolize. Also, metabolism produces waste molecules when we use nutrients for energy or other cellular functions. Urea, a nitrogen-containing waste, is an example of a waste product of amino acid metabolism. These wastes must be removed from the blood so that they do not build up to toxic levels. The kidneys and the urinary system not only provide this crucial service of waste removal, but they also contribute significantly to body fluid homeostasis. In this exercise, we will first examine the organs of the urinary system, and then we will focus on the process of urine formation.

Organs of the Urinary System

The Kidneys

The kidneys are bean-shaped organs about the size of your fist, and they are located near the posterior abdominal wall (Figure 18.1a). Like some other organs we have encountered in the study of the human body, the kidneys have an outer **cortex** and an inner **medulla.** The cortex is where most of the blood is filtered and enters the microscopic kidney tubule system, called the **nephron.** The medulla mainly contains tubules and ducts involved in urine formation and concentration (Figure 18.1b). The tiny collecting ducts of the medulla (Figure 18.1c) eventually dump the urine they produce into the **renal pelvis,** which collects and funnels urine into the ureter (Figure 18.1b).

The **ureters** from each of the two kidneys lead down to the **urinary bladder,** a saclike organ that stores urine. Both the ureters and urinary bladder have smooth muscle in their walls. In the ureters, the muscle produces wavelike contractions that push urine toward the bladder. In the urinary bladder, the smooth muscle contracts as part of a reflex, which helps to push urine through the **urethra** to the outside of the body (Figure 18.1a).

As you can see in Figure 18.2, the urethra is longer in males because it passes through the penis. In this figure, you can also see the two sphincter muscles that control urination. The **internal urethral sphincter,** which is near the base of the bladder, is involuntary. The **external urethral sphincter,** which is 1 to 2 inches farther down the urethra, is voluntary.

ACTIVITY 1

Observing the Organs of the Urinary System

Materials for this Activity

Fetal pig
Dissection equipment
Disposable gloves
Human torso models

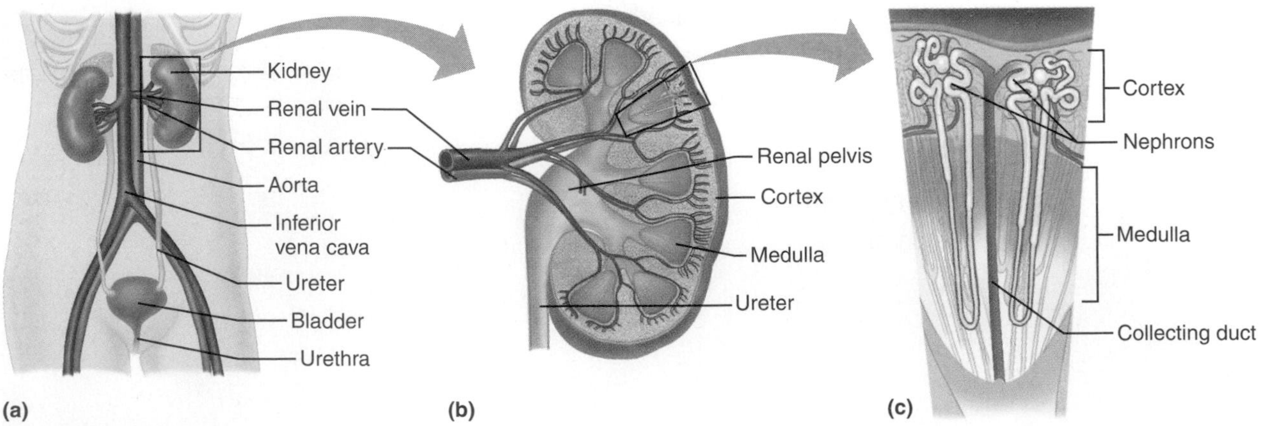

FIGURE 18.1 The human urinary system. (a) Structures of the urinary system within the body. (b) Internal structure of the kidney. (c) Nephrons within the cortex and medulla in relation to the collecting duct, through which urine produced by the nephron passes.

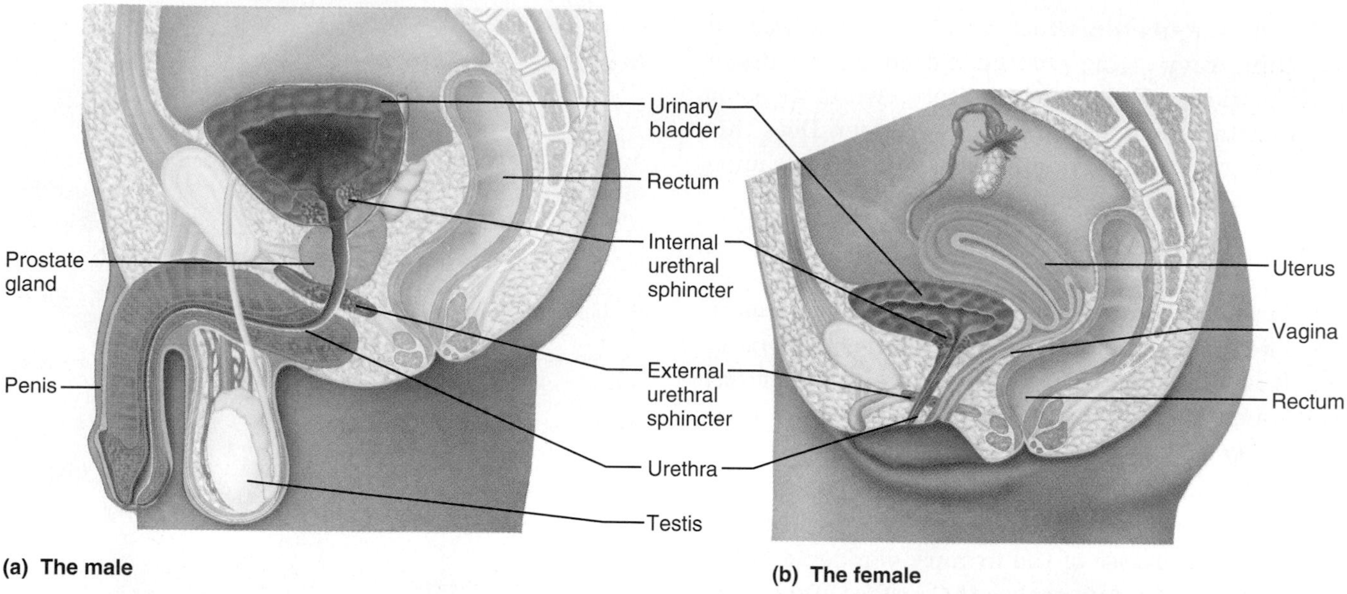

(a) The male

(b) The female

FIGURE 18.2 The bladder, the urethra, and associated organs in the human male and female. (a) The urethra is longer in men than in women. (b) Women generally have a smaller bladder capacity because their bladders are compressed by the uterus.

Although your fetal pigs should be open from previous exercises, you may need to cut farther down in the pelvic cavity to view the urethra. See Figure 18.3 for views of the male and female urinary systems of the fetal pig. Review the safety information provided by your instructor and the precautionary information for performing dissections presented in Exercise 13.

1. Move the digestive organs over to one side of the pig's abdominopelvic cavity, and look for the bean-shaped **kidney** located near the posterior abdominal wall. You may need to break and remove the plasticlike connective tissue that is covering the kidney to see it clearly. Repeat this process to find the kidney on the other side. Speculate as to why animals have two kidneys, as well as duplicates of other important organs as well.

____

____

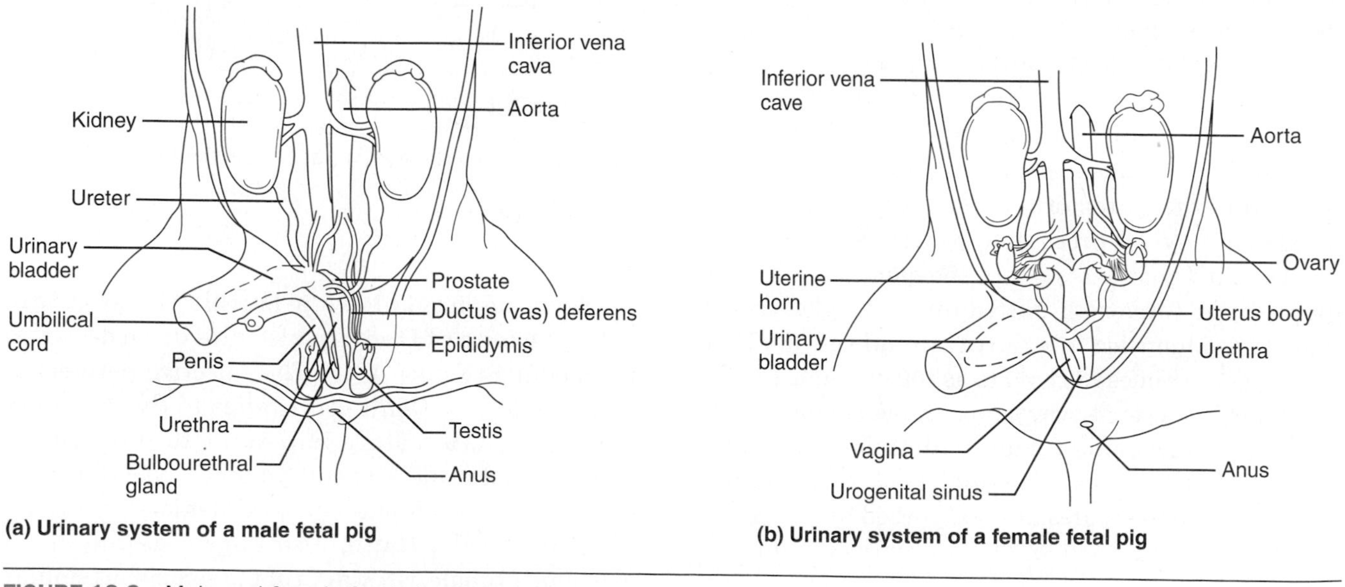

(a) Urinary system of a male fetal pig

(b) Urinary system of a female fetal pig

FIGURE 18.3 Male and female fetal pig urinary systems.

2. Find the **ureter,** which will look like a piece of thin, wavy string emerging from the middle of the kidney. Follow each ureter down to its point of attachment to the urinary bladder. Does this tube use simple gravity feed, or does it use muscle to push urine into the bladder?

   gravity
   ____

3. Find the **urinary bladder,** which looks more like a long, muscular tube than a saclike organ. It is located between the two umbilical arteries, which helps to make the urinary bladder relatively easy to find.
4. Use a scalpel to cut through the ventral body wall, slightly off center. Find the start of the urethra at the base of the urinary bladder, and follow it to where it reaches the surface of the skin. The urethra is the portion of the urinary system that is most different in males and females.
5. Using Figures 18.1 and 18.2 as references, find the following structures on the human torso models:
   - Kidney
   - Renal cortex
   - Renal medulla
   - Renal pelvis
   - Ureter
   - Urinary bladder
   - Urethra ■

The Nephron and Urine Formation

A kidney typically contains about 1 million nephrons, which are the microscopic tubule systems that do the work of the kidney. A knot of leaky capillaries called the **glomerulus** is surrounded by the nephron's **glomerular capsule,** which is also known as Bowman's capsule (Figure 18.4). The combination of the glomerulus and glomerular capsule is found in the cortex of the kidney, which is where blood is filtered for processing by the nephron.

The fluid that leaks out of the glomerulus, called **glomerular filtrate,** is picked up by the glomerular capsule and funneled into the **proximal tubule.** This tubule is sometimes referred to as the proximal *convoluted* tubule due to its wavy appearance (Figure 18.4). This is a hardworking tubule, with cells loaded with mitochondria to provide energy for active transport. Most valuable molecules are reabsorbed by the proximal tubule, and most of the water lost from the glomerulus is reabsorbed by the tubule as well.

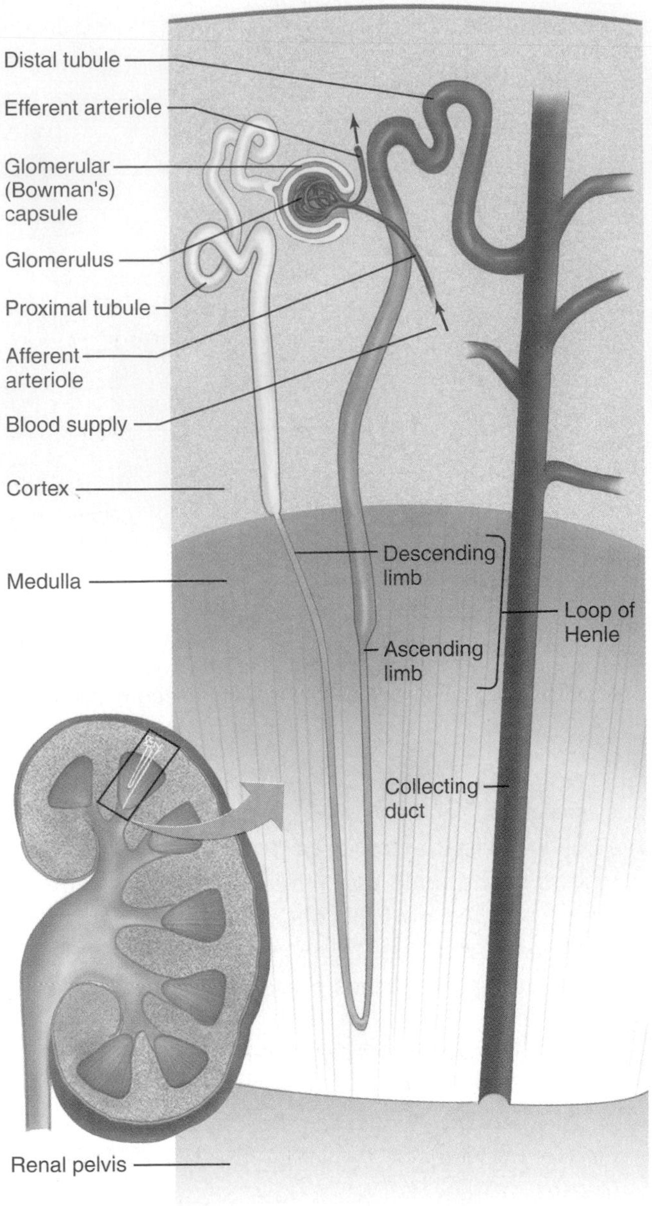

FIGURE 18.4 The tubular structure of a nephron. In this view, a portion of the glomerular capsule is cut away to expose the blood vessels that serve the nephron.

Some nephrons have a hairpinlike structure called the **loop of Henle,** which dips down deep into the medulla (Figure 18.4). The selective permeability of the loop of Henle contributes to the medulla's salt gradient: a feature important to producing a concentrated urine. Consult your textbook for more details on the work of the loop of Henle.

The **distal tubule,** also known as the distal *convoluted* tubule, is a wavy tubule somewhat similar in appearance to the proximal tubule (Figure 18.4). Like the proximal tubule, this tubule reabsorbs some important molecules. It also *secretes* molecules into the urine.

The **collecting duct,** or collecting tubule as it is sometimes called close to its point of attachment to the distal tubule, is the place where regulation of water reabsorption occurs (Figure 18.4). Collecting ducts are sensitive to a hormone, **antidiuretic hormone (ADH),** which was first discussed in Exercise 13, The Endocrine System. Without ADH, water remains trapped inside the collecting duct and is eventually delivered to the bladder. This action results in the production of a high volume of dilute urine. When ADH is present, water easily passes out of the collecting duct into the kidney tissue of the medulla, where it is picked up by capillaries. Thus, much less water is dumped into the pelvis for eventual delivery to the bladder. This action results in the production of a lower volume of concentrated urine.

ACTIVITY 2

Observing the Tubules of the Nephron

Materials for this Activity

Model of the nephron

Using Figure 18.4 as a reference, locate the following structures on a model of the nephron.

- Glomerulus
- Glomerular capsule
- Proximal tubule
- Loop of Henle
- Distal tubule
- Collecting duct

Which tubules seem to have a wavy appearance?

____ ■

ACTIVITY 3

Conducting a Urine Formation Experiment

Materials for this Activity

Urine specimen containers

Clinistix or other suitable testing strips that test urine pH and specific gravity

Disposable gloves

Drinking cups

Salted pretzels

Distilled water

Baking soda (sodium bicarbonate)

Caffeinated coffee, tea, or diet soda

Disinfectant solution

In this experiment, you will experience firsthand the process of urine formation and the factors that may affect urine formation. Substances that we eat or drink can affect the urine we produce as we dump wastes and excess materials into the urine. Similarly, *not* drinking water has a great impact on urine formation. We lose a little water every time we exhale, and if the water is not replaced by drinking, our body fluids become hypertonic (more concentrated).

Consuming salt is a fast way to mimic the effect of not drinking water for several hours because it quickly makes our body fluids hypertonic, just as water loss without replacement would do. This triggers ADH release and our thirst drive, mechanisms for maintaining homeostasis. Drinking a large amount of distilled water makes our body fluids hypotonic (more dilute), which inhibits ADH release. Sodium bicarbonate (baking soda) molecules, like many drugs and most other water-soluble substances, are simply dumped into the urine as a waste product. Finally, caffeine increases blood pressure in the glomerulus, increasing glomerular filtration rate. The following experiment will test the response of the urinary system to each one of these factors.

For this experiment, it will be necessary to test your urine after consuming one of four designated items. When handling urine or any body fluids, it is imperative that proper precautionary and sanitary procedures are followed. Urine samples should be disposed of in a restroom toilet. Any containers or items that have come in contact with urine should be properly disposed of or soaked for 1 hour in disinfectant solution and washed before re-use. Your instructor can advise you of the proper way to handle, test, and dispose of your samples. The single most important precautionary step is to handle *only* your own sample. Wear gloves, and use disinfectant solution to wipe down all work areas on which you have handled urine.

Note: For best results, it is important that you do not consume anything other than a small quantity of water for 2–3 hours before you begin this experiment. If you have recently consumed anything that could interfere with this experiment (such as salty foods or caffeinated beverages), you should empty your bladder and wait 1 hour before starting this experiment.

1. Work in groups of four, and decide which one of the four items each person in the group will consume. It is *not* advisable for a person with high blood pressure or other cardiovascular or urinary system disorders to consume either the caffeinated beverage or pretzels.
2. Consume *one* of the following *within 5 minutes* time:
   - One serving of pretzels (see the food label for serving size) and no more than 3–4 ounces of tap water
   - Three 8-ounce cups of distilled water
   - Two cups of distilled water with a tablespoon of baking soda mixed in each
   - Twelve ounces of caffeinated coffee, tea, or diet soda
3. Obtain a specimen container, and obtain a sample of your urine immediately after consuming the item in Step 2. Test this sample for pH and specific gravity, and enter that information as your baseline data in Box 18.1. Obtain and record the information for the other three test items in Box 18.1 as well.
4. Half an hour after the start of the experiment, attempt to obtain another sample. Test your sample, this time recording the volume as well. If you produce more urine than the sample container can hold, do your best to estimate the amount. If you cannot produce a sample at the ½-hour point (which is likely for at least one of the test items), write "no data" in the ½-hour column boxes and wait for the 1-hour interval.
5. One hour after the start of the experiment, obtain another sample. Test your sample for specific gravity, pH, and volume. Again, if you produce more urine than the sample container can hold, do your best to estimate the amount. Repeat at 1½ and 2 hours from start time and enter your data in Box 18.1.
6. Evaluate your data. In the following space, describe the changes to specific gravity and pH, and calculate the volume generated per hour by adding the volumes recorded during the experiment.

____

____

____

____

____

Which of the substances caused an increase in volume?

____

Which of the substances caused an increase in pH?

____

Which of the substances caused the specific gravity to decrease (approach 1.000)?

____

Salted pretzels made the subject's body fluids hypertonic and caused the release of ADH. The resulting more concentrated urine should have a higher specific gravity, and a very low volume of urine should be produced per hour. Did your group's salt/hypertonic fluid data match the expected results?

____

Summarize what was happening in the subject's kidneys during this experiment.

____

____

Distilled water made the subject's body fluids hypotonic and inhibited the release of ADH. The resulting more dilute urine should have a lower specific gravity (approaching 1.000), and a high volume of urine should be produced per hour. Did your group's distilled water/hypotonic fluid data match the expected results?

____

Summarize what was happening in the subject's kidneys during this experiment.

____

____

Bicarbonate ions dumped into the urine as a waste product tend to absorb hydrogen ions and raise pH above 7.0. Thus, a pH that increased at least one point above baseline would be a positive test for bicarbonate ions in the urine. Was your group's bicarbonate ion test positive?

____

Box 18.1 Urine Test Data

Test Item	Parameter	Baseline	½ Hour	1 Hour	1½ Hours	2 Hours	
Pretzels	Specific Gravity						
	pH						
	Volume	X					Total Vol.
Distilled Water	Specific Gravity						
	pH						
	Volume	X					Total Vol.
~~Baking Soda~~ OTHER	Specific Gravity						
	pH						
	Volume	X					Total Vol.
Caffeine	Specific Gravity						
	pH						
	Volume	X					Total Vol.

Would water-soluble drugs also pass into the urine, and could they be detected by an appropriate test?

____

Caffeine increased glomerular filtration rate and overloaded the nephrons' mechanisms to reabsorb substances, including water. Considering it is not just water reabsorption that was affected, urine output should have been relatively high; but you may find little change in specific gravity or pH. Did your group's caffeine data match the expected results?

____

Compare your results to the expected results and explain.

______________________________ ■

EXERCISE

19

THE REPRODUCTIVE SYSTEM

Objectives

After completing this exercise, you should be able to

1. Identify components of the female reproductive system.
2. Trace the route of an egg from ovary to vaginal opening.
3. Identify components of the male reproductive system.
4. Trace the route of a sperm from the testis to the urethral opening.
5. Compare the reproductive system of the fetal pig to that of a human.
6. Explain the role of condoms in preventing sexually transmitted infections.

Materials for Lab Preparation

Equipment and Supplies for Dissection

- ❑ Fetal pigs (male and female if possible)
- ❑ Dissection equipment
- ❑ Dissection tray
- ❑ Bone cutters
- ❑ Disposable gloves
- ❑ Autoclave bag for biologically hazardous material

Equipment and Supplies for Condom Experiment

- ❑ Latex condoms
- ❑ Natural membrane condoms (e.g., lambskin)
- ❑ Threads
- ❑ 10-ml or 25-ml graduated cylinders
- ❑ Dye solution in squeeze bottles (Dye solution: clothing dye dissolved in warm water so as to be sufficient to show as colored water)
- ❑ 200-ml beaker
- ❑ Paper towels

Models

- ❑ Human female pelvis (full, half, or dissectible)
- ❑ Human female reproductive organs
- ❑ Human pregnancy series (if available)
- ❑ Human male pelvis (full, half, or dissectible)
- ❑ Human male reproductive organs

Introduction

The human reproductive system consists of the tissues and organs required to create a new human being. Both male and female reproductive systems have specific functions that play complementary roles to achieve the goal of reproduction. The male reproductive role is to produce sperm and deliver them to the female reproductive tract. The female reproductive role, initially, is to produce unfertilized eggs. The sperm and egg may combine in the female reproductive tract to produce a fertilized egg. If fertilization occurs, the female uterus then provides an environment in which the embryo, later called a fetus, can safely develop.

Female Reproductive System

The reproductive structures of the female may be considered in terms of both *external* and *internal* organs. In this exercise we will be concerned principally with the three monthly cycles of the female reproductive system: the egg cycle, the ovarian cycle, and the uterine or menstrual cycle, and the internal structures that contribute directly to human reproduction.

The Egg Cycle

Eggs, or female sex cells, are produced in a process called **oogenesis.** This process is started in the first trimester of pregnancy, and by birth, the female baby has approximately one million egg cells, called **oocytes,** stored in the immature ovary. Here they remain in a state of suspended animation, which can last from 10 to 14 years. At puberty, approximately 400 thousand oocytes remain in the ovary.

Each month, one oocyte will be selected for **ovulation.** If more than one oocyte is ovulated, multiple births are possible. Fraternal or non-identical twins result from different oocytes being fertilized by different sperm. Identical twins result from a single oocyte being fertilized by a single sperm, followed by separation of the fertilized egg's daughter cells during development. During a woman's lifetime, fewer than 500 oocytes will be ovulated.

Speculate on why the female reproductive system contains so many more oocytes than will ever be released to begin the development process during a woman's lifetime.

___

___

___

Ovarian Cycle

The ovary has two functions: to store the oocytes, and to produce female sex hormones. In the ovary, a series of changes occurs over the course of a month, which is directly linked to the maturation of the oocyte. The main function of the ovarian cycle is to produce two sex hormones, **estrogen** and **progesterone.**

There are three phases in the monthly ovarian cycle (Figure 19.1a). The **follicular phase** takes about 14 days. During this time, the **follicles,** or ovarian cells around the developing oocyte, grow larger, mature, and begin to produce estrogen, the female sex hormone. The **ovulatory phase** is triggered by two hormones, called **luteinizing hormone (LH)** and **follicle stimulating hormone (FSH),** which is released from the pituitary, a small organ located in the brain. During **ovulation,** the ovary wall ruptures, and the oocyte is expelled into the peritoneal cavity.

What is the relationship between the pituitary, which is located in the brain, and the ovaries, which are located in the pelvic area?

___

___

Finally, in the **luteal phase,** which takes about 10 days, the ovarian cells around the now-expelled oocyte, called the **corpus luteum,** begin to secrete two hormones called **estrogen** and **progesterone.** If pregnancy does not occur, these ovarian cells begin to degenerate, and the hormonal output ends; if pregnancy does occur, these ovarian cells will continue producing hormones for about three more months.

Estrogen is the primary female sex hormone. It prepares the uterus for pregnancy; it causes girls to grow faster than boys during early adolescence; and it enhances secondary sex characteristics for females, such as larger breasts and hips (more fat), softer skin (increased skin hydration), wider hips (pelvic changes in preparation for pregnancy and delivery), and growth of axillary and pubic hair. **Progesterone** regulates the uterine cycle, reduces uterine movements during pregnancy, and prepares the breast for milk production.

Compared to boys, why do girls generally have wider hips and larger breasts?

___

___

___

Uterine or Menstrual Cycle

In the uterus, a monthly series of events occurs, which is also linked to the maturation of the egg. The function of the uterine or menstrual cycle is to provide safe housing and nutrition for the developing fetus.

There are also three phases to the monthly uterine cycle (Figure 19.1b). The **menstrual phase** consists of the sloughing off of the uterine lining, also called the **endometrium.** This process is accompanied by bleeding for three to five days and is referred to as **menstruation.** The **menstrual period** is the

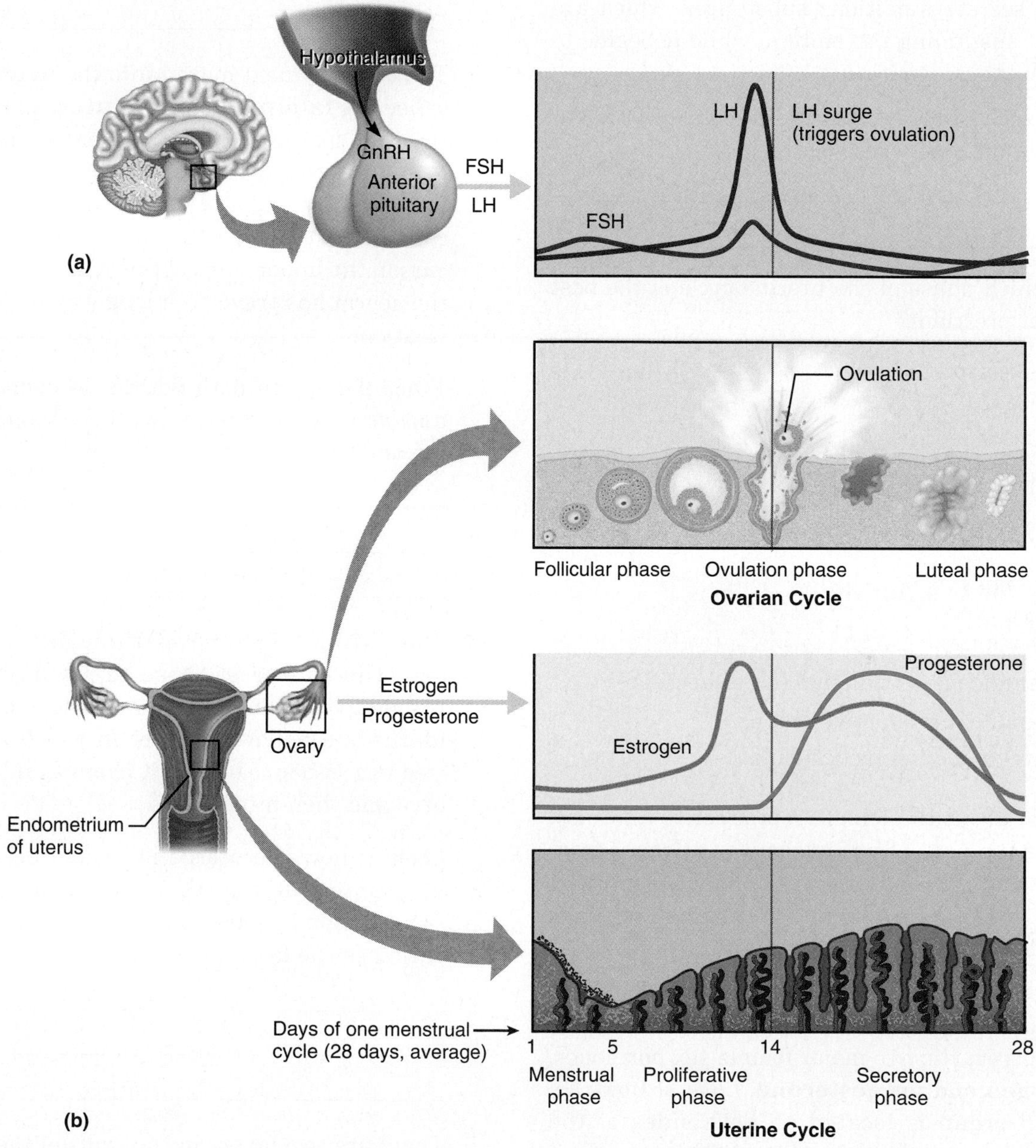

FIGURE 19.1 Ovarian and uterine cycles. (a) The ovarian cycle is the monthly changes in the ovary that are linked to oocyte maturation. Ovulation is triggered by surges of FSH and LH. (b) The uterine cycle is the monthly changes in the uterine lining, which responds to rising levels in estrogen and progesterone.

period of visible menstrual flow, which is a discharge consisting of shed tisues and blood flowing through the vagina, and out the vaginal opening. During the **proliferative phase,** which takes approximately 9 days and is triggered by rising levels of **estrogen,** the endometrium is rebuilt and blood vessels again proliferate. In the **secretory phase,** which takes approximately 14 days and is stimulated by **progesterone,** the endometrium becomes even thicker, and begins to secrete nutritious substances, which are capable of sustaining the embryo while it begins to implant in the endometrium.

During which phase of the uterine cyles is the endometrium the thinnest? The thickest?

____

____

During which phase of the uterine cycle is the best time to get pregnant?

____

ACTIVITY 1

Female Reproductive Anatomy

Materials for this Activity

Models

Human female pelvis (full, half, or dissectible)
Human female reproductive organs
Human pregnancy series (if available)

Let's learn about the female reproductive anatomy, by tracing the path of an egg, as it is released from the ovary and travels through the female reproductive tract. Use Figure 19.2 and Figure 19.3 together with the female reproductive models in your lab.

1. The **ovaries,** which are the primary female reproductive organs, produce eggs, called **oocytes,** and secrete the two major female sex hormones, **estrogen** and **progesterone.** Look at how this paired organ is located on both sides of the **uterus,** and is held in place, within the peritoneal cavity, with several strong ligaments.
2. After ovulation, the egg cell is released from the ovary into finger-like projections over the ovary called **fimbriae.** Look at how the fimbriae are draped over the ovary. The fimbriae contain thousands of cilia, which beat in unison to capture the oocyte and transport it to the next location.

   If the oocyte is not caught by the fimbriae, what are the chances for the oocyte to get to the oviduct?

   ____

   ____

3. The oocyte is next moved into the **oviduct,** also called the **fallopian tube,** or **uterine tube.** Notice that sperm is already present in the oviduct in Figure 19.3. **Fertilization** is the successful union of the egg and sperm, and the result is called a **fertilized egg.** Ideally, fertilization occurs in the upper third of the oviduct. Remember the sperm has traveled a long way to get to this location.

   Trace the sperm path within the female reproductive system, starting with the vaginal opening and ending with the oviduct.

   ____

   ____

   ____

4. The fertilized egg moves through the oviduct, over the course of three to six days, to the **uterus,** commonly known as the **womb.** The uterus is a hollow chamber in which the fertilized egg is nourished as it changes into an embryo and then into a fetus.

   Look at how the uterus is located between the large intestine and the urinary bladder. Give some reasons for frequent urination and constipation as the fetus grows bigger.

   ____

   ____

   ____

5. There are two layers in the walls of the **uterus.** The **endometrium,** the inner layer, is composed of epithelial and connective tissues. It is rich and thick in nutrients and will eventually become

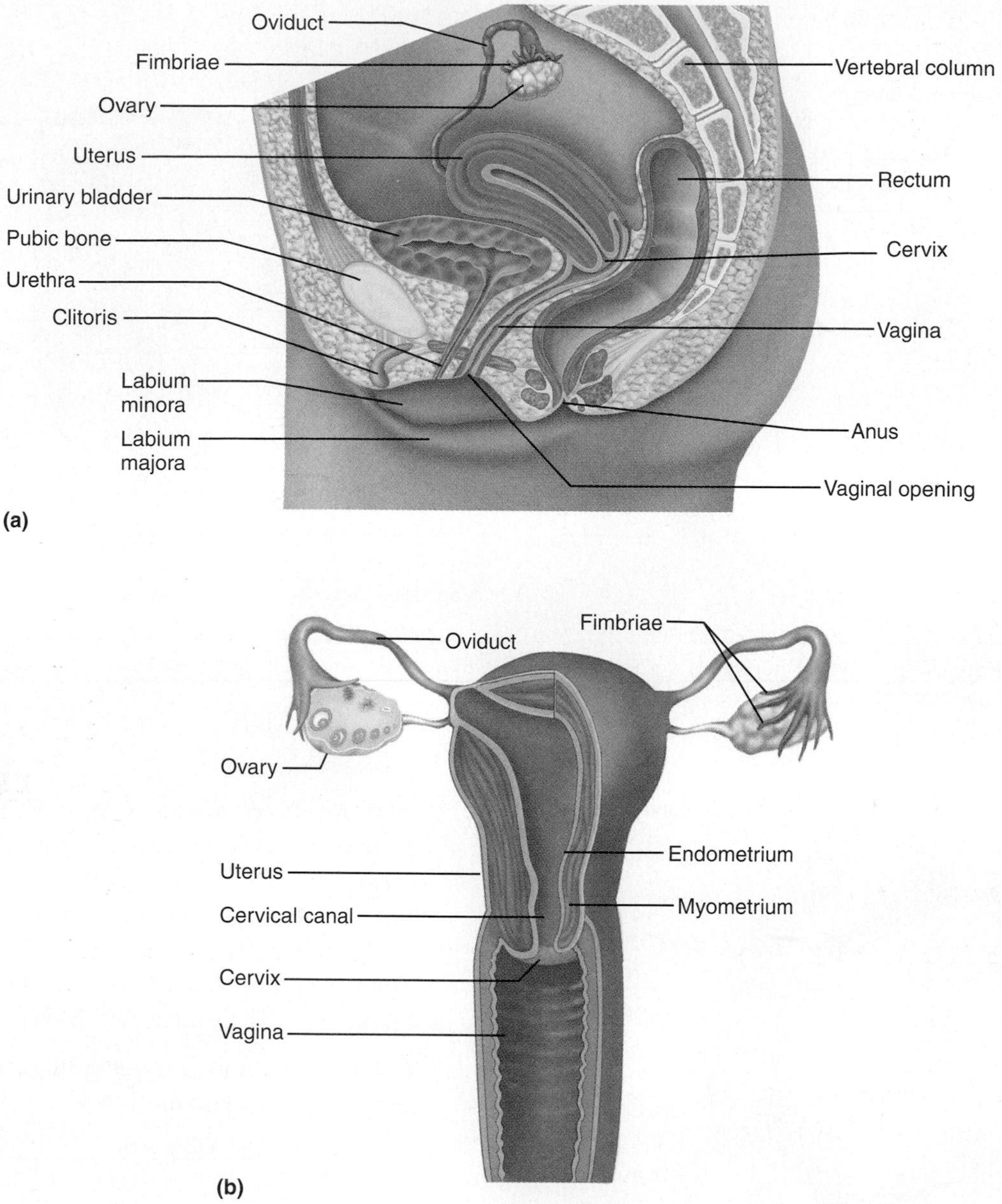

FIGURE 19.2 Female reproductive organs. (a) Midsagittal section showing the internal and external reproductive organs. (b) Anterior view of the internal reproductive organs.

the **placenta,** which provides shelter, nourishment, and waste material removal, for the growing fetus. The **myometrium** is the outer layer, and it is composed of thick layers of smooth muscles, called the **uterine muscles.**

If the embryo implants in the oviduct, this is called an ectopic pregnancy. Compare the structure of the uterus and the oviduct. Explain why an ectopic pregnancy is more likely to spontaneously abort than a regular pregnancy.

____

____

____

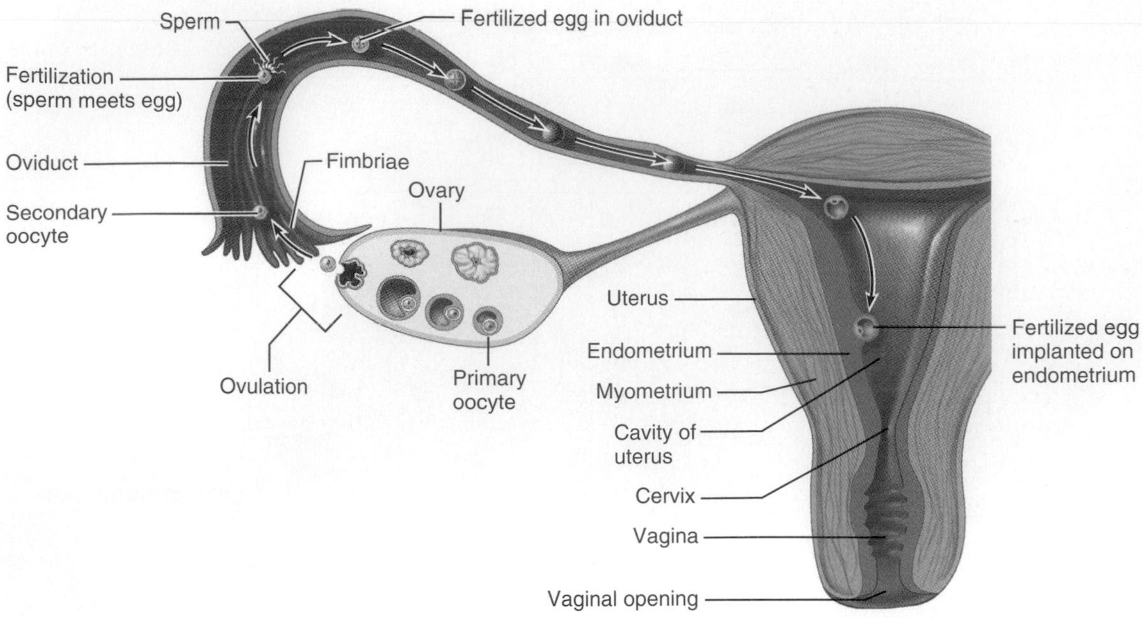

FIGURE 19.3 Path of an oocyte. The path of an oocyte begins with ovulation from the ovary and travels from the oviduct to the uterus.

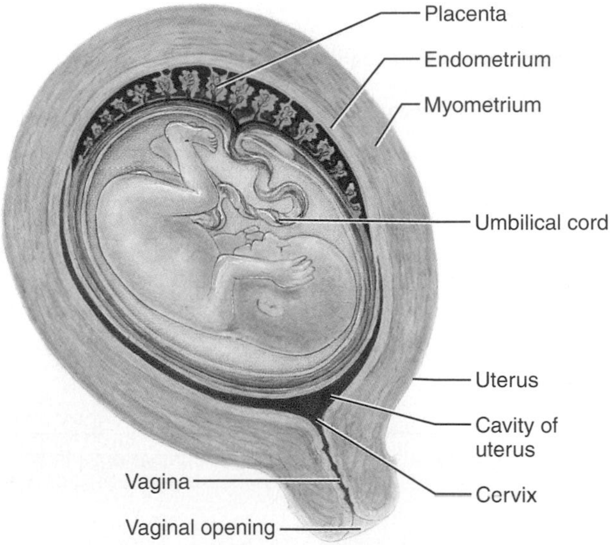

FIGURE 19.4 A full-term fetus inside the uterus.

6. During **labor,** the uterine muscles provide powerful uterine contractions to first expel the fetus thru the **cervix, vagina,** and out of the **vaginal opening.** Later, the contractions continue in order to detach the **placenta** from the uterus. To prevent post-partum bleeding and infection, the placenta and attached fetal membranes, commonly known as the **afterbirth,** must be completely removed. Compare the relatively empty uterus in Figure 19.3 and the almost full-term fetus in the uterus in Figure 19.4. Notice the extent of the "stretch" of the uterus, and how completely protected the fetus is inside the uterus. There is a limit to this stretch, of course, and at a certain size, the uterus will be torn and damaged.

   If a small woman carries a large baby, what are the dangers to the mother?

   ____

   ____

   ______________________________ ■

Male Reproductive System

The male reproductive system produces sperm, which must travel through the female reproductive system to potentially fertilize the egg released by the female ovaries. **Testosterone,** the male sex hormone, is produced in the testes (singular *testis*). This hormone produces secondary sexual characteristics such as facial and body hair and a deeper voice. In the context of reproduction, testosterone stimulates sperm formation by stimulating growth of the male reproductive structures—ducts, glands, and the penis.

ACTIVITY 2

Male Reproductive Anatomy

Materials

Models

Human male pelvis (full, half, or dissectible)
Human male reproductive organs

Let's learn about the male reproductive anatomy by tracing the path of a sperm as it is released from the testes and travels through the male reproductive tract. Use Figure 19.5 together with the male reproductive models for this lab.

1. The **testes** (singular *testis*) lie in the **scrotum,** a sac of skin and other membranes located outside the abdominopelvic cavity. The external location is both good and bad. The advantage is a cooler temperature, which is ideal for sperm production, or **spermatogenesis;** the disadvantage is the exposed location.
2. The testes are comprised of the **seminiferous tubules** where the actual sperm formation takes place. These tubules are tightly packed, and they total about 100 meters or 100 yards in length. Here, **spermatogonia** will undergo a series of cell divisions finally becoming **immature sperm.**

   What is the advantage of such long lengths in the seminiferous tubules? *Hint:* greater surface area.

   ____

   ____

3. The seminiferous tubules meet to form a single coiled 4-cm (1.5-in) duct called the **epididymis** located just outside the testis but still inside the scrotum. Sperm maturation begins in the epididymis.

   What is the disadvantage in storing the sperm in the seminiferous tubules?

   ____

   ____

4. The **vas deferens,** also known as the ductus deferens, is quite long, measuring about 45 cm (18 in). In the vas deferens, the sperm travels from the epididymis, which is outside the body, and enters the pelvic cavity in the interior of the body. The vas deferens ends in a widened area called the **ampulla** where the sperm may be temporarily stored.

   What is the advantage in storing sperm in the ampulla?

   ____

   ____

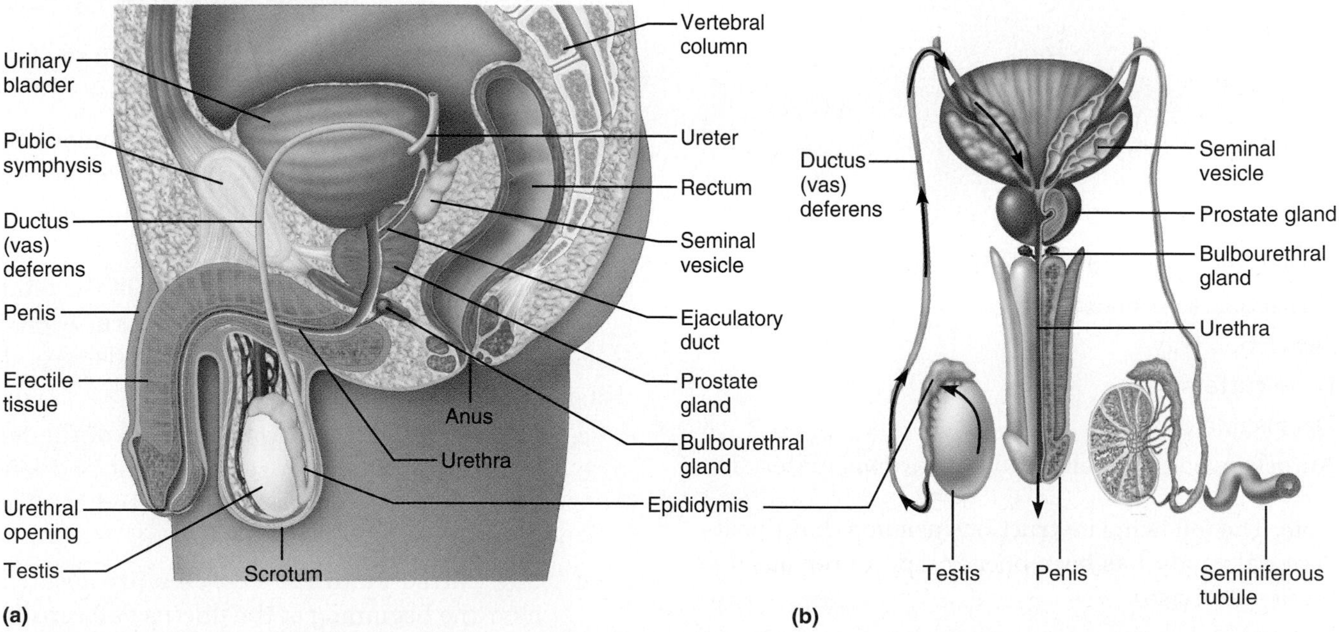

FIGURE 19.5 Male reproductive organs. (a) Midsagittal section. (b) Anterior view.

5. When a male is sexually stimulated and ejaculates, there are smooth muscles in the walls of the epididymis and vas deferens that contract expelling the mature sperm into the **ejaculatory duct.**
6. **Semen** is sperm combined with gland secretions from the **seminal vesicle** and **prostate gland.** The **bulbourethral glands,** also known as **Cowper's glands,** produce a mucous that neutralizes the acidic urine in the urethra and provides lubrication during sexual intercourse.
7. During a sexual climax, rhythmic muscle contractions will propel the sperm through the 20-cm (about 7-in) long **urethra.** Most of the urethra is located in the **penis,** which is the male organ of sexual intercourse. Its reproductive function is to deliver the sperm internally to the female as close to the cervix as possible.

   Notice that the urethra is used for both urine and semen movements. Name one disadvantage of this "common duct."

   ______________________________________ ■

Fetal Pig Reproductive System

We will dissect the male and female fetal pig in order to identify the major fetal pig reproductive structures and to compare the major differences between the fetal pig and human reproductive systems.

ACTIVITY 3

Dissecting a Fetal Pig

Materials

Fetal pigs (male and female if possible)
Dissection equipment
Dissection tray
Bone cutters
Disposable gloves
Autoclave bag for biologically hazardous material

Note: The following instructions assume that the abdominal cavity has been opened up in previous dissection exercises.

1. Obtain the listed supplies, and return to your lab area. For a few moments, study Figure 19.6a if you have a female pig or Figure 19.6b if you have a male pig. The focus of this dissection is the reproductive structures in the pelvic portion, but there are other tubular structures in the area (e.g., the vena cava, colon, urinary bladder, and umbilical vein). Review these ducts so you can differentiate between these ducts and those of the reproductive system.
2. If you are examining a female fetal pig, perform the following steps:
   a. Identify the **uterus,** which is Y-shaped with two horns. Such a design in the uterus allows the mother pig to produce litters.
   b. Below the kidney, identify the **uterine tube,** a funnel-shaped tube that curves around the **ovary,** which appears to be a round mass.
   c. Cut a median line through the pubic symphysis. (You may need bone cutters.)
   d. Follow the uterus towards the bottom of the fetal pig until you identify the **vagina.** Note that the urethra, which drains the bladder, and the vagina enter a common chamber.

How does this fetal pig structure compare with the female human structure?

____

____

____

If you are using a male fetal pig, perform the following steps:

a. Identify the urethral orifice just under the umbilical cord. Carefully cut through the skin of the orifice to identify the **penis,** which lies within a fold of skin.
b. Follow the penis toward the back of the fetal pig where it joins with the **urethra.** A pair of glands, the bulbourethral glands, are at this intersection.
c. Follow the urethra toward the top of the fetal pig, until you can see the right and left **ductus deferens,** encapsulated in the **spermatic cord.**
d. The paired **seminal vesicles** are located near the beginning of the ductus deferens.
e. The **prostate gland** is located between the bladder and **urethra,** but it is difficult to find.

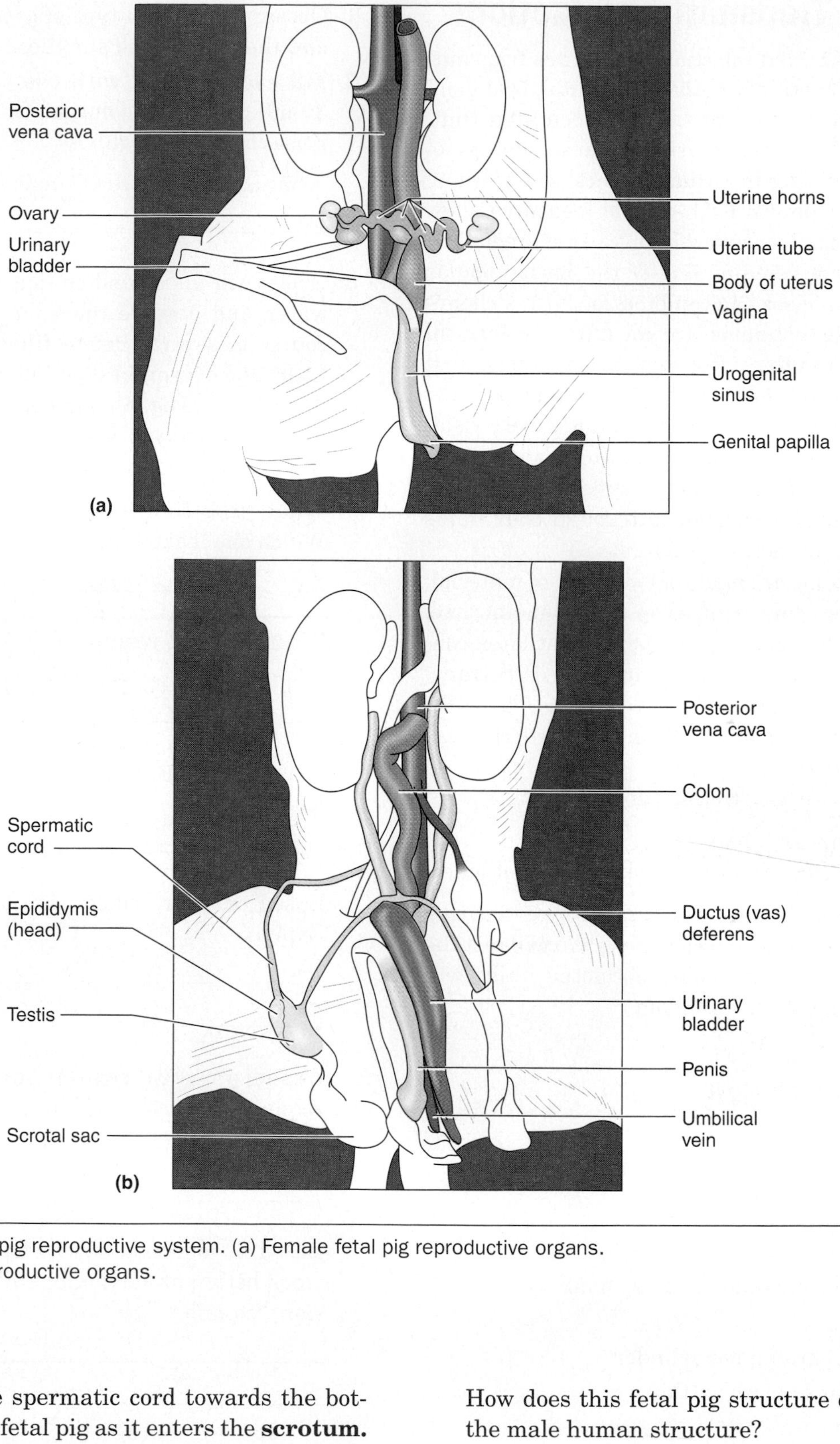

FIGURE 19.6 Fetal pig reproductive system. (a) Female fetal pig reproductive organs. (b) Male fetal pig reproductive organs.

f. Follow the spermatic cord towards the bottom of the fetal pig as it enters the **scrotum.**

g. Carefully cut through the scrotal sac and identify the **testis,** a round structure, and the **epididymis,** a fine tube running alongside the surface of the testis.

How does this fetal pig structure compare with the male human structure?

____

____________________________________ ■

Sexually Transmitted Infections

Sexually transmitted infections (STIs) are transmitted by sexual contact, whether by genital, oral-genital, or anal-genital contact. It is speculated that, worldwide, STIs are increasing because the age of people participating in sexual contact is decreasing and because improved birth control measures allow people to be sexually active without fear of pregnancy.

The condom remains one of the most popular methods for "safe sex." In combination with a chemical spermicide, condoms are an effective form of birth control, and they offer some protection against STIs because they form a physical barrier against the transmission of body fluids. Condoms are considered especially effective in preventing transmission of acquired immunodeficiency syndrome (AIDS), which is transmitted primarily through body fluids during sexual contact.

Male condoms are made of two types of material: latex, a synthetic material, or an animal membrane, such as lambskin. It is hypothesized that latex condoms provide more protection against AIDS transmission than natural membrane condoms. The AIDS virus is extremely small, so it can penetrate through very small pores.

Let's test two assumptions about condoms:

1. All condoms are alike.
2. A condom fresh from the package will not leak.

We will test the first assumption by comparing latex to natural membrane condoms. We will test the second assumption by seeing if colored fluids will leak out of a tightly-tied condom.

ACTIVITY 4

Comparing Condoms

Materials

Latex condoms
Natural membrane condoms (e.g., lambskin)
Threads
10-ml or 25-ml graduated cylinders
Dye solution in squeeze bottles
200-ml beaker
Paper towels

1. Obtain one of each type of condom, two beakers, and the other listed supplies.
2. Fill each condom with the same volume (20–30 ml) of the dye solution, tie them tightly, and rinse them thoroughly.

   What is the purpose of rinsing the condoms?

   ____

3. Place each filled condom in a separate beaker of water, and observe them for leakages over the course of several hours. Check them each half hour (and overnight if possible), and indicate the degree of leakage in a general manner (e.g., 1 = a little bit leaky, 2 = leaky, and 3 = very leaky). Record your observations in Box 19.1.

   What were the results within the first hour? Which one leaked first?

   ____

   What were the results after 3 hours? Which one did not leak?

   ____

   What were the results overnight? Which one leaked?

   ____

   Based on your results, are all condoms alike? Explain.

   ____

   ____

   Based on your results, do condoms leak? Explain.

   ____

   ____

   Based on your results, which type of condom provides better protection against AIDS transmission. Explain.

   ____

   ______________________________ ■

Box 19.1 Comparing Condoms

Time	Degree of Leakiness*	Comments
0.5 hr		
1 hr		
1.5 hr		
2 hr		
2.5 hr		
3 hr		
Overnight		

* Degree of Leakiness: 1 = a little bit leaky, 2 = leaky, and 3 = very leaky.

EXERCISE

20

GENETICS

Objectives

After completing this exercise, you should be able to

1. Describe the basic principles of inheritance.

2. Explain the connection between genetics, DNA, and chromosomes.

3. Draw and complete Punnett squares, and calculate simple genetic probabilities.

4. Describe the concept of linkage.

5. Interpret pedigree charts.

Materials for Lab Preparation

- ❑ Meiosis models
- ❑ Two coins
- ❑ Two small boxes
- ❑ Four prepainted tongue-depressor (or Popsicle) sticks
- ❑ Whitefish blastula microscope slide
- ❑ Microscope

Introduction

The study of the transfer of hereditary material from parents to offspring is the science of genetics. For asexual reproduction by mitotic division, the genetic makeup of the offspring would be fairly predictable, considering offspring are typically exact copies of the parent. Conversely, sexual reproduction allows genes from different organisms of the same species to mix, creating far more variety among offspring. However, a new form of cell division that cuts the chromosome number in half was necessary for sexual reproduction to work. **Meiosis** is that form of cell division that produces **gametes,** or sex cells, with one-half the usual number of chromosomes. The two gametes fuse to form a new cell with one full set of chromosomes. In this exercise, we will study meiosis and then the different patterns of inheritance made possible by the fusion of gametes.

Mitosis

Before discussing how sex cells are produced, and how they are related to the study of genetics, it is important to have a basic understanding of how cells divide in the first place. The process of cell division in human cells is called **mitosis.** Because the usual purpose of mitosis is to produce two daughter cells from one mother cell, mitosis usually involves three aspects: (1) copying of the DNA; (2) portioning of the DNA into separate areas where new nuclei will form around it; and (3) splitting the cell in two.

It is important to note that while all cells go through the first two steps some do not go through the third. For example, the megakaryocyte cells in bone marrow that produce platelets do not split and just become large, multinucleated cells. However, this is the exception, not the rule, as most cells undergo mitosis in order to divide.

A cell that is to undergo mitosis will receive some internal or external stimulus during **interphase,** the part of the cell cycle when it is otherwise busily doing its job. The DNA is replicated during the **S-phase** ("synthesis" phase) of interphase, and the cell then begins preparations for mitosis.

Mitosis begins with **prophase,** when the copies of the DNA molecules wrap themselves around proteins and coil up into **chromosomes.** The nuclear membrane disintegrates as the cell prepares to move the chromosomes. A **mitotic spindle** forms for the purpose of attaching to chromosomes and pulling them to opposite sides of the cell (Figure 20.1).

During **metaphase,** spindle fibers pull the chromosomes to the center or **equator** of the cell. **Anaphase** occurs when the spindle fibers snap the connection between double chromosomes or **sister chromatids,** and pull them to opposite sides of the cell. Finally, **telophase** is the reversal of prophase, where the chromosomes uncoil, the nuclear membrane reforms, and the cell *usually* splits in two. The process of cell splitting is called **cytokinesis.**

ACTIVITY 1

Observing the Stages of Mitosis

Materials for this Activity

Whitefish blastula microscope slide
Microscope

View a slide of the whitefish blastula, which goes through the process of mitosis somewhat similarly to human cells. The entire blastula (with hundreds of cells) may look like one cell under low power, so make sure to go up to at least 400× magnification to ensure you are viewing individual cells within the round blastula. Select examples of each phase of mitosis from your slide. Draw an example of each phase in Box 20.1. Refer to Figure 20.1 as necessary. ■

Meiosis

Meiosis makes the process of reproduction and inheritance significantly more complicated than asexual reproduction. Depending upon the number of chromosomes in the cells of the life form and how much variation there is within the species, there can be thousands or millions of genetically different offspring that are possible from each sexual reproduction event. Basically, meiosis starts out similarly to mitosis. After the chromosomes double, the cell prepares to divide up these chromosomes—one-half going to each new cell. The difference is that the cells divide a second time in meiosis, which then cuts the chromosome number in half. See your textbook for more information on mitosis and meiosis.

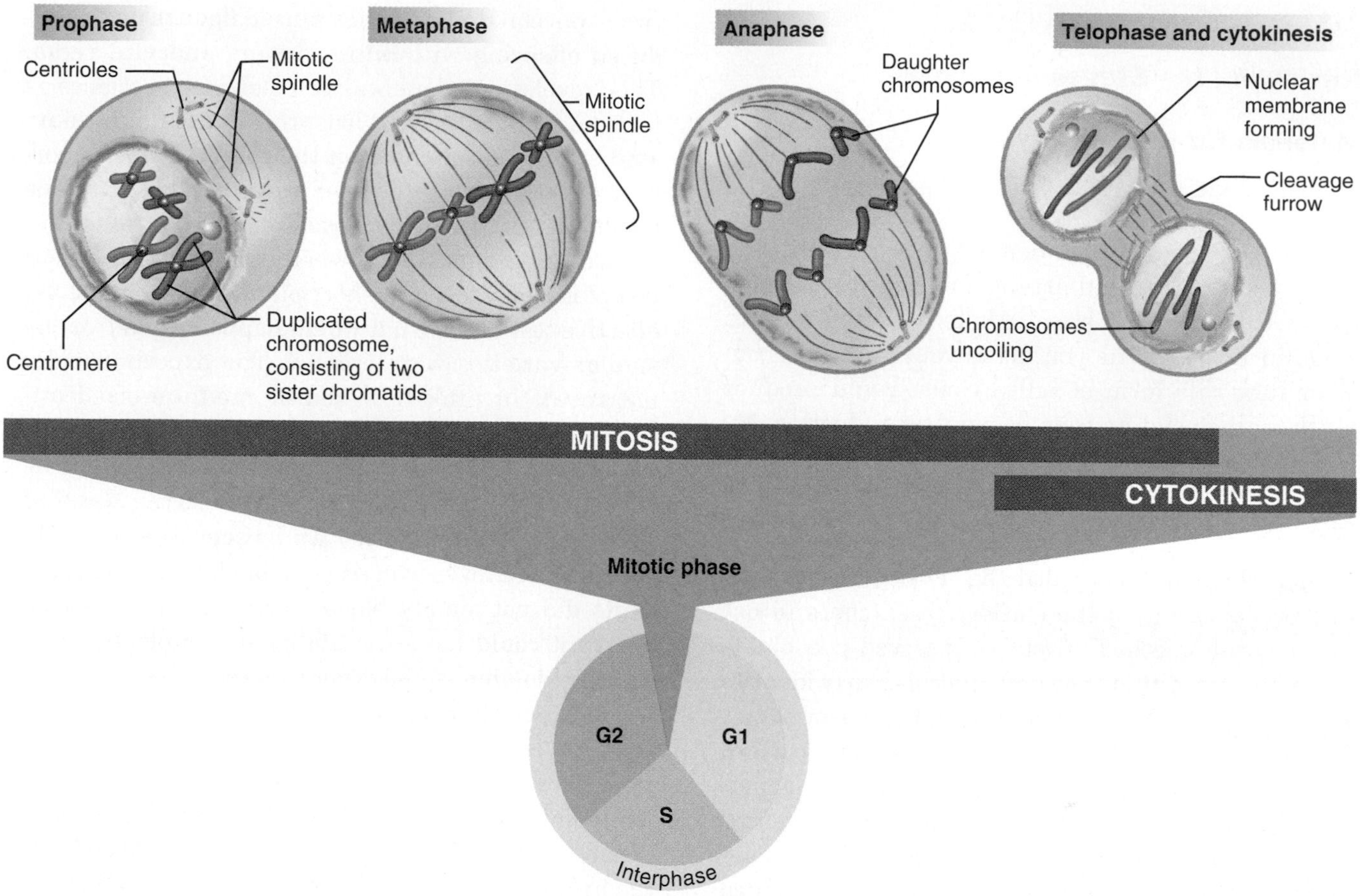

FIGURE 20.1 Mitosis. Mitosis comprises most of the mitotic phase of the cell cycle. The duplicated chromosomes become visible in prophase, line up in metaphase, separate in anaphase, and uncoil and are surrounded by nuclear membranes in telophase.

Box 20.1 Stages of Mitosis

Prophase	Metaphase	Anaphase	Telophase

ACTIVITY 2

Reviewing the Stages of Meiosis

Materials for this Activity

Meiosis models

Review the models that demonstrate meiosis, and follow the number and arrangement of the chromosomes through each stage. Make a sketch of each stage in Box 20.2. As you make your sketches, consider how this form of cell division could produce much more variation than asexual reproduction. ■

Basic Genetics

Gregor Mendel, often called the "father of genetics," was probably one of the earliest researchers to use the scientific method. First, he observed pea plants and determined that they had several clearly identifiable traits (e.g., flower color, pea shape, size of plant). He then wondered if there was a measurable pattern to the inheritance of these traits. He formed a generalization based upon his observations that purebred plants would produce offspring exactly like themselves through numerous generations. This seemed true—purebred plants with purple flowers *never* produced offspring with white flowers and vice versa. This "revelation" alone was of little value to science as farmers already knew at least this much, but his next step is what set Mendel on the road to some important discoveries. He decided to collect observations about a cross between opposite purebred strains.

Crossing a purple-flowered and white-flowered pea plant, or a tall and short plant, was a simple yet effective test to see what would happen when two dissimilar varieties were crossed. The experiment did not result in pink-flowered or medium-sized offspring. All offspring of such purebred crosses had purple flowers and were tall. From this experiment he formed a hypothesis, later referred to as the **"law"** or **principle of dominance and recessiveness.** Apparently, whatever was responsible for causing these traits did not simply blend them together; instead, one trait could have the ability to completely overshadow (dominate) the other (the recessive trait).

Mendel then crossed these offspring among themselves. Once again, no blending occurred (no pink flowers or medium-sized plants), but a few white flowers and a few short plants were produced. Whatever was responsible for causing these traits in the hybrids did not become lost or assimilated in the

Box 20.2 Stages of Meiosis

Prophase I	Metaphase I	Anaphase I	Telophase I

Prophase II	Metaphase II	Anaphase II	Telophase II

dominant trait. He referred to this phenomenon as the **"law"** or **principle of segregation.** Mendel did not know about genes and DNA, so we will now leave his story for another time and move forward into modern genetics.

Genes are the segments of DNA on a chromosome responsible for producing a particular trait such as hair color. However, not all hair color genes are identical. Each variety of a gene for a particular trait is called an **allele.** For example, everyone has hair color genes, but some have blond alleles for that gene, some have brown alleles, and so on.

The **phenotype** is the observable trait expressed, such as blue or brown eyes. The **genotype** describes the alleles present in the offspring. For example, people can have freckles because they have two identical alleles of the freckles gene (FF). Or they may have no freckles because they have two identical alleles of the nonfreckles gene (ff). There is a third possibility: people can have freckles because they have one of each allele (Ff). Because having freckles is dominant, they only need to have *one* freckles allele to display that phenotype. Because we bring two of these alleles together to form a single cell or "zygote," the suffix **zygous** is used to describe the genotype. When describing genotype in words (not letters as in "FF," "Ff," or "ff"), the terms **homozygous** (same alleles) or **heterozygous** (different alleles) are used to describe purebred and mixed alleles respectively. For example, "FF" means homozygous dominant (with freckles); "Ff" means heterozygous dominant (with freckles); and "ff" means homozygous recessive (without freckles).

Using the Punnett Square

A **Punnett square** is a visual demonstration of simple genetic probabilities in chart form. It is a simple graphical device where each of the parent's alleles that could end up in the gametes are placed along the top and side of a box. The box is divided into columns and rows, depending upon the number of possible gametes for each trait to be studied. In each empty box, you simply enter the letters representing the alleles for that intersection as you look up to the top of the column and then to the side of the row. This method can be demonstrated with a Punnett square showing a cross of two people who are "hybrids" (heterozygous) for freckles (Table 20.1). Note that three out of the four boxes have at least one capital "F" and would represent a person with freckles. Only one box has two lowercase "f's" and would represent the recessive, nonfreckled person. This **3:1 ratio** is typical of a cross of two hybrid genotypes.

Table 20.1 ***Punnett Square Demonstrating Two Individuals Who Are Heterozygous for Freckles***

Parent 1 → Parent 2 ↓	F*	f*
F	FF	Ff
f	Ff	ff

*F = freckles; f = no freckles

ACTIVITY 3

Exploring Genetic Probabilities Using the Punnett Square

Materials for this Activity

Two coins

Part 1: Coin-Toss Experiment

Although coins may seem quite different from chromosomes, flipping two coins is somewhat analogous to what happens to chromosomes during meiosis. Just as a person and his or her mate will randomly contribute *one* of their pairs of each chromosome, each of the two coins may randomly land on *either* heads (H) or tails (h). Thus, you can complete the Punnett square for heads and tails in Box 20.3.

Box 20.3 Coin-Toss Punnett Square

Coin 1 → **Coin 2 ↓**	**H**	**h**
H		
h		

Work in pairs, and *carefully* flip two coins simultaneously. (You and your partner each flip a coin at the same time.) Record your data for 20 tries in Box 20.4 and on the laboratory blackboard.

Box 20.4 Coin-Toss Work Sheet

	Your data	Entire class data
Both heads (HH)		
One head, one tail (either Hh or hH)		
Both tails		

Just like the freckles example, one-fourth of the boxes in the coin-toss Punnett square (Box 20.3) would be HH, one-fourth would be hh, and half should be mixed (either Hh or hH). Your data should approximate this distribution as well.

Did your results come within 5% of the expected results (25%–25%–50%)?

____

If not, did the data for the entire class come closer?

____

Explain any potential sources of experimental error.

____

____

Part 2: Hairy Hands Experiment

Having hair on the back of the hands is a dominant trait, compared to the absence of hair on the back of the hands. Thus, a person would have to have both recessive alleles (hh) to have hairless hands, considering even one of the dominant alleles would produce hair on the back of the hands. Fill in the Punnett square presented in Box 20.5, which represents the mating of two people who have hairy hands but are "hybrids" in that they carry the recessive allele for hairless hands. People carrying a recessive allele without displaying the trait are called **carriers.**

Box 20.5 Hairy Hands Punnett Square 1

	H	h
H		
h		

Your Punnett square should match the coin-toss Punnett square (see Box 20.3).

Of the four boxes, how many have at least one capital H? _______________

How many have two lowercase h's? _______________

Is it possible for two people with hairy hands to produce a child without hair on the backs of his or her hands?

____

Would it be possible to produce a child without hair on the backs of his or her hands if one of the parents had two dominant alleles?

____

Prove your answer by completing the Punnett square presented in Box 20.6.

Box 20.6 Hairy Hands Punnett Square 2

	H	H
H		
h		

Does your Punnett square match your prediction? _______________

Explain your answer.

_______________ ■

Probabilities

Although Punnett squares provide great ways of visually demonstrating the possible outcomes of a genetic cross, **probabilities** are often better ways of determining results for more complicated crosses. Probabilities are expressed as a fraction of *one,* like the "whole" pie in a pie chart. The simple rule as it applies to genetics is that probabilities of events that must happen simultaneously (like mixing of egg and sperm in mating/crossing) are *multiplied.* Nonsimultaneous events and alternate probabilities are *added.*

For a simple example, let us work through the coin-toss "cross" (Hh × Hh) as a probability. What would be the probability of obtaining a toss where both coins are tails (hh)? The chance of coin 1 landing as tails is 50%, or one-half, and the chance of coin 2 landing as tails is also 50%, or one-half. According to the procedure, getting both coins to land as tails is something that must happen at the same time, so we must multiply these two numbers together. Thus, one-half *multiplied* by one-half would be *one-fourth,* which nicely matches the one out of four boxes that were "hh" in your coin-toss Punnett square.

There are *two* ways to get a coin toss that is mixed: One toss could be Hh (coin 1 = heads, coin 2 = tails), and another toss could be hH (coin 1 = tails, coin 2 = heads). The probability of each of these different tosses would be the same one-fourth that we calculated in the previous example. But because there are two independent, nonsimultaneous ways that this can happen (Hh or hH), we *add* them: $\frac{1}{4} + \frac{1}{4} = \frac{2}{4}$ or $\frac{1}{2}$.

How many of the four boxes in your coin-toss Punnett square represent a mixed toss?

____

Does this match the probability calculated here?

____

Explain your answer.

____

____

ACTIVITY 4

Calculating Probabilities

Without using a Punnett square, answer the following questions using the probabilities technique explained in this section. (You may compare your results to a Punnett square to check your work.)

1. What is the probability of tossing two coins and getting two heads?

2. What is the probability of a hairy-handed father who is a carrier for the hairless allele (Hh) and a mother who has no hair on the back of her hands (hh) producing a child with hairy hands? ■

ACTIVITY 5

Calculating Probabilities Using Two Methods

1. Being able to roll one's tongue is a dominant trait, and being unable to roll the tongue is a recessive trait. Using both the probability method and then a Punnett square, determine the probability of two carrier parents (Rr × Rr) producing a child who cannot roll his or her tongue.

2. Having unattached earlobes is a dominant trait, and having attached earlobes is a recessive trait. Using both the probability method and then a Punnett square, determine the probability of a carrier parent with unattached earlobes and his mate with attached earlobes (Uu × uu) producing a child with unattached earlobes. ■

Sex Determination, Linkage, and Pedigree Charts

A different kind of inheritance pattern involves the only mismatched chromosome pair in sexually reproducing organisms, namely the **X** and **Y chromosomes.** The Y chromosome apparently carries the genes needed to produce a male, and it is smaller than its X-chromosome partner. A person with two X chromosomes is female, and a person with an XY combination is male (Table 20.2).

However, the sex chromosomes do have genes for traits other than sex determination. These "other genes" are said to be **linked** to the sex chromosome because they are obligated to travel with it when the gametes come together to produce a fertilized egg. Most **sex-linked** traits are linked to the X chromosome because the smaller Y chromosome has very few genes that are not related to sex determination.

This XY combination creates a potential problem for males, who do not have the safeguard of a complete second sex chromosome with an allele which would protect them against expressing a (usually recessive) mutant gene. The truncated Y chromosome is missing many of the alleles found on the X chromosome. So, in a male, there is only a single allele on his X chromosome, with no matching allele on the Y chromosome. This is why X-linked genetic disorders occur far more frequently in males. When a male receives a "diseased" X chromosome from his mother, the Y chromosome from his father has no matching, healthy allele to compensate for it.

Table 20.2 ***Punnett Square Demonstrating Sex Determination by the X and Y Chromosomes***

Parent 1 → Parent 2 ↓	X	X
X	XX Female	XX Female
Y	XY Male	XY Male

ACTIVITY 6

Using Pedigree Charts to Determine Patterns of Inheritance

Part 1

Patterns of inheritance can be studied by looking at a person's genetic history using a device called a pedigree chart (Figure 20.2). Commonly referred to as a "family tree," a pedigree chart has boxes and circles that designate males and females, respectively, and its rows indicate each generation. Because these charts are typically used by medical professionals or geneticists to trace a genetic disease or trait of interest, a solid box or circle is used to designate a person with the phenotype in question, and a lighter or colorless box or circle is used to indicate a person without the trait. Carriers, who are heterozygous for the allele, are either shown with a different color or a box or circle that is half solid.

Figure 20.2 displays a chart typical of an X-linked disorder such as color-blindness. Using the terminology explained previously, describe the individual labeled as No. 1 in the figure.

____

Describe the individual labeled as No. 2 in the figure.

____

Describe the individual labeled as No. 3 in the figure.

____

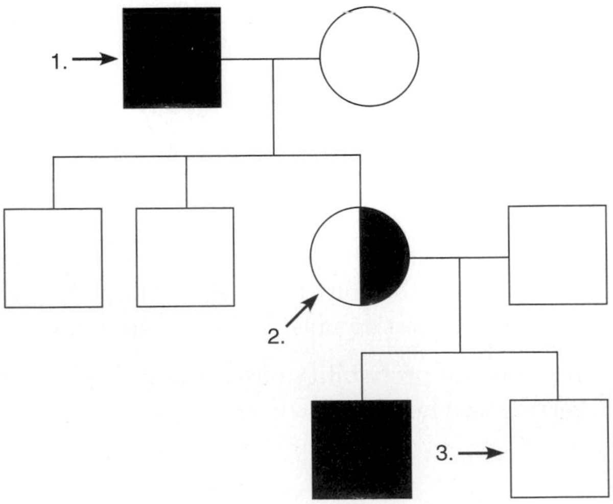

FIGURE 20.2 A simple pedigree chart displaying a typical X-linked disorder.

Part 2

Review the pedigree chart in Figure 20.3. Complete the chart by adding the following offspring to the couple in the chart: 2 sons—one color-blind and one with normal vision; 1 daughter.

Is the mother or the father the source of the color-blindness gene?

____

Consider your answer to the previous question, and darken in half of the box for the appropriate parental unit chosen.

What is the probability that the daughter is a carrier?

____

It is important to note that only some linked genes are on the sex chromosomes. The other 22 pairs of chromosomes contain thousands of genes, many of which are linked and transmitted together to the offspring. ■

ACTIVITY 7

Analyzing Patterns of Inheritance Using the Experimental Cross Method

Materials for this Activity

Two small boxes

Four prepainted tongue-depressor (or Popsicle) sticks

Work in pairs, and obtain two small boxes labeled "experimental cross male" and "experimental cross female." These boxes each contain two tongue-depressor sticks that have been painted on each side according to the following color schemes:

- **Sex (side one):** Blue = Male alleles (Y chromosome); Pink = Female alleles (X chromosome)
- **Color-blindness (side two):** Red = Normal color vision allele; Black = Recessive color-blindness allele

Both of the sticks in the female box are pink on one side, but on the other side, one is red and the other is black. This box represents a female who is a *carrier* for color-blindness (X^BX^b). In the male box, one stick is pink on one side and red on the other side. The other stick is blue on one side and unpainted on the other side. This box represents a healthy male (X^BY).

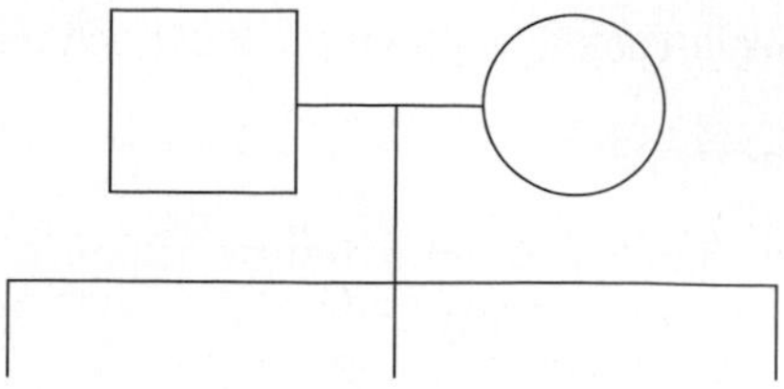

FIGURE 20.3 A blank pedigree chart.

Why is the blue stick unpainted on its reverse side?

____

1. *Without looking,* randomly take one stick out of each box. This combination represents the first of your hypothetical offspring. First determine the sex of your child. Two pink sticks would be a female, and one pink and one blue stick would be a male. Then, determine if the child has color-blindness. Recall that color-blindness is recessive, so even one normal allele, represented by the red stick, results in a child with normal color vision. Record the results of this cross in Box 20.7.

Box 20.7 Color-Blindness Punnett Square 1

	X^B	Y
X^B		
X^b		

2. Return the sticks to their respective boxes and shake the contents. Repeat the process described in Step 1 11 more times, tallying the results in Box 20.7.

   Did you obtain any color-blind females? ______________

Propose a reason why this result occurred.

____

In the space provided below, draw a pedigree chart for your family.

3. Next, complete the Punnett square in Box 20.8. This Punnett square will provide you with the *theoretical* ratios for this experimental cross.

   Based on the Punnett square, what proportion of your females *should be* color-blind?

   ____

   What proportion of your males *should be* color-blind? ______________

4. Based upon your *experimental* results in Box 20.7, what proportion of your females were color-blind?

   ____

Box 20.8 Color-Blindness Punnett Square 2

	X^B	Y
X^B		
X^b		

Does this match the *theoretical* results of your Punnett square in Box 20.8?

____

Explain your answer.

____

Based upon your *experimental* results, what proportion of your males were color-blind?

____

Does this match the *theoretical* results of your Punnett square?

____

Explain your answer.

____

5. Next, pool your data with that of the other groups in the class.

   Does your percentage of color-blind offspring more closely match the Punnett square results?

   ____

   If so, propose a reason for this result.

   ____

Section ______________________ Name ______________________

Date ______________________ Instructor ______________________

EXERCISE 20

GENETICS

Review Questions

1. What advantage does sexual reproduction have over asexual reproduction?

____

____

____

2. Compare and contrast meiosis and mitosis.

____

____

____

3. Match the term to its best description:

____	Alleles	a. the observable trait expressed by an organism
____	Phenotype	b. the genes for a trait present in an organism
____	Genotype	c. the different varieties of a gene for a particular trait
____	Homozygous	d. two genes traveling together on the same chromosome
____	Linkage	e. the state of having two identical alleles for a particular trait

4. Having unattached earlobes is a dominant trait over attached earlobes. Calculate the probability of two parents who are heterozygous for unattached earlobes having a child with attached earlobes. Perform a Punnett square to check your answer *after* first using the probabilities method to answer this question.

5. Who is responsible for sex determination in humans, the male or female parent?

____

Explain.

____

____

6. Briefly explain sex linkage and linkage.

____

____

Why do X-linked genetic disorders occur more frequently in males?

____

____

EXERCISE

21

DNA TECHNOLOGY AND GENETIC ENGINEERING

Objectives

After completing this exercise, you should be able to

1. Explain the connection between nucleotide sequences, DNA structure, and genes.
2. Extract DNA from plant cells and describe how it looks to the naked eye.
3. Explain how the structure of DNA allows us to identify individuals through DNA fingerprinting.
4. Perform the procedure for DNA fingerprinting.
5. Understand how genetically modified plants are created and explain their potential benefits and drawbacks.

Materials for Lab Preparation

Note to Instructors: Although we recommend the kits below, your laboratory setting may necessitate a different procedure. We recommend that you read through the activities in this exercise to ensure that you have all the needed supplies prior to starting the activity.

Equipment and Supplies

- ❑ Transparent tape
- ❑ Scissors
- ❑ Blender
- ❑ PowerPac power supply
- ❑ Safety glasses
- ❑ Heat-resistant vinyl gloves
- ❑ Whole onions
- ❑ Tissues (to wipe eyes)
- ❑ Knife
- ❑ 250-ml beakers
- ❑ 100-ml graduated cylinder
- ❑ 37°C and 65°C water baths or beaker and hot plate
- ❑ Thermometer
- ❑ Ice
- ❑ Wooden splints
- ❑ Cell lysis solution
- ❑ DNA precipitation solution
- ❑ Conical funnel and filter paper
- ❑ Mini-Sub Cell GT
- ❑ 1.5-ml micro test tubes
- ❑ Test tubes and test tube racks
- ❑ Gel staining trays
- ❑ Gel support film
- ❑ Micropipettes and tips
- ❑ Computer with Internet access (optional)
- ❑ Printer (optional)

Kits

- ❑ Carolina Biological Supply Kit # WW-15-4704 (or equivalent)
- ❑ Bio-Rad Laboratories Kit 166-0007EDU (recommended)

Introduction

DNA (deoxyribonucleic acid) is found in all living cells, and it carries each cell's genetic information. **Genes** are the segments of DNA that carry the code for making a molecule or regulating a cellular function. When scientists discovered the structure and nature of DNA just over half a century ago, it was an enormous leap forward in the field of biology. The growing knowledge of DNA function and advances in technology that allow manipulation of DNA molecules (DNA technology) have empowered biologists to manipulate the biochemistry of cells. In recent decades, the world has been greatly changed by the application of DNA technology to produce medical and commercial products. The ability to alter the genetics of organisms through DNA technology is called **genetic engineering.**

Chemical Characteristics of DNA

First, it is important to understand the basic structure of the DNA molecule. DNA is a double-stranded molecule composed of building-block molecules called **nucleotides.** Nucleotides are made up of a phosphate group, a 5-carbon sugar molecule and a ringed nitrogen-containing molecule referred to as a **base** or **nitrogenous base** (Figure 21.1).

There are four varieties of nucleotides in DNA; each variety has a different base but they all have the same sugar and phosphate portions. Each nucleotide takes the name of the base that makes it unique: adenine, thymine, cytosine, or guanine. In a DNA molecule, the sugar of one nucleotide is covalently bonded to the phosphate of the next nucleotide, leaving the base poking inward. The opposing strand is set up similarly, and the inward-facing bases connect with each other through hydrogen bonding, giving DNA the "twisted ladder" shape referred to as a **double helix.** The sugar-phosphate backbones make up the sides, and the connected bases make up the rungs of the twisted ladder (Figure 21.2).

The bases of the nucleotides match up in the center in a specific pattern. Adenine (A) and thymine (T) nucleotides always pair with each other, as do guanine (G) and cytosine (C) nucleotides. This is called **complementary base pairing.** You will never find As or Ts matched with Gs or Cs in normal DNA molecules.

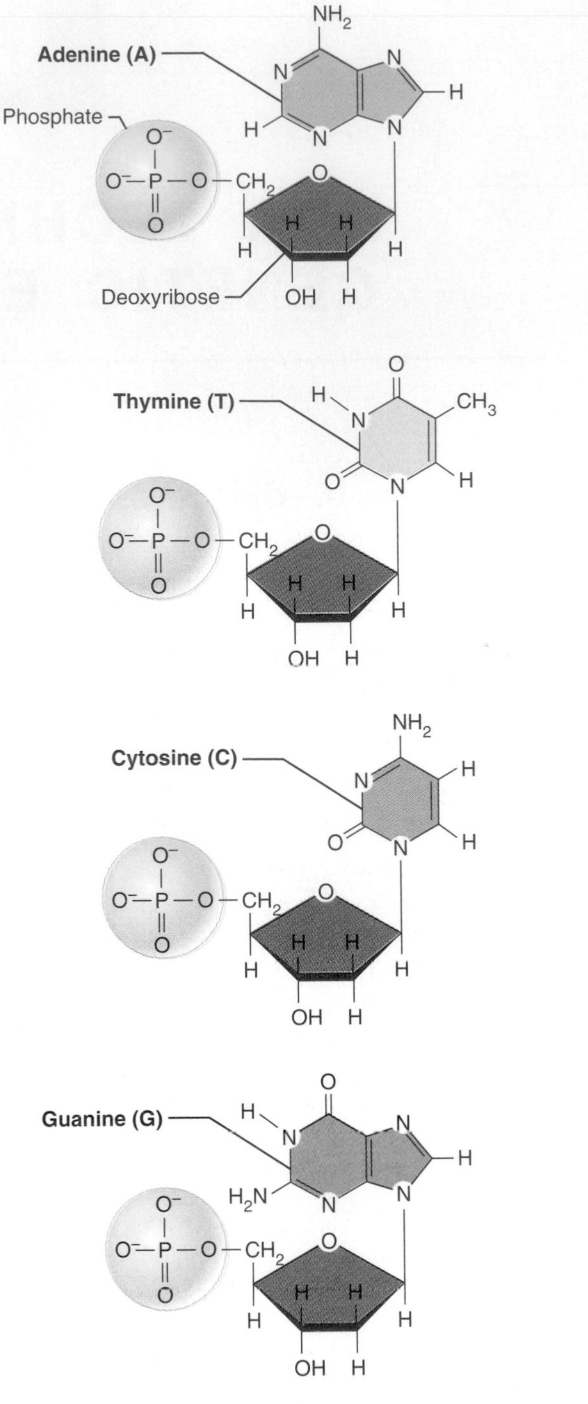

FIGURE 21.1 The four nucleotides that comprise DNA. The phosphate and sugar groups are identical in all four nucleotides.

ACTIVITY 1

Discovering the Structure of DNA

Materials for this Activity

Transparent tape
Scissors

Part 1: Structure of DNA

1. Cut out the nucleotide molecules on page 253 by cutting along the dashed lines.
2. Connect four molecules in the sequence G-C-A-C (from top to bottom) by taping the rectangular sugar of one molecule to the plus-shaped phosphate of the next. This will be the *left strand* of your DNA molecule. Lay this out flat on your table with the G at the top and C at the bottom.

   With which nucleotide will you start when constructing the right side strand from the top?

   ____

3. To construct the right side of your DNA, take the cutout of the nucleotide you will place at the top, and try matching up the bonding "arms" of the nitrogenous base (which represent hydrogen bonding areas) with the bonding arms of the G nucleotide on the left strand. You should find that they don't seem to line up. What can you do to get those bonding arms to match up with each other? (*Hint:* Be creative, and leave no stone unturned when examining the possibilities!)

   ____

   ____

4. Once you have figured out the trick, line up (but do not tape) the nucleotides together at the central arms, and repeat this process for the remaining three nucleotides. Also remember the sugar-phosphate covalent bonds by lining up (but not taping) those molecules as you build the right strand.

   In this activity, you have essentially walked in the footsteps of James Watson, who was equally puzzled at first about the way these molecules could fit together.

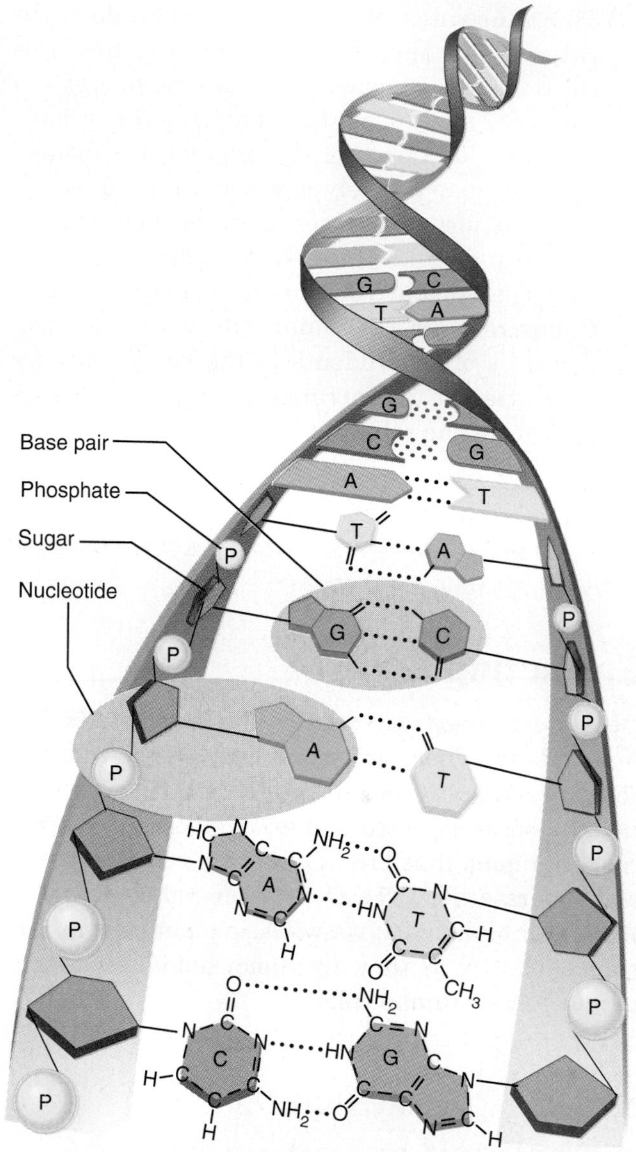

FIGURE 21.2 The double helical structure of DNA.

Part 2: Replication of DNA

1. Remove the nucleotides of your untaped right strand, leaving the left strand in place.
2. Now, reattach the loose nucleotides to the left strand in a new order, but be sure to maintain the appropriate bonds with the right strand.

   Can you do it? ______________________________

   Why or why not?

   ____

   ____

This illustrates how the DNA molecule replicates itself: the strand splits, and new nucleotides are attached one at a time to create a complete strand again. Complementary base pairing allows each of the original "backbones" to produce an entirely new sister strand identical to its former partner. Enzymes help to split the strands and place the complimentary nucleotides—just as your hands did in this activity. Compare your DNA molecule with that produced by other students in the lab. What does your attempt to rearrange the nucleotides suggest about the reliability of DNA replication?

___

___ ■

Extracting DNA

DNA must be extracted before it can be studied or manipulated. There are several ways to do this, but the process that typically works best in an educational laboratory setting is extraction using chemical techniques that break down the cell membrane and separate the DNA from other cell contents. In the following activity, we will use plant cells as our source of DNA, as they are somewhat easier to manipulate than human cells.

ACTIVITY 2

Extracting DNA from Onion Cells

Materials for this Activity

Carolina Biological Supply Kit # WW-15-4704 (or equivalent)
Cell lysis solution (e.g., 10% sodium dodecyl sulfate)
DNA precipitation solution (concentrated ethanol)
Test tubes
Filter paper
Wooden splints
Safety glasses
Heat-resistant vinyl gloves
Whole onions
Tissues (to wipe eyes)
Knife
Blender
250-ml beakers
100-ml graduated cylinder
65°C water bath or beaker and hot plate
Thermometer
Ice
Conical funnel

Note to Instructors: The directions that follow are provided for guidance; your kit or other source of supply may come with more detailed instructions.

1. Prepare a 65°C water bath. A beaker and hot plate may be used in place of a commercial laboratory water bath. Place a 250-ml beaker of DNA precipitation solution (concentrated ethanol) on ice to ensure it will be cold when needed.
2. Cut a medium onion (or half a large onion) into several pieces, and place in blender. Measure 65 ml of water in a graduated cylinder and add to the blender. Secure blender lid, blend at the *low* setting for 30 seconds, and then at the *liquify* setting for 10 seconds. Repeat 10 seconds of blending at the liquify setting until the onion is thoroughly liquified. Transfer the liquified onion into a clean beaker.
3. Add some of the liquified onion to a 15-ml test tube until it is almost half full. Add 7 ml of cell lysis solution and cap the tube. Invert and right the tube to mix contents (do not shake), and place in the 65°C water bath for 15 minutes.
4. After 15 minutes have elapsed, remove the tube from the water bath. Place a small conical funnel with filter paper in a 50-ml test tube, and pour the contents of your onion extract tube into the funnel. Allow the filtrate to drip for 10 to 15 minutes.
5. When most of the extract has filtered, remove the funnel and *slowly* add the cold ethanol while tilting the test tube about 45 degrees to minimize mixing and to allow the alcohol to float on top of the extract. Add an amount of ethanol about equal to the extract in the tube. If you have done this carefully, you should see a fairly clear layer of ethanol above a pinkish layer of extract.
6. The DNA is typically visible as a cloudy layer in between the two liquids (if not, add another 1–2 ml of cold ethanol). Slowly insert the wooden

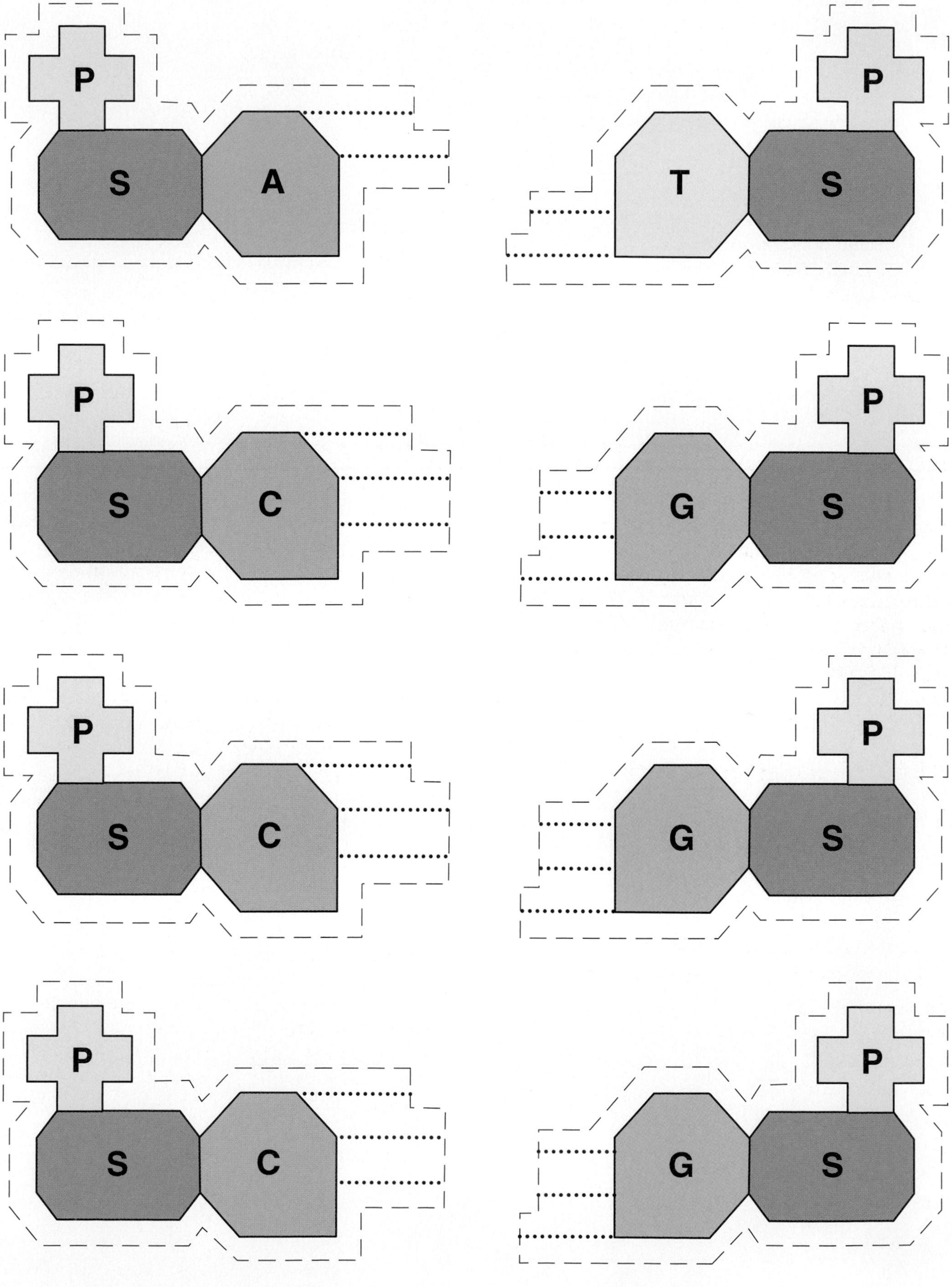
P
S
A
P
T
S
P
S
C
P
G
S
P
S
C
P
G
S
P
S
C
P
G
S

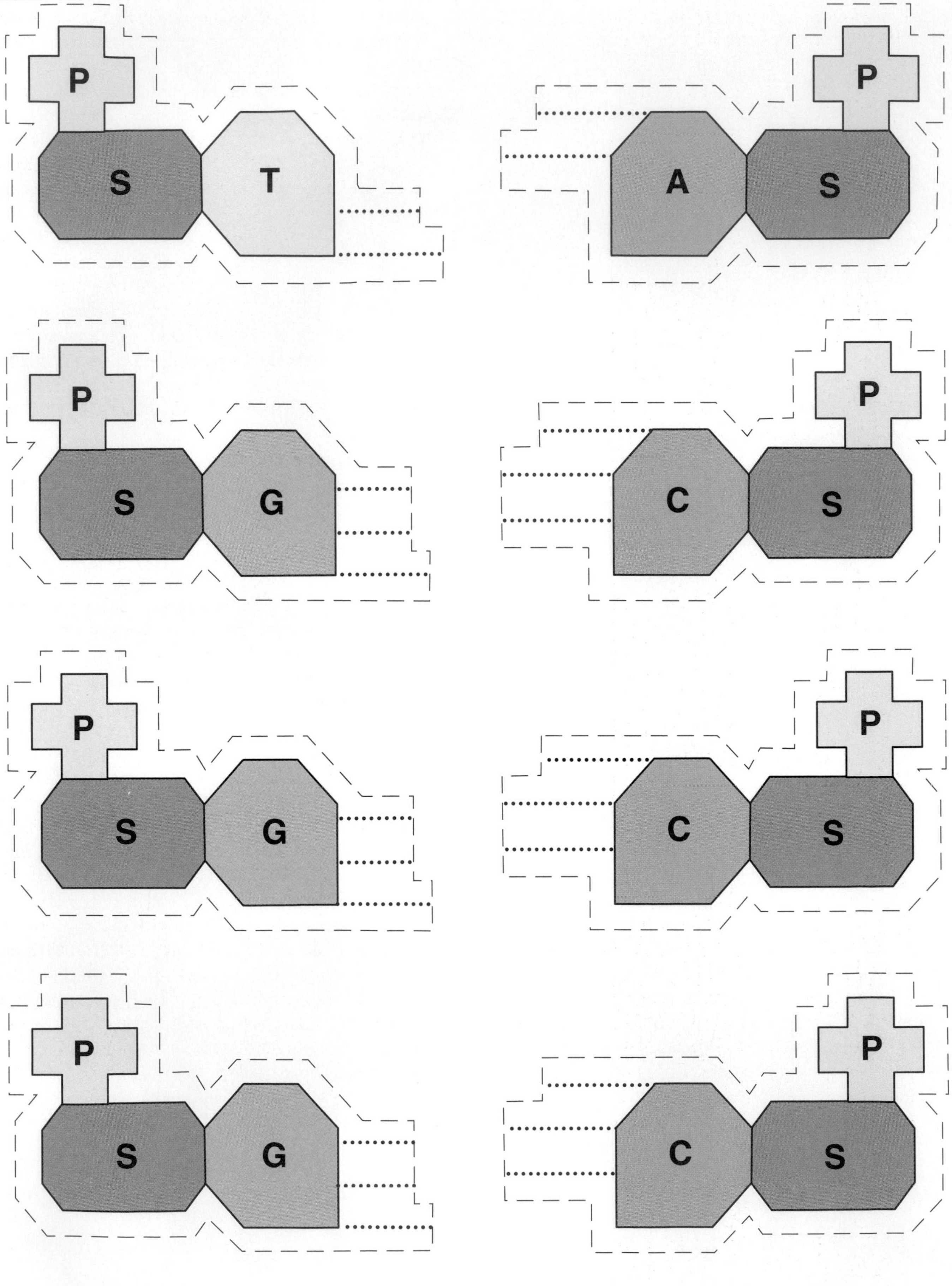
P
S
T
P
A
S
P
S
G
P
C
S
P
S
G
P
C
S
P
S
G
P
C
S

splint into the tube. Gently turn the splint as you slowly move it up and down to spool the DNA around it.

Describe the color and shape of the DNA on your splint.

____

Considering that the extraction/lysis solution was mostly water, speculate whether DNA is soluble in water:

____

In ethanol:

____

Test your answer above by placing your splint with DNA into a small, clean test tube half filled with distilled water and stir. Does it dissolve?

____

Before testing this water with pH paper, predict whether DNA would be acidic, basic, or neutral.

____

Record your pH test results:

____

Check your results against your prediction. What chemical properties of DNA influence pH?

____

______________________________________ ■

DNA Fingerprinting

A great deal of DNA is the same from person to person, assuring that each of us is usually born with one head, one four-chambered heart, and all the other things that make us human. But plenty of DNA varies greatly and can produce a reasonably unique pattern, or "fingerprint," when separated and analyzed.

DNA fingerprinting is accomplished by obtaining a sample of DNA and using **restriction enzymes** to cut up the DNA. Restriction enzymes are special enzymes that chop up DNA molecules at certain locations (Figure 21.3a). Because even slight differences in DNA sequences may result in DNA that is broken up into very different fragments, each person's DNA sequences will chop up differently, and the pieces will exhibit different patterns when they are separated. The most common method of separating these chopped up pieces is **gel electrophoresis,** a process that separates DNA fragments by size and weight and represents them as bands (Figure 21.3b).

When the DNA technology of gel electrophoresis was first developed, the equipment was both somewhat difficult to use and prohibitively expensive for the average educational laboratory. Refinements in equipment design, and relatively easy-to-use kits for educational purposes have made it possible to include DNA fingerprinting in undergraduate laboratory activities.

Like most separation techniques, gel electrophoresis is based on the principle that molecules move through a medium at different speeds depending on physical and chemical characteristics. As a general rule, small molecules will be drawn through a medium faster than larger ones. In gel electrophoresis, an electrical current pulls the negatively charged DNA fragments from wells on one side of the apparatus toward the other positively charged side of the gel. When the process is terminated, the fragments are seen as bands at various distances from the well in which the sample was originally placed. The smaller fragments are observed as having moved farthest away from the well, while the larger fragments form bands a shorter distance from the well (Figure 21.3c). Identical DNA samples will be chopped up into the same sized fragments and will move through the gel in exactly the same way (Figure 21.3d).

With rare exception, the separated bands of DNA fragments in a gel produce patterns that are considered sufficiently unique that the odds of two unrelated samples matching would be extremely unlikely. And, unlike partial fingerprints, a very small sample of DNA is often sufficient to produce a very clear DNA fingerprint. For these reasons, DNA fingerprinting is often used in crime scene investigations where blood, skin cells, or bodily fluids have been found.

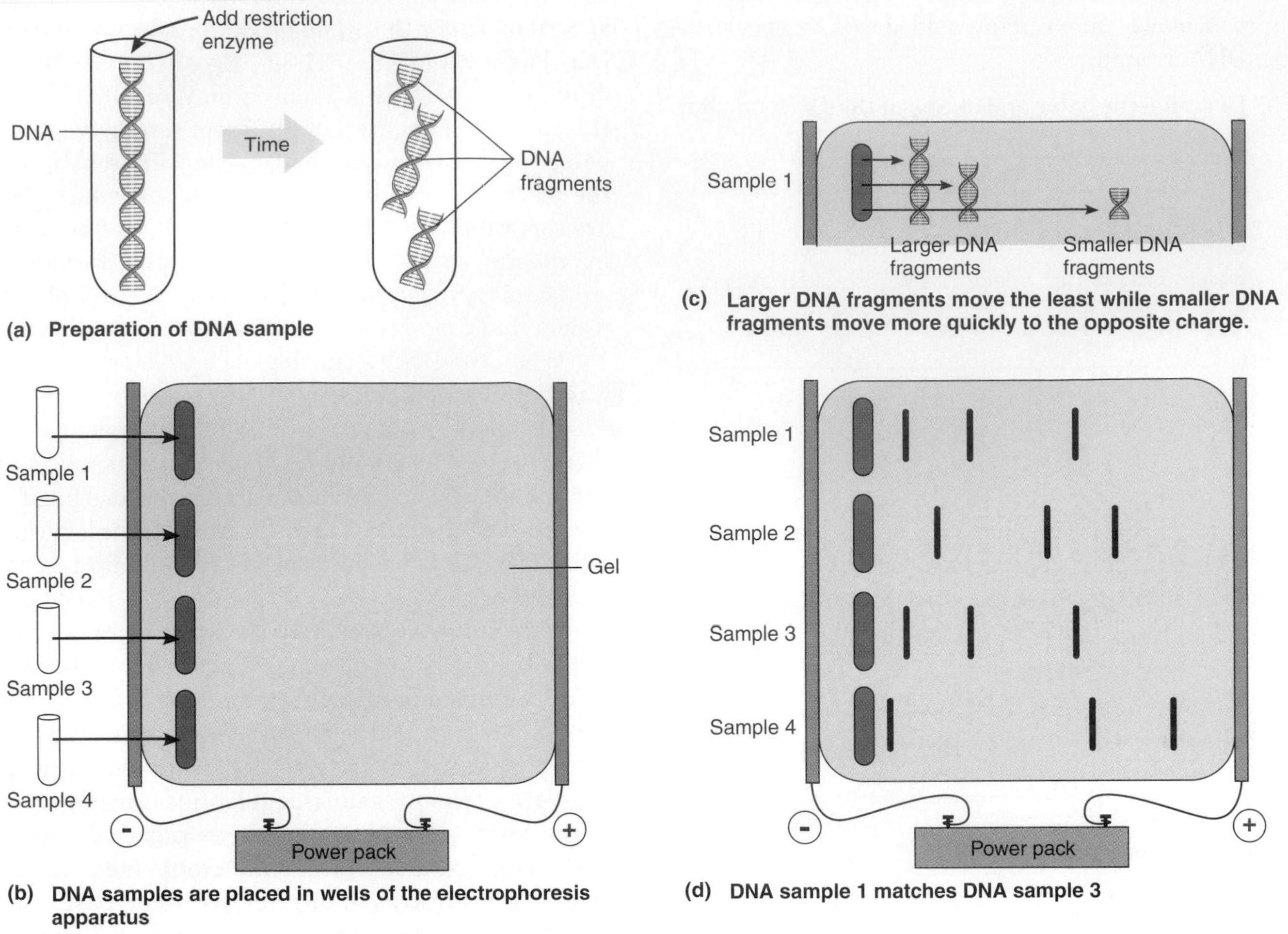
Add restriction enzyme
DNA
Time
DNA fragments
(a) Preparation of DNA sample
Sample 1
Larger DNA fragments
Smaller DNA fragments
(c) Larger DNA fragments move the least while smaller DNA fragments move more quickly to the opposite charge.
Sample 1
Sample 2
Sample 3
Sample 4
Gel
Power pack
(b) DNA samples are placed in wells of the electrophoresis apparatus
Sample 1
Sample 2
Sample 3
Sample 4
Power pack
(d) DNA sample 1 matches DNA sample 3

FIGURE 21.3 Gel electrophoresis. (a) Preparation of DNA for gel electrophoresis. (b) Set-up for gel electrophoresis. (c) Movement of DNA fragments across a gel plate. (d) DNA fingerprints showing one set of identical fingerprints.

In the following activity, you will perform DNA fingerprinting on substitute blood samples in order to analyze which suspect was at a crime scene. DNA from each of several suspects is provided.

ACTIVITY 3

DNA Fingerprinting and Crime Solving

Materials for this Activity

Bio-Rad Laboratories Kit 166-0007EDU (recommended)
DNA samples
DNA size standard
DNA sample loading buffer
Electrophoresis buffer
Agarose powder
DNA stain
Mini-Sub Cell GT
PowerPac power supply
1.5-ml micro test tubes
Test tube holders/racks
Gel staining trays
Gel support film
Micropipettes and tips
37°C water bath

Note to Instructor: Although we recommend the kit from Bio-Rad Laboratories, your laboratory setting and equipment may necessitate a different source of supply or procedure. The directions that follow are provided for guidance; your kit or other source of supply may come with more detailed instructions.

Part 1: Preparation of DNA Samples

1. Place the container of the restriction enzyme mix on ice.
2. Label the DNA sample test tubes, and add the DNA samples from the stock solutions according to the directions in your kit or as directed by your instructor. Use a new micropipette tip for each sample.
3. Mix the contents of each tube by flicking the bottom of the tube with a finger, and tap the bottom on your table top to collect the liquid at the bottom of the tube. Place the tubes in a 37°C water bath for 45 minutes. (*Note:* If time does not permit completion of this activity in one class meeting, tubes may be placed in a refrigerator after this incubation period until the next lab period; or to finish in one period, the instructor or staff may prepare steps 1 through 3 in advance so samples are ready to process.)

Part 2: Processing the Samples Using Gel Electrophoresis

1. Remove tubes and add loading dye solution to each tube according to the directions in your kit or as directed by your instructor.
2. Prepare your gel electrophoresis apparatus by adding the agarose gel and buffer, making sure the gel wells are near the black (negative) electrode and the base is near the red (positive) electrode.
3. Add a portion of the DNA samples from the tubes you prepared in Part 1 to each well according to the directions in your kit or as directed by your instructor. Use a new micropipette tip for each sample.
4. Place the lid on the electrophoresis chamber and plug it in to power supply. Set the power at 100V and allow to run for 30 minutes.
5. After 30 minutes, turn off the power and remove the lid of your electrophoresis apparatus. Carefully remove the gel and place in staining tray. Stain and destain the gel according to the directions in your kit or as directed by your instructor.

Which suspect sample matches most closely with that found at the crime scene?

___

What does this test suggest about the samples from the remaining suspects

___

___ ■

Recombinant DNA and Transgenic Organisms

One of the ways that DNA technology has been useful to society is the placement of beneficial genes from one organism into the cells of other organisms. Organisms that have had foreign DNA segments or genes inserted into their cells, in other words, have undergone genetic engineering, are called **transgenic organisms.** Transgenic organisms can be created from bacteria or other single-celled organisms to produce specific chemicals. Hormones and medications have been made by this method. A somewhat controversial use of the process is the production of transgenic plants. These genetically altered plants can be made resistant to pests or infused with more nutrients and vitamins than they would normally produce.

Transgenic plants can also be made pest resistant, which reduces or eliminates the need to spray them with potentially toxic pesticides. This can increase yields in areas of the world where people do not have access to farm chemicals. Food safety issues with transgenic plants, sometimes called **genetically modified organisms (GMOs),** are somewhat less clear, creating controversy within the industry.

ACTIVITY 4

Internet Research: Genetically Modified Foods

Materials for this Activity

Computer with Internet access
Notebook or printer to record information

This activity can be performed during the laboratory period if students have access to computers with Internet connections or as a homework assignment to be completed and discussed in the next class meeting.

1. Form groups of three or four students. Perform an Internet search for "Genetically Modified Food." Select several web sites to investigate, choosing an equal number of those that seem to be for and against genetically modified foods. Attempt to find at least one site that seems to provide balanced coverage of risks and benefits.

   Using the information you find, summarize the process by which a GMO is produced.

   ____

   ____

   ____

2. Discuss the information you find with members of your group. Fill in Box 21.1, and discuss these risks and benefits. Attempt to reach a consensus about whether the risks outweigh the benefits or if the reverse seems true. Your instructor may wish to coordinate a class discussion during which each group presents its conclusions and rationale.

   Weighing these risks and benefits, do you feel genetically modified foods are relatively safe or relatively unsafe? Why?

   ____

   ____

   ____

Box 21.1 Benefits and Risks of Genetically Modified Food

Benefits of Genetically Modified Food	Risks of Genetically Modified Food

Section ______________________ Name ______________________

Date ______________________ Instructor ______________________

EXERCISE

21

DNA TECHNOLOGY AND GENETIC ENGINEERING

Review Questions

1. Summarize the relationship between the two opposite strands of theDNA molecule and the way in which DNA is copied.

2. Which portions of the DNA molecule are connected by covalent bonds?

   Which portions are connected by hydrogen bonds?

3. What purpose was served by the blender and the cell lysis solution in the DNA extraction activity?

4. Why did you add the cold ethanol solution to the liquefied onion in the DNA extraction activity?

5. What are restriction enzymes, and what role do they play in DNA fingerprinting?

6. Explain the term electrophoresis, and describe its role in DNA fingerprinting.

7. What is the difference between genetically modified plants and hybrid plant varieties, which are produced by selective breeding?

8. Summarize the risks and benefits of genetically modified plants.

EXERCISE

22

EVOLUTION

Objectives

After completing this exercise, you should be able to

1. Understand the time line for human evolution compared to the evolution of other life forms on Earth.
2. Discuss the similarities and differences between humans, other mammals, and other primates.
3. Understand major trends in human evolution over the past 5 million years.

Materials for Lab Preparation

- ❑ Skeletal models of a cat, monkey, and human
- ❑ Colored pencils
- ❑ Current map of Africa

Introduction

Charles Darwin laid the foundation for the study of human evolution in his landmark book *The Origin of Species,* published in 1859. He outlined the concept of natural selection, which proposes that organisms adapt and change through long periods of time in response to environmental conditions. Tantalizingly, there was only one sentence in the book that related to humans; in the conclusion Darwin merely stated, "Light will be thrown on the origin of man and his history."

This was the beginning of more than a century of intensive investigation of human origins, and breathtaking discoveries have indeed shed much light on the human evolutionary path.

In recent decades, several lines of evidence have converged that provide information about the human lineage. First, numerous fossils from several sites indicate that the first humans emerged in Africa. These fossils mainly consist of bones and teeth and give us information about what humans looked like, what they ate, how they moved, and what their environment was like (Figure 22.1). The first humans looked similar to present-day apes such as chimpanzees, but they were unique in their locomotion. All hominids (modern humans and our upright ancestors and cousins) walk on two legs, a form of locomotion called bipedality. Bipedality profoundly shaped human anatomy and behavior, for example, freeing the hands for uses other than supporting body weight.

FIGURE 22.1 Fossilized human skeleton.

Human fossils are interpreted in the context of comparative anatomy. Molecular data indicate that humans are closely related to monkeys and apes, and we are most closely related to chimpanzees. (We share with them nearly 99% of our DNA!) Early human fossils are similar in anatomy to chimpanzees, and therefore, studies of present-day primates offer clues about early human behavior. In addition to fossils, geochronology has provided key information through techniques such as radiometric dating. These techniques have allowed scientists to precisely date the fossils that have been found and give us a firm chronology for events in human evolution. Together, the evidence indicates that humans and chimpanzees shared a common ancestor some 5 million years ago, at which time their evolutionary pathways diverged, and the family Hominidae set out on its own unique human journey (Figure 22.2).

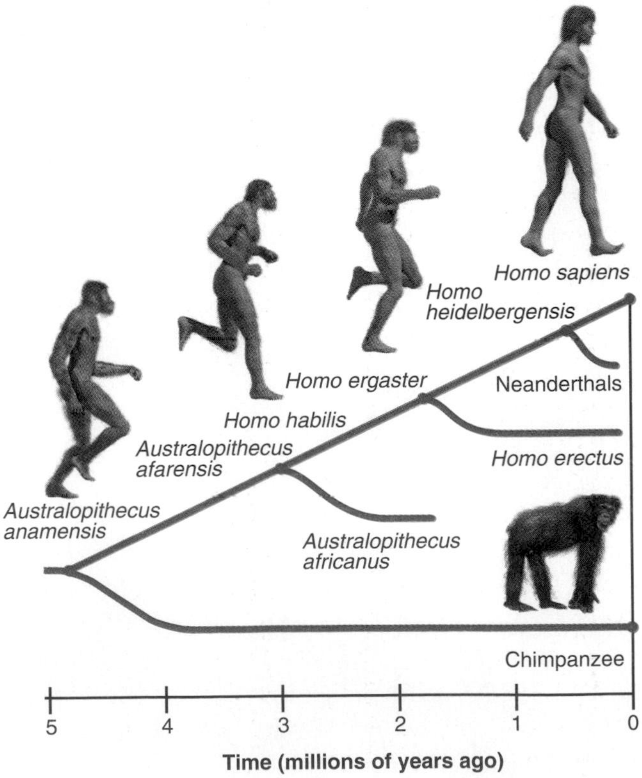

FIGURE 22.2 Proposed evolutionary pathway for the family Hominidae.

A Brief Evolutionary History of Life on Earth

Humans have been on Earth a short time compared to the history of life on Earth. One-celled organisms from South Africa and Australia show life originated 3.8 billion years ago. Hominids have existed a mere 5 million years! In the next activity, we will compare the evolution of our human lineage with that of other life forms. First, however, you will need some basic information.

Major Evolutionary Events

Life began at least *3.8 billion years ago* with the appearance of **prokaryotic cells** (single-celled organisms) such as **cyanobacteria. Eukaryotic cells** appeared *1.8 billion years ago.* Eukaryotic cells differ from earlier prokaryotic cells by having a true nucleus that contains the genetic material, DNA.

By *1 billion years ago,* eukaryotic organisms diversified due to **sexual reproduction** that paved the way for new combinations of genetic material. **Multicellular organisms** appeared approximately *700 million years ago.* These had specialized cells and tissues and complex body forms. These **invertebrates** exhibited an astonishing array of body forms, and some developed outer shells, which are well preserved in the fossil record.

The first **vertebrates** appeared about *500 million years ago.* These were ancient fish that had an internal skeleton with a backbone. Vertebrates continued to diversify over the next several hundred million years. Eventually, amphibians, reptiles, birds, and mammals emerged.

Mammals appeared approximately *200 million years ago.* The first mammals were small, and they revolutionized reproduction. Mammalian mothers mobilize their own internal resources to produce milk for their infants once they are born. Mammalian mothers expend tremendous energy for each offspring, with the result that the infants have a much greater chance of surviving and reaching adulthood. One branch of mammals that evolved were the **placental mammals.** Modern examples include cats, dogs, and primates. Placental mammals carry their fetus inside their bodies, providing constant warmth, nutrition, and protection during the early stages of development.

A widespread extinction that included the dinosaurs, among other organisms, occurred about *65 million years ago* and paved the way for mammals to diversify. All of the modern orders of mammals have their roots in this time, among them the **primates,** order Primata, which includes monkeys, apes, and humans.

Primates continued to diversify, and monkeylike fossil discoveries in Africa indicate that **higher primates** (also known as anthropoids), which share features with modern-day monkeys, appeared about *35 million years ago.*

Apes appeared about *20 million years ago.* They are distinguished from monkeys primarily because of features of their dentition, and they probably had a different locomotor pattern.

The earliest humans appeared *5–6 million years ago.* They belong to a group called australopithecines ("southern ape"), which is a general word for the genus *Australopithecus.* They are distinguished from apes because of their bipedal (upright) locomotion, and they show the beginnings of brain enlargement, which is a key feature of the human lineage.

ACTIVITY 1

Putting Human Evolution in Context

This activity is intended to give you a better understanding of how humans fit, with other life forms, in the evolutionary time line from Earth's formation.

Refer to Box 22.1, on page 264, which is a matching activity that features a time line on the left, and a series of life forms illustrated by icons on the right. Match the appropriate life form on the right with the evolutionary time period in which it appeared by connecting the dots. Refer to the section, "Major Evolutionary Events" for help. Answer the following questions based upon your completed time line.

1. When did the earliest life evolve?

____

2. What is the significance of the evolution of sexual reproduction?

____

____

____

3. What are features of the earliest vertebrates?

____

____

____

Text continues on page 265.

Box 22.1 An Evolutionary Time Line for Life on Earth

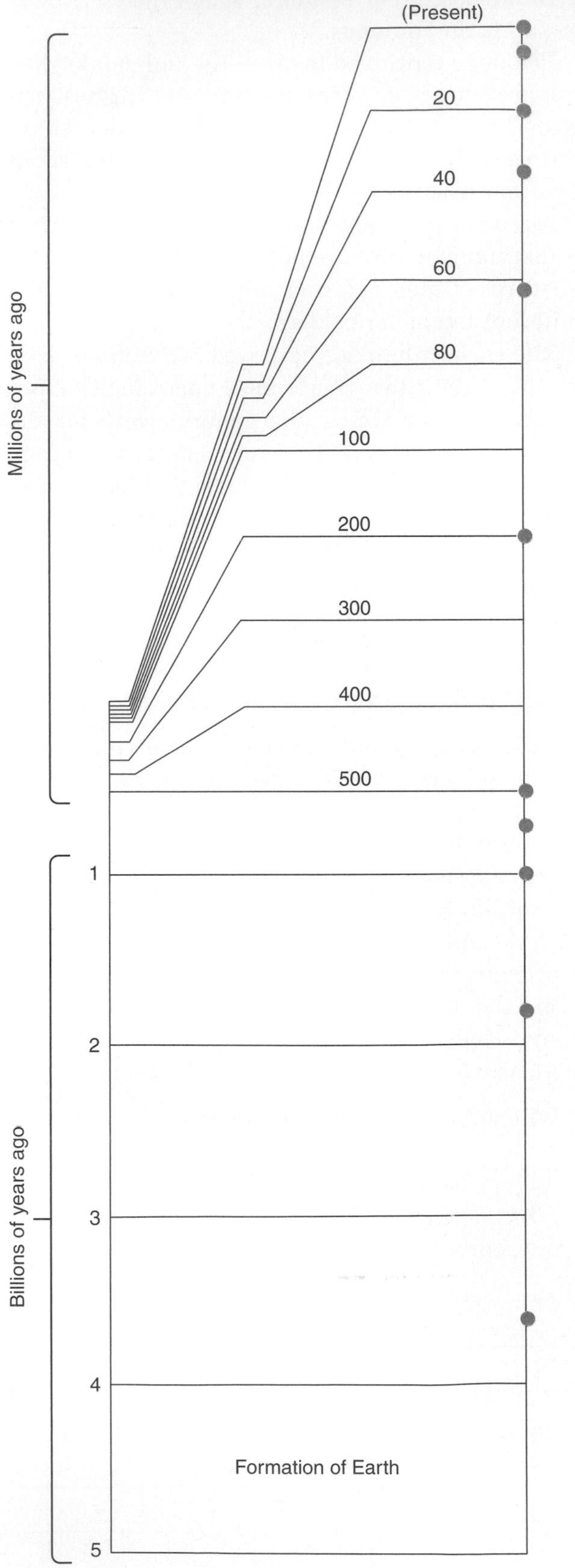

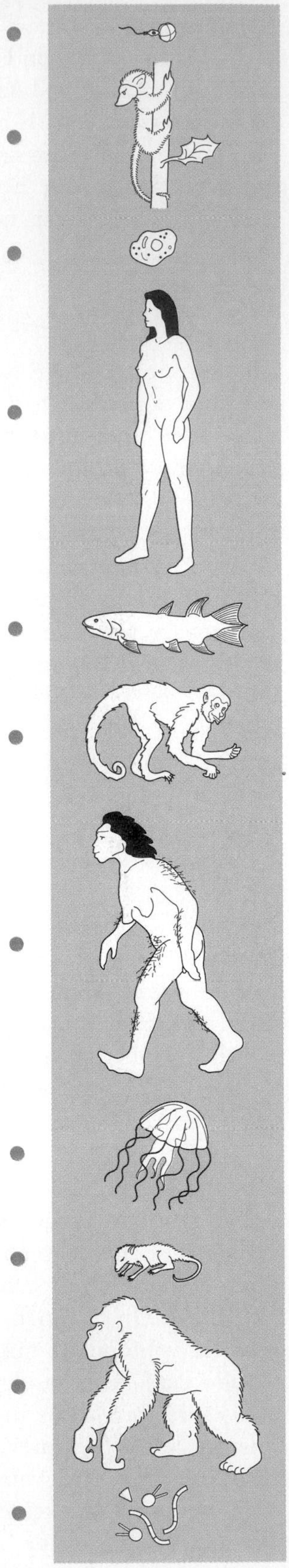

4. What distinguishes early mammals from other vertebrates?

____

____

____

5. When did the dinosaurs become extinct? What is the importance of this event for the evolution of mammals?

____

____

____

6. What distinguishes the earliest humans (hominids) from apes?

____

____

____ ■

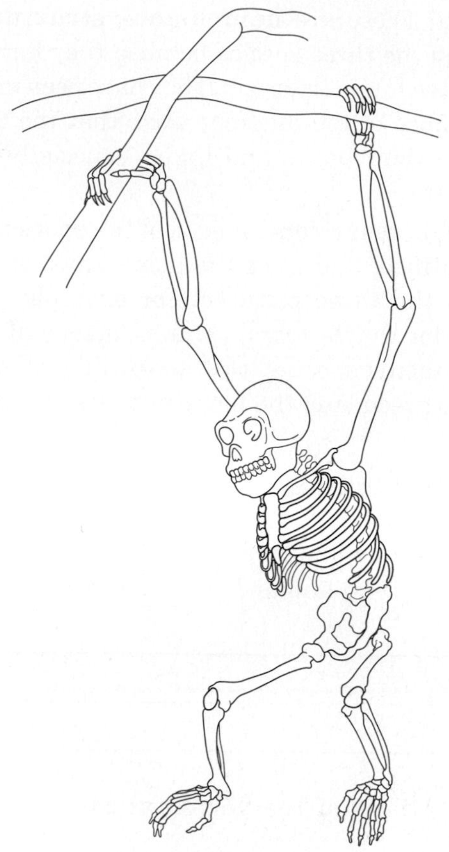

FIGURE 22.3 Generalized hominoid skeleton. Illustrates significant primate differences from other mammals.

ACTIVITY 2

Comparing Cats, Monkeys, and Humans

Materials for this Activity

Models

Cat skeleton
Monkey skeleton
Colored pencils

As pointed out in Activity 1, humans belong to the order Primata, one of the orders of mammals that diversified after the extinction of the dinosaurs 65 million years ago. Early primates resembled rodents; they were tiny, had a diverse diet, and lived in trees. Living primates include prosimians (which resemble some of the earlier primates from around 50 million years ago, but today they include the bush babies of Africa and the lemurs of Madagascar), as well as monkeys, apes, and humans.

Primates are characterized by grasping hands and feet (except humans who no longer have grasping toes) with opposable thumbs, a reliance on vision with forward-facing eyes and an ability to see color, and large brains compared to mammals of similar size (Figure 22.3).

Large brains provide the opportunity for learning, behavioral flexibility, and complex social interactions. Primates live in social groups, and their offspring are dependent on adults and take a long time to mature.

A comparison of a mammal (cat), a higher primate (a monkey), and a human will allow us to focus on several important features that help put human evolution further into context:

- Features shared by all mammals
- Features shared by primates
- Features that are unique to humans

In assigned groups, examine the cat, monkey, and human skeletons, and answer the following questions. *Refer to Figure 22.4a–c if you do not have cat or monkey models in your lab.* Prepare to explain your answers to the rest of the class.

1. Identify the vertebral column, scapula, humerus, ulna, radius, hand and wrist bones, pelvis, femur, tibia, fibula, and foot bones on each of the three

skeletons. These are **homologous structures** shared by the three species because they have a common evolutionary origin. They have been modified because of the different ways that the animals use their backs and limbs, especially for locomotion.

2. Choose different colors for each of the bones that you identified, and color the homologous structures in the three pictures. For example, you might color the vertebral column in each of the three pictures blue, the scapula red, the humerus green, and the forearm bones orange.

Structural Similarities

1. What are three similarities in the limbs of the cat, monkey, and human?

   a. ______________________________

   b. ______________________________

   c. ______________________________

2. What features of the vertebral columns are alike?

   ____

   ____

3. How are the skulls similar?

   ____

   ____

   ____

4. Explain why the animals have many similarities (or homologies) in their skeletons.

   ____

   ____

   ____

Structural Differences

While cats, monkeys, and humans share many features because they are mammals and they have a common origin, they have different evolutionary histories and different ways of life. These differences mean they have distinct anatomical features that have developed over the millions of years since they diverged from one another.

Hands and Feet

1. Refer again to the skeletal models or Figure 22.4. Examine the hands and feet. What are differences in the proportions of the wrist bones (carpals), metacarpals, and phalanges?

   ____

   ____

   ____

2. What part of the hand bears weight when the animal moves?

   ____

3. What are the differences in the fingers (especially note the ends of the digits)?

   ____

   ____

   ____

4. How do differences in the hands and feet relate to the animal's way of life (e.g., locomotion, feeding, social activities, tool use)?

   ____

   ____

   ____

Vertebral Column

1. Examine the vertebral column. What are the differences in the cervical region?

   ____

   ____

   ____

   ____

2. Can you explain the differences? *Hint:* Think about the way a cat uses its head and neck to eat, grasp prey, and perform other maneuvers, versus how a monkey might use its head. How does a monkey or human eat compared to a cat, and how might that behavior translate into structural differences?

   ____

   ____

   ____

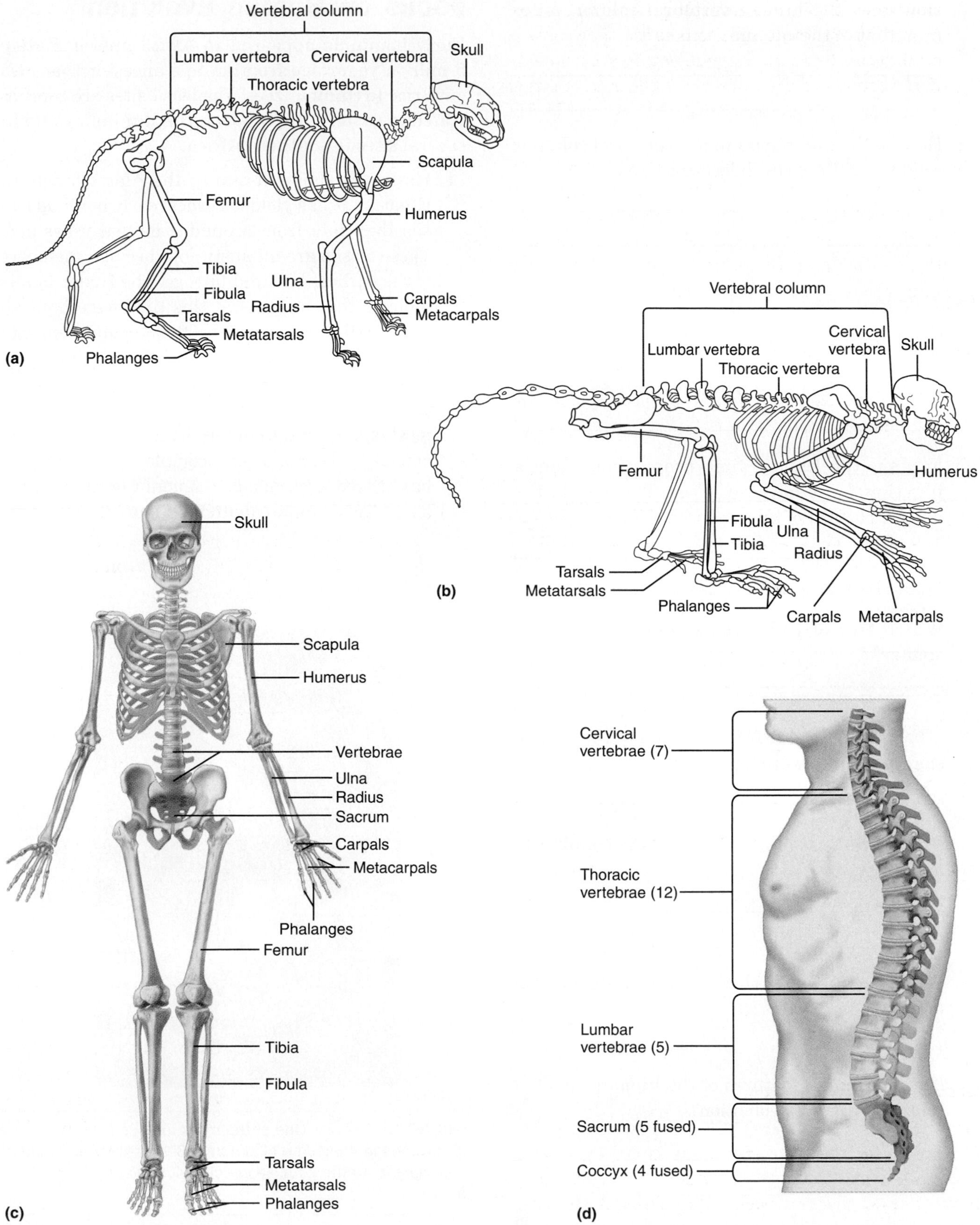

FIGURE 22.4 Comparison of mammal and primate skeletons. (a) Cat skeleton. (b) Old World monkey (langur). (c) Modern human skeleton. (d) Human vertebral column.

3. How does the human vertebral column differ from that of the cat and monkey?

4. How do the differences in the vertebral columns relate to differences in locomotion?

Limb Proportions

1. Compare the lengths of the forelimbs and hindlimbs of the cat and monkey to the human. Which two skeletons are most similar?

2. Explain the human limb proportions from a functional perspective. *Hint:* The key is the unique pattern of human locomotion.

3. Which two of the three skeletons are more similar?

   What anatomical features led you to the conclusion that these were not similar?

   Propose at least one reason why such similarity may exist.

4. What features of the human skeleton are unique?

5. How do unique features of the human skeleton relate to human locomotion?

Focus on Human Evolution

Early hominids appeared in Africa approximately 5 million years ago from an ape ancestor that also gave rise to chimpanzees. The fossil sites are concentrated in eastern Africa along the Rift Valley and in several sites in southern Africa.

1. **Hadar.** Hadar is a site in the Afar triangle in Ethiopia. It has yielded several early hominid fossils that date from around 3 million years ago. These fossils are all australopithecines, members of the earliest hominid lineage. The Hadar fossils indicate that humans at this time were bipedal, had small brains (about 400–450 cubic centimeters, compared to an average of 350 cubic centimeters for chimpanzees), and large teeth that were probably mostly adapted for eating plant food. The most famous find from the site is known as "Lucy," officially AL-288, a 40% complete skeleton that has offered a tremendous amount of information about early human anatomy (Figure 22.5).

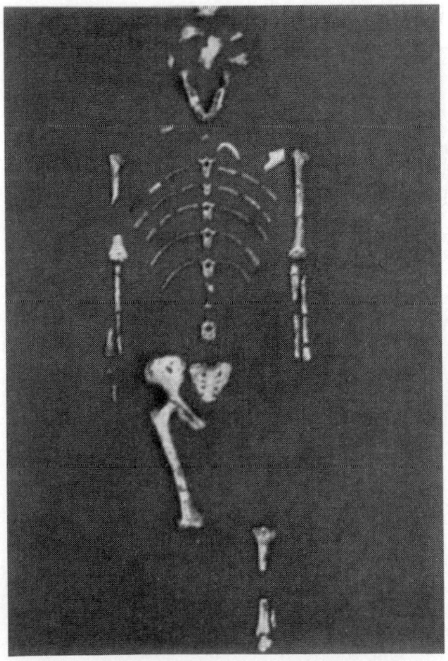

FIGURE 22.5 Lucy. One of the most famous fossil discoveries in Africa.The skeleton is 3.18 million years old and represents the hominid species *Australopithecus afarensis*.

2. **West Lake Turkana.** The western shores of Lake Turkana in Kenya have yielded some of the earliest undisputed hominids that date to about 4 million years ago. In addition, in the mid-1980s, researchers excavated a nearly complete skeleton of a teenage individual (affectionately known as the "Turkana Boy," KNM-WT 15000) that dates to 1.6 million years ago. This individual has been classified as a member of our own genus, *Homo,* and has a markedly larger brain than earlier australopithecines.
3. **Koobi Fora.** Located on the eastern shore of Lake Turkana, Koobi Fora has been explored under the leadership of Richard Leakey since the 1970s and has produced more than 200 hominid fossils. Most are fragmentary and range in age from about 2 million to 1.5 million years ago. One of the most remarkable is a skull known as KNM-ER 1470, which has a cranial capacity of more than 700 cubic centimeters and dates to about 2 million years ago. The brain size is significantly larger than australopithecines and indicates that the trend toward a larger brain was well underway. KNM-ER 1470 is assigned to our own genus, *Homo,* but the exact species to which it belonged is not clear.
4. **Olduvai Gorge.** Olduvai Gorge is located in Tanzania, and the findings there have had a profound impact on our understanding of human evolution. Louis and Mary Leakey discovered human fossils here beginning in the 1950s and introduced them to the world. One of the most famous was an extraordinarily robust australopithecine skull with enormous molar teeth and a ridge at the top of the skull for the attachment of chewing muscles. Together, the features indicate that these specialized australopithecines were adapted for chewing tough, fibrous plant food. The specimen, OH 5, was named "Zinjanthropus" by the Leakeys (which means East African man). Olduvai also produced other hominid specimens and some of the oldest stone tools, which date back 2 million years. The sequence of fossils and tools at Olduvai is well dated and provides crucial information about human evolution.
5. **Laetoli.** Not all fossils are bones and teeth (Figure 22.6). Some 3.7 million years ago, three australopithecines walked over a layer of volcanic ash that was moistened by a light rain. The ash hardened, and the preserved footprints were discovered by Mary Leakey, at Laetoli, a site near Olduvai Gorge in Tanzania. The footprints demonstrate that hominids were fully and efficiently bipedal by this time.
6. **Sterkfontein.** Sterkfontein is one of several cave sites located in South Africa that have fossils that date to approximately 1 to 3 million years ago and that were described beginning in the 1920s. Fossils from Sterkfontein include some rather complete crania, including Sts 5, affectionately known as "Mrs. Ples." Also discovered was part of a vertebral column and a nearly complete pelvis, which proved that these early australopithecines were competent bipeds but had small brains that were not much larger than chimpanzees. The South African sites continue to bring forth hominid fossils and stone and bone tools that deepen our understanding of human evolution.

FIGURE 22.6 Fossilized footprints. Found by Mary Leakey in 1978, it provided the first evidence of early hominid upright posture and locomotion.

ACTIVITY 3

Tracing Fossil Sites for Early Hominids

Materials for this Activity

Current map of Africa

1. For each fossil site previously presented in this text, complete the following columns in Box 22.2: Fossil age, Fossil example, and Fossil information. Note that the first item has been completed as an example.
2. Compare a current map of Africa to Figure 22.7. On Figure 22.7, draw the current national boundaries, label each nation, and draw in the Nile River.
3. Write the current country of origin for each fossil site in Box 22.2. Note that the number of each fossil site in Box 22.2 corresponds to a number on the map of Africa in Figure 22.7.
4. Note that most of the countries of origin are in Eastern Africa. Speculate on the relationship between the Nile River and the locations of most of the fossil sites.

____

____ ■

Box 22.2 Human Fossils

Fossil site	Fossil age	Fossil example	Fossil information	Country of origin
1. Hadar	3 mil. yrs.	Lucy, Al-288	bipedal, small brain, large teeth	
2. West Lake Turkana				
3. Koobi Fora				
4. Olduvai Gorge				
5. Lactoli				
6. Sterkfontein				

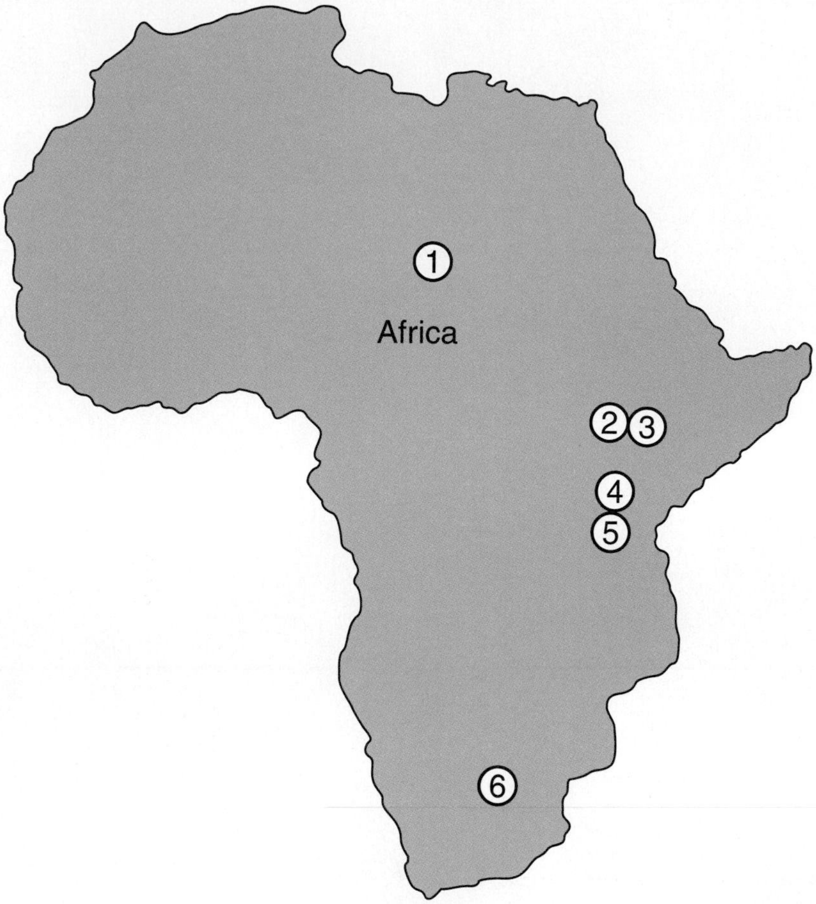

FIGURE 22.7 A map of the African continent, the site of many major fossil discoveries.

Section ____________ Name ____________

Date ____________ Instructor ____________

EXERCISE 22

EVOLUTION

Review Questions

1. Approximately when did the eukaryotic cells evolve? How were they distinct from earlier life forms? What was the significance of that feature for human evolution?

2. How do differences in skeletal anatomy of the cat, monkey, and human relate to differences in behavior (e.g., locomotion, feeding, hand use)?

3. What were some anatomical features of the earliest humans? What distinguishes them from apes?

4. List several sites where australopithecine fossils have been found. Give examples of the kinds of fossil evidence discovered.

5. What are some of the anatomical features of australopithecines?

6. Where are sites where *Homo* fossils have been found? What distinguishes our early *Homo* ancestors from the *Australopithecus* line?

7. Where were the fossil footprints found? How old are they? Why are they important?

8. Define at least one anatomical feature that changed as humans evolved.

9. What features most clearly distinguish primates from other mammals? Which feature do humans no longer share with other primates?

10. Which human fossils demonstrated bipedalism? Which fossils demonstrated larger brain capacity?

EXERCISE

23

HUMAN ECOLOGY

Objectives

After completing this exercise, you should be able to

1. Define human ecology.

2. Describe the lifestyle of hunter-gatherers, horticulturalist, and industrialized people.

3. Describe the impact of human lifestyles on the environment.

4. Describe human population growth trends in the last 2,000 years.

Materials for Lab Preparation

There are no special materials for this laboratory exercise.

Introduction

Human ecology is the study of the relationship between people and their environment. This broad discipline explores how people interact with other species such as plants and animals, how they interact with each other in a complex social environment, and how they deal with the effects of the inorganic environment (e.g., soil, climate). The study of how people interact with their environment has become critically important as we enter the new millennium. In this modern age, we are facing unprecedented world population levels and many environmental crises associated with human behavior, such as global warming and increased pollution.

Humans and Our Environment

From the past, we've learned that there is more than one approach to the way humans interact with the environment to meet their needs of energy use, reproduction, population growth, and impact on the land. As we explore three different ways of human life—hunter/gatherer, horticulturalist, and industrialized people—we can learn important lessons about the impact of human lifestyles on our environment and the solutions offered by non-industrialized ways of life.

In this final exercise, we'll explore the concept of **homeostasis,** or a state of dynamic balance between the different parts of a system and between humans and our environment. Today, our environment is experiencing a state of homeostatic imbalance as a result of many factors, including human population growth and pollution. Evidence of this imbalance is seen in diminishing biodiversity, global warming, rain forest losses, and an unsustainable future on planet Earth.

ACTIVITY 1

Human Lifestyles and Their Ecological Impacts

Your instructor will assign you one of three groups—hunter-gatherer, horticulturalist, or industrialized society. Use the detailed information about your assigned lifestyle to fill in Table 23.1 on page 279. Afterward, each group will participate in a class discussion, using the group discussion questions on page 278 as a guide.

!Kung Hunter-Gatherers

Traditionally, the !Kung people lived in small groups (15–50) that changed according to seasonal resources. Their diet consisted of mostly plant food (60% of calories were derived from more than 200 species of plants) and lean animal meat, including small animals such as rodents and reptiles and larger animals such as antelope (Figure 23.1). The !Kung used baskets, carrying bags, and digging sticks for gathering, and poison darts for hunting animals. The rich diversity of the diet meant that the !Kung had no observed vitamin or mineral deficiencies, with low levels of cholesterol and sodium intake. Food was shared with everyone in the group.

Anthropologists who studied the !Kung in the 1960s and 1970s were surprised to find that they worked only about 2.5 days per week to meet their subsistence needs. The rest of the time was spent socializing, eating, playing games, telling stories, and performing other activities. The !Kung had no concept of private property and had egalitarian relationships between men and women, with no defined leaders such as chiefs.

The age at menarche, or first menstruation among the traditional !Kung was 16.5 years, on

FIGURE 23.1 Hunter-gatherer lifestyle. A !Kung woman and her daughter crack gathered nuts on a stone.

average. Women gave birth to their first child at age 19 and had an average of 4.7 children over their lifetimes. Each child was breast-fed intensively (on demand, both day and night), and during this time ovulation was suppressed. Therefore, birth intervals were long, an average of 4.1 years. Age at menopause is more difficult to determine, but it seems that women reached menopause earlier than industrialized women, at about 40 years of age. This reproductive pattern means that !Kung women spent most of their lives either pregnant or lactating and not having ovulatory or menstrual cycles.

Infant mortality rates were high—as much as 20%—and consequently, overall life expectancy was short (about 30 years of age). However, individuals who did not succumb to accident or infectious disease in childhood could expect to live to old age. The primary causes of death were accident, complications of childbirth, or infectious disease.

FIGURE 23.2 Horticultural society. Three Totonac men farm a hillside cornfield in Veracruz, Mexico.

Horticulturists

Horticulturists live in settlements that vary in size but may number in the thousands. Traditional horticulturists may grow more than one crop in their gardens so that they have some variety in their diets, but they have much less diversity of nutrients than hunter-gatherers (Figure 23.2). Typically, horticulturists may also eat meat, either from domesticated animals or from hunting or trading with hunting peoples. For example, the Lese horticulturists from the Ituri Forest in Africa trade plant food and other goods with the Efe hunter-gatherers (sometimes called Pygmies) for the meat that the Efe procure from the forest.

Horticulturists have simple metal tools such as hoes that are used by people to cultivate plants. They work more than hunter-gatherers, particularly in certain seasons of the year such as seed planting and harvest. They settle villages and have private family spaces and possessions.

Among the Lese women of the Ituri Forest who have been extensively studied by human ecologists, the age at menarche is around 16 years old, and first birth typically occurs between age 18 and 19. Age at menopause is not well known. Lese women have three children over their lifetimes, a lower number than women bear in many simple agricultural societies. Women in such societies usually have four to eight children, with an average of six infants during their lives. Horticultural women typically breast-feed for shorter periods of time than the !Kung, and the interval between births is about 2 years. Infant mortality rates among traditional horticulturists are high—as much as 20–30%. Death is most often due to accident, complications of childbirth, or infectious disease.

Industrialists

Modern industrialized people (Figure 23.3) may live in large groups; some cities have more than a million people! Rapid technological advances are taken for granted—think of the changes in technology over the 20th century when we went from the horse and buggy to rocket ships, jet planes, and reliance on computers. People in industrialized society have access to lots of goods but work hard to attain them. In the United States, a 40-hour workweek is the norm.

Women in industrialized societies reach menarche at an average of 12 to 12.5 years of age. Menopause occurs at about 50 years of age. A woman's age at the birth of her first child is often postponed with the use of contraceptives, and the size of the family may be planned. The average number of children over a woman's lifetime is 2. Length

FIGURE 23.3 Industrialized society. A city planning group gather around a model city.

What are implications for women's health?

______

3. What are the dietary features of your group?

______

How would this diet influence the health of the people?

______

4. What is the major cause of death in your group, and how does it relate to lifestyle?

______

5. How does the lifestyle of your group impact upon its specific environment?

______

of lactation is extremely variable; some women do not breast-feed at all but feed their babies a prepared formula in bottles. Even if women do choose to breast-feed, they are often separated from the infant while they work, so the frequency and quality of lactation is different than that of women in nonindustrialized societies.

Life expectancy in developed countries has risen to about 76 years of age, thanks to low infant mortality (about 96% of children survive to be teenagers) and well-developed medical technology. Industrialized people die of diseases that develop as a consequence of lifestyle—high-fat diets, use of tobacco and alcohol, and lack of vigorous physical activity contribute to heart disease (the leading cause of death), cancers, stroke, and emphysema.

Group Discussion Questions

1. What are the implications for energy resources extracted from the environment from the perspective of your group's lifestyle?

______

2. Summarize the female reproduction pattern of your group.

______

What are the implications for population growth?

______

Human lifestyle has changed dramatically in the past 12,000 years, from the time that groups of people made the transition to reliance on domesticated plants and animals, to the breathtaking changes that have occurred in the past few hundred years with industrialization. The study of human ecology, specifically of the different patterns of life of nonindustrialized people such as hunter-gatherers and horticulturists, gives us insight into the impact of human behavior on the environment.

1. How have changes in human lifestyle contributed to the environmental challenges that we currently face?

______

2. What are some of the ecological problems faced by people as we enter the 21st century?

______

3. What can we learn from people with nonindustrialized ways of life that may offer solutions for some of our problems?

______ ■

World Human Population

As the environment improved for human beings over time, the world human population has steadily increased. Remember the concept of homeostasis: a dynamic balance between different forces within a

Table 23.1 Characteristics of Three Human Lifestyles

	Hunter-gatherers	Horticulturists	Industrialists
Group size			
Diet			
Tools/technology			
Hours worked per week			
Attitude toward private property			
Infant mortality			
Primary causes of death			
Age at menarche			
Age at first child's birth			
Number of children per woman (over a lifetime)			
Length of birth interval			
Approximate age at menopause			
Reproductive span*			
Number of years spent pregnant (over a lifetime)			
Number of years spent lactating			
Number of years of ovulatory/ menstrual cycles			

*Reproductive span = age at menarche to age at menopause.

system, in this case, human beings and the **ecosystem** of Earth. The maximum rate of growth for any population under ideal conditions is called **biotic potential.** The opposing forces, called **environmental resistance,** are environmental factors such as limited nutrients, energy, and land. **Carrying capacity** represents the balance between biotic potential and environmental resistance; it is a steady level of population of a given species that the ecosystem can support at a given point in time.

About 250 years ago, the world human population began to increase rapidly. Prior to this time, people lived in small, widely dispersed groups, with limited ability to shape the environment, so the human carrying capacity was quite low. Since 1750, the world human population has steadily grown, until there is now an exponential rate of growth. Our population growth curve has a **J-shape,** which is very troubling because it resembles the **exponential growth curve** of an organism with a population growth rate tending toward the biotic potential, suggesting that there are unlimited environmental resources.

We are approaching our human biotic potential. From reading the text, what are some of the environmental changes that contributed to this population explosion?

Speculate on the earth's carrying capacity for humans.

Speculate on the effects of the human population from of a major disaster such as the icebergs melting during global warming.

ACTIVITY 2

Population Explosion

Plot the changes in population over the past 2,000 years using the graph in Figure 23.4. The x axis represents time (the number of years), while the y axis represents population (numbers of people).

1. From the graph, approximately when did the growth curve increase the most? *Hint:* Look for the area of the curve with the most slope ("rise over run").

2. From the graph, if there was a major food problem in the years 1000, 1500, or 1900, which would have been the most devastating? Why?

3. What are the environmental implications of an explosive human population growth?

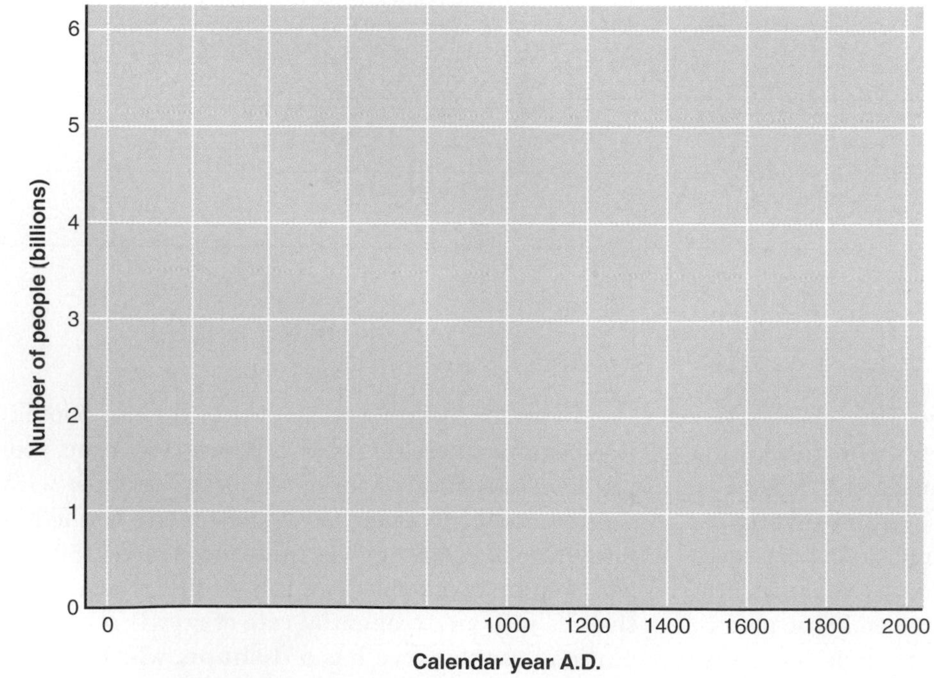

World Human Population

0 A.D.	250 million
1650	500 million
1900	1 billion
1950	2 billion
1985	4 billion
2000	6 billion

FIGURE 23.4 Explosion growth curve.

Section Name

Date Instructor

EXERCISE

23

HUMAN ECOLOGY

Review Questions

1. What is the discipline of human ecology? How can studies of nonindustrialized people offer insight into current environmental problems such as pollution and global warming?

2. What are the main features of a hunter-gatherer lifestyle? What insight does the study of !Kung hunter-gatherers offer for understanding human biology?

3. How has the human lifestyle changed in the past 12,000 years? What impacts have the changes had on human biology? On the environment?

4. Summarize human population growth over the past 2,000 years. Why has the human population increased so dramatically over the past century? What are some environmental effects of increasing population? Speculate on some of the consequences of the population increase for human biology.

APPENDIX A

THE METRIC SYSTEM

Measurement	Unit and abbreviation	Metric equivalent	Metric to English conversion factor	English to metric conversion factor
Length	1 kilometer (km)	= 1000 (10^3) meters	1 km = 0.62 mile	1 mile = 1.61 km
	1 meter (m)	= 100 (10^2) centimeters	1 m = 1.09 yards	1 yard = 0.914 m
		= 1000 millimeters	1 m = 3.28 feet	1 foot = 0.305 m
			1 m = 39.37 inches	
	1 centimeter (cm)	= 0.01 (10^{-2}) meter	1 cm = 0.394 inch	1 foot = 30.5 cm
				1 inch = 2.54 cm
	1 millimeter (mm)	= 0.001 (10^{-3}) meter	1 mm = 0.039 inch	
	1 micrometer (μm) [formerly micron (μ)]	= 0.000001 (10^{-6}) meter		
	1 nanometer (nm) [formerly millimicron (mμ)]	= 0.000000001 (10^{-9}) meter		
	1 angstrom (Å)	= 0.0000000001 (10^{-10}) meter		
Area	1 square meter (m^2)	= 10,000 square centimeters	1 m^2 = 1.1960 square yards	1 square yard = 0.8361 m^2
			1 m^2 = 10.764 square feet	1 square foot = 0.0929 m^2
	1 square centimeter (cm^2)	= 100 square millimeters	1 cm^2 = 0.155 square inch	1 square inch = 6.4516 cm^2
Mass	1 metric ton (t)	= 1000 kilograms	1 t = 1.103 ton	1 ton = 0.907 t
	1 kilogram (kg)	= 1000 grams	1 kg = 2.205 pounds	1 pound = 0.4536 kg
	1 gram (g)	= 100 milligrams	1 g = 0.0353 ounce	1 ounce = 28.35 g
			1 g = 15.432 grains	
	1 milligram (mg)	= 0.001 gram	1 mg = approx. 0.015 grain	
	1 microgram (μg)	= 0.000001 gram		
Volume (solids)	1 cubic meter (m^3)	= 1,000,000 cubic centimeters	1 m^3 = 1.3080 cubic yards	1 cubic yard = 0.7646 m^3
			1 m^3 = 35.315 cubic feet	1 cubic foot = 0.0283 m^3
	1 cubic centimeter (cm^3 or cc)	= 0.000001 cubic meter = 1 milliliter	1 cm^3 = 0.0610 cubic inch	1 cubic inch = 16.387 cm^3
	1 cubic millimeter (mm^3)	= 0.000000001 cubic meter		
Volume (liquids and gases)	1 kiloliter (kl or kL)	= 1000 liters	1 kL = 264.17 gallons	1 gallon = 3.785 L
	1 liter (l or L)	= 1000 milliliters	1 L = 0.264 gallons	1 quart = 0.946L
			1 L = 1.057 quarts	
	1 milliliter (ml or mL)	= 0.001 liter	1 ml = 0.034 fluid ounce	1 quart = 946 ml
		= 1 cubic centimeter	1 ml = approx. ¼ teaspoon	1 pint = 473 ml
				1 fluid ounce = 29.57 ml
			1 ml = approx. 15–16 drops (gtt.)	1 teaspoon = approx. 5 ml.
	1 microliter (μl or μL)	= 0.000001 liter		
Time	1 second (s)	= ¹⁄₆₀ minute		
	1 millisecond (ms)	= 0.001 second		
Temperature	Degrees Celsius (°C)		°F = ⁹⁄₅ °C + 32	°C = ⁵⁄₉ (°F − 32)

ART AND PHOTOGRAPHY CREDITS

Exercise 1
1.1 Courtesy of the National Cancer Institute
1.2 Shirley Bortoli
1.3 Shirley Bortoli
1.4 Shirley Bortoli
1.5 © Benjamin Cummings/Elena Dorfman/Shirley Bortoli
1.6 Shirley Bortoli

Exercise 2
2.1 Courtesy of Leica Microsystems, Inc./Shirley Bortoli
2.2 Shirley Bortoli

Exercise 3
3.1 Adapted from Johnson, *Human Biology,* 3e, Pearson Education, 2006
3.2 Adapted from Johnson, *Human Biology,* 3e, Pearson Education, 2006
3.3 Adapted from Johnson, *Human Biology,* 3e, Pearson Education, 2006
3.4 © Benjamin Cummings. Photo by Alan Bell, University of New England/Shirley Bortoli
3.5 © Bert Atsma/Shirley Bortoli
3.6 © Eric Graves/Photo Researchers, Inc./Shirley Bortoli
3.7 © Ed Reschke, Peter Arnold, Inc./Shirley Bortoli

Exercise 4
4.1 Adapted from Johnson, *Human Biology,* 3e, Pearson Education, 2006
4.2 Shirley Bortoli
4.3 Shirley Bortoli
4.4 Shirley Bortoli
4.5a © Keith R. Porter, Science Source/PhotoResearchers, Inc.
4.5b © David M. Phillips, Science Source/PhotoResearchers, Inc.
4.5c Courtesy of Dr. Mohnadas Narla, Lawrence Berkeley Laboratory

Exercise 5
5.1 © John D. Cunningham/Visuals Unlimited/Shirley Bortoli
5.2 © Benjamin Cummings, photo by Alan Bell, University of New England/Shirley Bortoli
5.3 © Benjamin Cummings, photo by Alan Bell, University of New England/Shirley Bortoli
5.4 © Benjamin Cummings, photo by Alan Bell, University of New England/Shirley Bortoli
5.5 © Bert Atsma/Shirley Bortoli
5.6 © Ed Reschke, Peter Arnold, Inc./Shirley Bortoli
5.7 © Ed Reschke, Peter Arnold, Inc./Shirley Bortoli
5.8 © Ed Reschke, Peter Arnold, Inc./Shirley Bortoli
5.9 © Benjamin Cummings, photo by Alan Bell, University of New England/Shirley Bortoli
5.10 © Eric Graves/Photo Researchers, Inc./Shirley Bortoli
5.11 © Ed Reschke, Peter Arnold, Inc./Shirley Bortoli
5.12 © Benjamin Cummings, photo by Alan Bell, University of New England/Shirley Bortoli
5.13 © Ed Reschke, Peter Arnold, Inc./Shirley Bortoli

Exercise 6
Table 6.1 Adapted from Marieb, *Human Anatomy and Physiology,* 7e, Benjamin Cummings, 2005
6.1 Adapted from Marieb, *Human Anatomy Laboratory Manual,* 7e, Benjamin Cummings, 2005
6.2 Shirley Bortoli
6.3 Adapted from Johnson, *Human Biology,* 3e, Pearson Education, 2006
6.4 Adapted from Marieb, *Human Anatomy Laboratory Manual,* 7e, Benjamin Cummings, 2005
6.5 © Jenny Thomas, Addison Wesley Longman/Shirley Bortoli

Exercise 7
7.1 Adapted from Johnson, *Human Biology,* 3e, Pearson Education, 2006
7.2 Adapted from Johnson, *Human Biology,* 3e, Pearson Education, 2006
7.3a–d © Ed Reschke, Peter Arnold, Inc./The Left Coast Group
7.4 Adapted from Marieb, *Human Anatomy and Physiology,* 7e, Benjamin Cummings, 2005

Exercise 8
8.1 Adapted from Johnson, *Human Biology,* 3e, Pearson Education, 2006
8.2 Adapted from Johnson, *Human Biology,* 3e, Pearson Education, 2006
8.3 Adapted from Johnson, *Human Biology,* 3e, Pearson Education, 2006
8.4 Adapted from Johnson, *Human Biology,* 3e, Pearson Education, 2006
8.5 Adapted from Johnson, *Human Biology,* 3e, Pearson Education, 2006
8.6 Adapted from Johnson, *Human Biology,* 3e, Pearson Education, 2006
8.7 Adapted from Johnson, *Human Biology,* 3e, Pearson Education, 2006
8.8 Adapted from Johnson, *Human Biology,* 3e, Pearson Education, 2006

Exercise 9

9.1a © R. Calentine, Visuals Unlimited
9.1b Adapted from Johnson, *Human Biology,* 3e, Pearson Education, 2006
9.2a, b Adapted from Johnson, *Human Biology,* 3e, Pearson Education, 2006
9.2c © K. G. Murtil, Visuals Unlimited
9.3 Adapted from Johnson, *Human Biology,* 3e, Pearson Education, 2006
9.4a Adapted from Johnson, *Human Biology,* 3e, Pearson Education, 2006
9.4b © Fred Hossler, Visuals Unlimited
9.5 Adapted from Johnson, *Human Biology,* 3e, Pearson Education, 2006
9.6 Adapted from Johnson, *Human Biology,* 3e, Pearson Education, 2006
9.7 Adapted from Johnson, *Human Biology,* 3e, Pearson Education, 2006

Exercise 10

10.1 Adapted from Johnson, *Human Biology,* 3e, Pearson Education, 2006
10.2 Adapted from Johnson, *Human Biology,* 3e, Pearson Education, 2006
10.3 Adapted from Marieb, *Human Anatomy and Physiology,* 7e, Benjamin Cummings, 2005
10.4 © Benjamin Cummings. Photo by Victor Eroschenko/Shirley Bortoli
10.5 Adapted from Johnson, *Human Biology,* 3e, Pearson Education, 2006

Exercise 11

11.1 Adapted from Johnson, *Human Biology,* 3e, Pearson Education, 2006
11.2 Adapted from Johnson, *Human Biology,* 3e, Pearson Education, 2006
11.3 Adapted from Johnson, *Human Biology,* 3e, Pearson Education, 2006
11.4 Adapted from Marieb, *Human Anatomy and Physiology,* 7e, Benjamin Cummings, 2005
11.5 © Manfred Kage/Peter Arnold, Inc./Shirley Bortoli
11.6 © Martin Rotker/Phototake/Shirley Bortoli
11.7 Adapted from Johnson, *Human Biology,* 3e, Pearson Education, 2006
11.8 © Benjamin Cummings. Photo by Dr. Sharon Cummings, University of California, Davis/ Shirley Bortoli/Precision Graphics
11.9a Adapted from Marieb, *Human Anatomy and Physiology Laboratory Manual, Pig Version,* 8e, Benjamin Cummings, 2005
11.9b © Benjamin Cummings. Photo by Elena Dorfman
11.10 Shirley Bortoli

Exercise 12

12.1 Adapted from Johnson, *Human Biology,* 3e, Pearson Education, 2006
12.2 Adapted from Johnson, *Human Biology,* 3e, Pearson Education, 2006
12.3 Adapted from Marieb, *Human Anatomy and Physiology,* 7e, Benjamin Cummings, 2005
12.4 Adapted from Johnson, *Human Biology,* 3e, Pearson Education, 2006
12.5 Adapted from Johnson, *Human Biology,* 3e, Pearson Education, 2006
12.6 Adapted from Johnson, *Human Biology,* 3e, Pearson Education, 2006
12.7 Adapted from Johnson, *Human Biology,* 3e, Pearson Education, 2006
12.8 Adapted from Johnson, *Human Biology,* 3e, Pearson Education, 2006
12.9 Adapted from Marieb, *Human Anatomy and Physiology,* 7e, Benjamin Cummings, 2005

Exercise 13

13.1 Adapted from Johnson, *Human Biology,* 3e, Pearson Education, 2006
13.2 Adapted from Johnson, *Human Biology,* 3e, Pearson Education, 2006
13.3 Shirley Bortoli
13.4 Shirley Bortoli

Exercise 14

14.1 Shirley Bortoli
14.2 Adapted from Johnson, *Human Biology,* 3e, Pearson Education, 2006
14.3 Adapted from Johnson, *Human Biology,* 3e, Pearson Education, 2006
14.4 Adapted from Johnson, *Human Biology,* 3e, Pearson Education, 2006
14.5 Adapted from Johnson, *Human Biology,* 3e, Pearson Education, 2006
14.6 © Benjamin Cummings. Photo by Jack Scanlan, Holyoke Community College/Shirley Bortoli

Exercise 15

15.1 Adapted from Johnson, *Human Biology,* 3e, Pearson Education, 2006
15.2 Adapted from Johnson, *Human Biology,* 3e, Pearson Education, 2006
15.3 Adapted from Johnson, *Human Biology,* 3e, Pearson Education, 2006
15.4 Adapted from Johnson, *Human Biology,* 3e, Pearson Education, 2006
15.5 © Benjamin Cummings. Photo by Wally Cash, Kansas State University/Shirley Bortoli
15.6 Adapted from Johnson, *Human Biology,* 3e, Pearson Education, 2006
15.7 Adapted from Johnson, *Human Biology,* 3e, Pearson Education, 2006

Exercise 16

16.1 Adapted from Johnson, *Human Biology,* 3e, Pearson Education, 2006
16.2 Adapted from Johnson, *Human Biology,* 3e, Pearson Education, 2006; photos © CNRI/Phototake

16.3 Adapted from Johnson, *Human Biology,* 3e, Pearson Education, 2006
16.4 Shirley Bortoli
16.5 Adapted from Johnson, *Human Biology,* 3e, Pearson Education, 2006
16.6 Adapted from Johnson, *Human Biology,* 3e, Pearson Education, 2006
16.7 Shirley Bortoli

Exercise 17
17.1 Adapted from Johnson, *Human Biology,* 3e, Pearson Education, 2006
17.2 Adapted from Johnson, *Human Biology,* 3e, Pearson Education, 2006
17.3 Adapted from Johnson, *Human Biology,* 3e, Pearson Education, 2006
17.4 Adapted from Johnson, *Human Biology,* 3e, Pearson Education, 2006
17.5a Adapted from Marieb, *Human Anatomy and Physiology Laboratory Manual, Pig Version,* 8e, Benjamin Cummings, 2005
17.5b © Benjamin Cummings. Photo by Elena Dorfman

Exercise 18
18.1 Adapted from Johnson, *Human Biology,* 3e, Pearson Education, 2006
18.2 Adapted from Johnson, *Human Biology,* 3e, Pearson Education, 2006
18.3 Shirley Bortoli
18.4 Adapted from Johnson, *Human Biology,* 3e, Pearson Education, 2006

Exercise 19
19.1 Adapted from Johnson, *Human Biology,* 3e, Pearson Education, 2006
19.2 Adapted from Johnson, *Human Biology,* 3e, Pearson Education, 2006
19.3 Adapted from Marieb, *Human Anatomy and Physiology*, 7e, Benjamin Cummings, 2005
19.4 Adapted from Marieb, *Human Anatomy and Physiology,* 7e, Benjamin Cummings, 2005
19.5 Adapted from Johnson, *Human Biology,* 3e, Pearson Education, 2006
19.6 Adapted from Marieb, *Human Anatomy and Physiology Laboratory Manual, Pig Version,* 8e, Benjamin Cummings, 2005

Exercise 20
20.1 Adapted from Johnson, *Human Biology,* 3e, Pearson Education, 2006
20.2 Shirley Bortoli
20.3 The Left Coast Group, Inc.

Exercise 21
21.1 Adapted from Johnson, *Human Biology,* 3e, Pearson Education, 2006
21.2 Adapted from Johnson, *Human Biology,* 3e, Pearson Education, 2006
21.3 Shirley Bortoli
Base Pairs Shirley Bortoli

Exercise 22
Box 22.1 Shirley Bortoli
22.1 Kenneth Garrett
22.2 Adapted from Johnson, *Human Biology,* 3e, Pearson Education, 2006
22.3 Shirley Bortoli
22.4ab Shirley Bortoli
22.4cd Adapted from Johnson, *Human Biology,* 3e, Pearson Education, 2006
22.5 Cleveland Museum of Natural History
22.6 John Reader/Science Photo Library/Custom Medical Stock Photography
22.7 Shirley Bortoli

Exercise 23
23.1 © Anthony Bannister
23.2 © Tony Stone Images, Inc.
23.3 © Prentice Hall, Inc.
23.4 Shirley Bortoli

INDEX

Page references followed by *fig* indicates an illustrated figure; followed by *t* indicates a table; followed by *b* indicates a box.